rock • paper • pixels

The purpose of this book is to avail students of the many different reasons and back-stories equated with the advent of interaction design and its effect on our world. Since the beginning of recorded history, humankind has desired to keep track of its events and accomplishments and to find a way to properly represent those occurrences to the rest of the world. Now with the use of interaction design and new technologies it has become possible to present the story through current and diverse platforms.

As the new generation comes of age, a new method is being utilized in all areas of communication. This new generation – Generation i – desires its communication at a different rate than its predecessors. This new Gen-i wants its information Internet-ready and interactive. This generation has become involved in the immediate and it is not going away. Interactive media is here to stay – from the halls of famous museums to the conference rooms of corporations to the entertainment centers in our homes. A new vision of information distribution has taken root. The main goal of this book is to help you take your best steps in that new direction. To help you understand the effects of this new platform and to be aware of its abilities and its dangers.

As a friend once said, "no matter how thin you slice it, there are always two sides."

Patrick Aievoli started his career in 1978 as a designer for local editorial and advertising companies. In 1984, he became a promotional designer at McGraw-Hill. Professor Aievoli has been a full-time academic since 1988 when he left his position as senior designer, print promotion, at the McGraw-Hill Book Company for a teaching position at SUNY Farmingdale. During his time at McGraw-Hill, Patrick helped in the creation of McGraw-Hill's first interactive CD-ROM "Encyclopedia of Science and Technology" in 1987. In 1990–1996, Professor Aievoli completed his thesis on "The Use of New Media in Higher Education," culminating in an interactive art history CD-ROM featuring core and dynamic content along with a simplified suite of online learning tools. In 1998, he became a full-time faculty member at LIU Post in Brookville, New York. He became the director of the campus' Interactive Multimedia Arts graduate program in 1999 and built the program from the start to its closing in 2018. He still teaches in the undergraduate design program. Professor Aievoli has authored four books prior to *Rock • Paper • Pixels*. Although he is a dedicated academic, Professor Aievoli is still involved in the online arena and has consulted for some of the NYC metro area's largest media companies.

rock • paper • pixels

How Our Need to Communicate Created a New Economy, Society, and Individual

What will be the Effects of Artificial Intelligence and Quantum Computing?

Patrick Aievoli

CRC Press
Taylor & Francis Group
Boca Raton London New York

CRC Press is an imprint of the
Taylor & Francis Group, an **informa** business

Designed cover image: © Patrick Aievoli

First edition published 2025
by CRC Press
2385 NW Executive Center Drive, Suite 320, Boca Raton FL 33431

and by CRC Press
4 Park Square, Milton Park, Abingdon, Oxon, OX14 4RN

CRC Press is an imprint of Taylor & Francis Group, LLC

Library of Congress Cataloging-in-Publication Data
Names: Aievoli, Patrick, author.
Title: Rock paper pixels : how our need to communicate created a new
economy, society, and individual: what will be the effects of artificial
intelligence and quantum computing? / Patrick Aievoli.
Description: Boca Raton, FL : CRC Press, 2025. | Includes bibliographical
references and index.
Identifiers: LCCN 2024011932 (print) | LCCN 2024011933 (ebook) |
ISBN 9781032518916 (hbk) | ISBN 9781032518923 (pbk) | ISBN 9781003404347 (ebk)
Subjects: LCSH: Technological innovations. | Artificial intelligence. |
Multimedia communications.
Classification: LCC HC79.T4 A39 2025 (print) | LCC HC79.T4 (ebook) |
DDC 303.48/34--dc23/eng/20240612
LC record available at https://lccn.loc.gov/2024011932
LC ebook record available at https://lccn.loc.gov/2024011933

ISBN: 978-1-032-51891-6 (hbk)
ISBN: 978-1-032-51892-3 (pbk)
ISBN: 978-1-003-40434-7 (ebk)

DOI: 10.1201/9781003404347

Typeset in Times
by SPi Technologies India Pvt Ltd (Straive)

This book is dedicated first and foremost to my parents, Patsy and Jerry, without whose love and support nothing would have been possible. To my sister Andrea for putting up with me for a very long time. Also to my family, Adele, Allison, Philip, Carly, and Ryan.

I would also like to thank all of my students over the years, especially Janice LaBarbera, Richard Neal, and Larry Wolfe, truly a great team. We started all of this back before anyone even knew what we were saying. I just got to write about it.

I would also like to welcome the next generation, one I like to call Gen-i, generation interactive, Internet, immediate, information, impulsive, etc.

But this is dedicated especially to my beautiful, wonderful, fearless granddaughter Avery Mae Aievoli.

The ball is in your court, baby girl; Grumpy says, "Game on!"

Contents

1 Bang! Boom! and Bang Again! *The Second Big Bang*

The who, why, when, how, where, and cost of all this advancement and chaos.

1.1 BANG! BOOM! AND BANG AGAIN

1.1.1 The Second Big Bang!

First it was churches, then palaces, then office buildings, and today it is the cell towers on top of those buildings. Since the mid-1980s our society has gone through a series of changes we as a whole have not experienced since the Age of Enlightenment. Gone are the times of waiting for days for a response, to any form of communication between individuals or businesses. Today everything is instant. Messages flow back and forth in real time from the furthest parts of the world. We communicate with friends and family in that instance but now that instance has become the actual reality. The walls of time and space have collapsed and everything is immediate. We can study anywhere in the world with the best faculty and colleagues. We can hear our favorite music just as soon as it is released. Watch endless episodes of our favorite show regardless of space or time. Attend concerts from our living rooms. Get food delivered to our door. In essence we never have to leave the sanctuary of our homes. All of this convenience has brought society to a new consciousness, a new acceptance of how life should be lived. How life should be experienced. The cheapest prices, the fastest delivery, the ease of almost non-involvement. All of these luxuries are for the taking – but at what price?

This first chapter is meant to set the stage for how we got here from the viewpoint of our need to communicate as a species and the steps toward the creation and absorption of technology. Connectivity is the key to this movement (IoT – the Internet of Things). But this connectivity goes further than the hardware and software; it goes toward the ripple effect of these changes. The Industrial Revolution of the early 1900s brought about a shift from an agrarian society to a citified industry-driven one. This Digital Revolution – an Incunabula if you will – as cliche of a term as that may seem is actually correct. It has revolutionized and rebirthed how we work, live, and interact with each other.

The next chapters will give examples of these changes, both the good and bad of "how our need to communicate generated a new economy, society, and individual." Subsequent chapters will chronicle how this shift from analog to digital has been a long time coming but with the recent events in technology it has gained speed going downhill faster and with more strength than many could imagine, quickening Moore's Law – "technology doubles every eighteen months" (Padua, 2011).

1.2 ECONOMY, SOCIETY, AND INDIVIDUAL

This section is broken down into economy, society, and individual in order to give a primer to how this all happened and where it is going. So how has this massive change flowed down to affect each one of these areas?

DOI: 10.1201/9781003404347-1

Economy – how have these new innovations affected the economy?
Society – how has society had to respond?
Individual – how have these changes affected the individual?

1.2.1 Economic Changes – Starting Now

The ability to communicate worldwide in a nano-second has tremendous potential for industry, media, education, information, and the arts. This is reflected immediately in the global economy. Entire industries have been created by this evolution into a new era of communication and the speed by which these industries move has also quickened.

Careers that did not exist five or ten years ago are now mainstays and purposeful needs in diverse industry categories. The management of social media profiles and media compliance are now high on priority lists from corporations to single-person shops to global leaders to local politicians. Everyone from individuals to entire nations have found that the need to present themselves in the proper form within these global town halls has come to the forefront. As the satirist Mark Twain once said, "a lie will fly around the whole world while the truth is getting its boots on…." This has never been as true as it is today.

Example: On May 22, 2023, a "newsbreak" shook that day's news cycle. The "newsbreak" showed an image of billowing smoke coming from an administrative-looking building that was near the Pentagon. This post reverberated across the Internet, caused disruption in the stock market, and was picked up by numerous news outlets. And reached the Russian news service RT.

The reality was that this was an AI-generated image and, therefore, not real. However the effects were real. Stock market disruptions, news service bought in, and even nations as far away as Russia were more than happy to show how America was falling apart. Through the power of generative AI (Marcelo, 2023).

This is a perfect example of how a specific new technology – AI (Artificial Intelligence) – coupled with a delivery platform – X (formerly Twitter) – disrupted the day's economy. Stock markets dipped when the "news" was announced. It was a short time event but, nonetheless, it affected stock trading globally for that day and deepened the belief that the world was in peril.

Has all of this disruption been good or has it taken a turn downward during its coming of age? These effects have changed the landscape in some cases for better but at this time in many cases for worse. While many new opportunities have been born, entire industries have been removed, deleted from our society, lives have been changed forever. Career opportunities of the old method have faded. Families have been affected and future careers changed and altered forever. Even our way of simply communicating with each other has been ramshackled, falling into a new existence. Gone is the ability to actually deal with a situation, to be civil, and is replaced with an attempt to hide behind the screens of our devices. No longer heroes in the street but keyboard warriors in our cars and basements. This is not the perspective of a few Luddites – this is the perspective of many who have been in the communications and digital industry for decades.

The fragmentation of marketplaces and society in general has had ripple effects that have demolished industries and obviously those events have set society on a course toward chaos. However, there are also many benefits which we will discuss later.

1.2.2 Media Fragmentation: Must See TV – Gone Forever?

The traditional media landscape has been upended like never before. The audience has been splintered and fractured to a point where it is no longer "broadcasting" but now "narrowcasting." Media companies are scurrying around to piece together ad revenues to balance back to a time when there were many more limited sources and streams. This fragmentation has also destroyed the "shared experience" of a society into smaller and smaller numbers. No longer is there what was called a "watercooler" moment. No longer do groups of people meet up in the office to talk about last night's episode. This shared real-life experience has now gone "tribal." Due to the overly comfortable environments created by these online communities, social discourse has given way to these echo-chambers, and bubbles. Bubbles of like-minded individuals that have no time for checks and balances and no time for critical thinking and discourse. This move toward a tribal mentality has fragmented society. The question is how do we hold onto this new amazing advancement and make sure the future curves toward benevolence over destruction?

1.2.3 Birth of New Economies – Disruption via Destruction

Schumpeter's theory of Creative Destruction has borne out some of the greatest achievements of the last 100 years. This theory states that each innovation has an arc of time. During this time these explosive disruptions or innovations destroyed what came before and hence created a new blueprint for advancement in the field (Adler, 2019). The term "disruption" is used like "grande latte" today – it is heard every place you turn. As these disruptions occur, the core of society is also disrupted whenever an existing model is forced into being completely reworked – jobs, relationships, lifestyles change. This affects the economy as well as families, and communities. Just look at the rust belts of the Northeast. A perfect and simple example is our means of transportation.

The internal combustion engine ended the dominance of the horse as transportation but opened up the world to greater and more comfortable means of transportation. Most people were appreciative of this change but the blacksmiths were not so happy. Now electric cars are disrupting the oil industry. Car manufacturers are adapting beautifully. But the global oil conglomerates are fighting this disruption tooth and nail. And bringing about wars across the face of this planet scouring Third World nations for their resources to use in this new industry.

This has occurred time and time again throughout our history. Inventions, innovations, and changes occur and spur tremendous opportunities for many but the few that resist – as Schumpeter explains – actually suffer worse damage than if they tried and failed to innovate. These hesitant ones somehow feel that they can weather this storm of progress and come out from the other side unscathed. Or they hope that they can get to retirement before it actually affects them, this is a very rare occurrence. Few are the numbers of industries and individuals that have seen this play out well for them. This arc of change has left many on the side of the road wondering what had happened. Resistance is futile. Traditional media and its networks have suffered the most in this fragmentation. But also have the surrounding businesses – print is almost completely dead as an industry and then so have the support systems for that industry. No print industry means lumber is down, trucking is down, ink production is down, warehousing is down, office space is down. Big effect – big ripples. This is felt through many forms of industry, not just communications, and with this comes a new form of communication – the echo chamber.

1.2.4 Society – Echoing Back

This fragmentation has affected and increased tribalism not just through common hatreds but also through common likes. When you fragment media and cause tribalization, you reinforce the strength of the separate camps due to the lack of reasonable discussion. Everyone goes to their corner and nothing goes forward to rational thought. It all just stops cold. We have always depended upon the social collective, the shared experience to go forward. This primarily occurred in the media once the "Fairness Act" was repealed. The "Fairness Act" required a news organization to show the two sides of an issue fairly. This was repealed during President Reagan's administration.

Web and app design, ecommerce initiatives, and global online educational possibilities have opened markets to maybe the best of the best? Or possibly rumor disguised as fact? Obviously, the print industry has felt these changes. But again we have to ask what effect has economic gain brought about into society?

The use of technology in some cases has affected society as well as industry in ways that shake it to its core. What was once a typical structure of commuting to work or school has given way to the world of an online/remote work arrangement. This has changed how we live and interact with each other. The concept of quality over quantity has broken through and with it structures have been reformatted. We no longer go to an office and derive our objectives from a hierarchy but have now given the freedom to create our schedule and make our choices. This can be a great thing for those of us who have a form of self-discipline and the ability to manage time properly. Technology has given us these choices and these freedoms. But at what cost? Have we lost the means to understand a civil perspective or in general consideration for others due to this egocentric choice paralysis? More time at home in our comforts can be a very good thing. Especially when you consider commutation time is now non-existent. But what is the backlash of such isolation?

Ironically, all of this new change comes from Sir Isaac Newton's *Principia*, 1728. And in basic motion (physics), as Newton's Third Law states, "for every action (force) in nature there is an equal and opposite reaction." What will be and what has been the "reaction" to all of this motion? Has it all been benevolent or have we fallen to our weakest and most basic instincts? Do we actually need these instant interactions? Is it a necessity for our continuation as a lifeform? As a species we do need to gather and communicate to survive? (Newton, 2016).

According to an article by Michael Platt, PhD – in Dana Foundation's Report from Neuroscience 2019 – "Human beings are wired to connect" (Sukel, 2019).

How will the advent of these new technologies and specifically Artificial Intelligence (AI) become benevolent or malevolent to these interactions? Will AI pull us further away from the human interaction and morality that we seem to be abandoning? Recent articles show this fear has infiltrated the current psyche with some of the innovators expressing their fears. AI researcher and developer Geoffrey Hinton declared he would be stepping away from his position at Google due to these concerns. Sam Altman (founder of OpenAI – now owned by Microsoft) responded to questions from Congress by imploring them to regulate this new invention. Altman and other experts in the field, as we will discuss, have been loud critics of this new technology running wild and creating greater havoc than social media ever has done (ChatGPT Creator Issues Stark Warning on AI, 2023).

To this end Congress has shown concern with how to regulate the imposition of AI on the workforce and its effect on overall global economics. Asa Raskin and Tristan Harris in a recent presentation for the Centre for Humane Technologies exposed how we need to get in front of the curve on AI or we are doomed to suffer the effects similar to social media. This is real and currently on the mind's of many of today's thought leaders. Later on we will discuss the advantages and disadvantages of the coming of AI along with the strength of quantum computing and their effects on the global marketplace and the worldwide economic outlook.

1.2.5 Individual – a Newer AI – "Anxiety Interruptus"

Many more cases of anxiety and depression are being reported today than ever before. Is that because we have lost the stigma surrounding admitting such conditions or has the effect of this self-created loneliness reached into a deeper level of the individual? Is this a side effect of the self-prescribed "Anti-Social Social Club"? Or has the isolation and self-idolizing of the small screen led to a form of dehumanization of this generation? A generation that prides itself on identifying clearly its agenda and position seems to be the one most affected by this venture into a "metaverse-like" being. These are the opinions of the actual individuals who built this Golem. Tristan Harris, founder of the Centre for Humane Technologies, has been warning the analog and digital community that this new technology is affecting and atrophying our own abilities to deal with logic and reason.

On May 4, 2023, an opinion article in the *New York Times* stated that case clearly. The article positions our society as one where we are hypercautious in our parenting which is damaging our youth to believe in an exaggerated sense of danger in their lives. Throughout the article this hypercautious tendency, this anxiety-ridden, overtly depressed generation is tied directly to the outcomes of the financial collapse of 2008 and the general sense of fear perpetrated by the current political climate (Brooks, 2023).

But although this seems to have always happened during times of growth and evolution in business and media, it appears to have created an all-new level of human expense and depth today. And this experience seems to have had long-lasting and expansive effects on this population. This is the perfect product cycle – like drug addiction – a "drug-app." Facebook whistleblower Frances Haugen's interview on *60 Minutes* with Scott Peller October 3, 2021, is an example of that feedback cycle.

In the interview Ms. Haugen details the procedures used and recorded in internal documents at Facebook of their process of tracking dopamine cycle to be coordinated with the timing of "like" cycles on Facebook and also noted that these internal documents warned of the effects of Instagram (another Meta product) on young women. Noting that they detailed the actual harm inflicted on these young women by Instagram as being far greater than most other forms on social media (Pelley, 2021).

We use the "drug-app" then we crave the drug. The difference is there is no outside source – it is all internal. Users create the content that generates the internal stimulant that we crave and the "pusher" gets paid by huge corporations to run ads based on the data of user interaction. Then the cycle continues – more need for serotonin – we go back online to get the "hit," the "like," and the collective "we" generates more main ingredients and produces more "drug" and we go back for more. All the while creating the sleep deprivation cycle that causes the need for more "drugs" to make us feel better. Documentation on these effects from the National Institute of Health detail their potency.

According to the National Institute of Health, sleep disturbances are part of the top ten caution signs of suicides and overall mental problems within adolescents. Concluding that using electronic devices interferes with sleep therefore causing damage by shortening sleep hours, and the oversleep quality experienced by these individuals (Pirdehghan et al., 2021).

Patterns such as these create long-term effects in our brains. Repeated processes are the means by which neural pathways are constructed especially in young people. The National Center for Biotechnology Information cites in a recent article the effects of Internet and video-game usage on the brain especially in adolescents and younger children. Some of these effects were adverse neurocognitive changes, with significant reductions in gray matter of the orbitofrontal cortex. These are the areas in the brain dealing with impulse control and decision making. This is very concerning in the early development of children, typically some of the most avid users of video games and overall Internet utilization (Firth, 2019).

There would appear to be some incredible facts here. This could be likened to the "fake news" narrative that is spreading like wildfire throughout our society. The question that needs to be asked here is – who benefits from this narrative? The Internet platforms or the National Institute of Health? Joking – not joking.

If we changed the topic to cigarettes and their effect on health, how many would say this is "fake news"? Would anyone say this was generated to minimize the damage caused by tobacco? Or is it because this device and the platforms hold our "memories," our persona, and our identity that we choose not to believe its potential damage? Is it just too valuable to lose for us to doubt its dangers?

Oddly, while writing this, the current surgeon general released a statement saying this should be viewed as an addiction. Dr. Vivek Murthy wrote, "There are ample indicators that Social Media can also have a profound risk of harm to the mental health and well-being of children and adolescents" (Richtel, Pearson, and Levenson, 2023).

Also, others are comparing it directly to tobacco addiction. "Social media has been described as more addictive than cigarettes and alcohol," Shirley Cramer, chief executive of the Royal Society for Public Health, said in a statement. "It is no longer possible to ignore it when talking about young people's mental health issues" (MacBride, 2017).

We will discuss how to thwart these ill effects and turn it into a benevolent outcome. This will not be easy and will require a global intervention if at all possible. This choice will be explored further in this book but for now let's talk about how we got here in the sense of technology.

1.2.6 We've Got the Technology: Okay What the Hell Do We Do with It Now?

This book will present the arc of occurrences from the start of man's record keeping and on to the historical changes that created a new economy, society, and individuals. The need to communicate to survive has always been at the center of this change even though we have clothed that need with ornamentation. The ornamentation has been necessary to cut through the noise and clutter of multiple messages, to attract the right viewers. Today we clothe this need with fake dances or silly bad magic tricks, fake jumps over cars, or even worse – real jumping over cars. Challenges are set up online that reach across continents. Rivalries are sometimes fake, sometimes fun. Sometimes they are kind of unreal expectations, sometimes they are dead real. Sometimes they lead to radical progressive change, sometimes to lethal permanent stagnation and to a new form of isolationism that has been used before in history.

This isolation inward to technology has been preyed upon by what has been called "The Weaponization of Loneliness." In a recent article in *The Atlantic*, former Secretary of State Hillary Clinton quotes the surgeon general, "Where Dr. Murthy states that this "epidemic of loneliness and isolation threatens Americans' personal health and also the health of our democracy" (Clinton, 2023).

This "epidemic" (as it has been called) has been brought to life through the screen of these handheld companions.

All of this ornamentation, this guise of belonging and acceptance has been to create a following, a community, a group of similar-minded people – an Internet-based society. The form of movement and conditioning can be seen as murmuration. The way birds, sometimes entire flocks, follow a leader without thought or rationale. This murmuration shown in a human form is best seen through social networks. And the award for best name in social media goes to "Twitter" now "X."

1.2.7 IMHO – Best Name Ever

Twitter was the most appropriately named social platform ever – IMHO. The issue is where are they going and is it of their own mind? All social media platforms bear this burden or process and progress. The question is – has this always been the case with all forms of media? Typically, media has been portrayed through a passive environment. You watch but don't generate or contribute. Maybe the ratings or viewing time but not to the direct conversation. Now we – the audience – create the conversation, the content, the direction. Or do we? – we will discuss that further. Obviously, user-generated content has become the basis for this new world of communication. As stated, before it becomes a feedback cycle. The algorithm is controlling the direction of the conversation. But as it becomes much more prominent with the development and integration of advanced computer science – what is the value-added benefit? One such benefit is the speed of search. The intuitive nature of your Google search based on your history. Is this good or is it like "Clippy" from early MS Word days? Does it lead you on a path that minuses exploration and the very lonely self-journey of what if? Will neural pathways become slick with no indents?

At our core the goal has always been survival. Survival has always been the driving force behind the need for progress. How do we survive whether it is career-based or social-based or for the individual – is at our core. This need is now manifested through the means with which we communicate. Yep – the medium has completely become the message. What has been the origin of these tools, how did we get here, how have we used them, what has been the basis that has propelled us in this journey, and where will they bring us next?

1.3 FROM OPPOSABLE THUMBS TO FAKE FINGERS

1.3.1 Lascaux to Gutenberg to ChatGPT

This section is meant as a primer for Chapter 2, which discusses our path from early writing and recordkeeping to the advancements of today's supercomputing AI – ala "ChatGPT" and "Dall-E."

How did we get here, and where will it bring us? Why is it necessary to understand how we got here? Isn't that old news? Well not according to Asa Raskin and Tristan Harris from the Centre for Humane Technologies. Their message seems to be the old adage – "if we ignore history we are destined to repeat it." In their recent YouTube presentation, they both stated that we need to understand the past to best regulate and be forewarned about the effects of AI. We ignored social media as just fun and cat videos and now we see how this has destroyed our youth via anxiety, depression, and ultimately suicide.

Our need to communicate is documented as far back as the cave walls of Lascaux. Was survival at the base of this need? Steven Heller in his book *Graphic Design History* notes that at the base of these cave drawings were broken arrowheads accompanied by chips in the paintings of prey on the cave walls (Heller and Ballance, 2001).

So as we see here, the need and purpose to survive may be at the core of all of these works. Bigger than the need for art in the case of Lascaux is how it worked as design – something that serves a purpose. This purpose being food for survival. But this has always been our core purpose – to simply survive. From accepting this principle, we need to investigate whether or not what we are embracing will actually allow us to survive or merely exist?

But it seems we need to embrace this new lifestyle if we are to maintain our status in the tribe. It is definitely warmer next to the fire. And if you wanted to stay next to the fire and keep warm you better play nice with the tribe or learn to make fire yourself. This has been the impetus for most of our technological advancements over time starting with these early symbol drawings to our next area of concern. From the spiritual to the necessary.

Well where did our need and love of technology start? Basically like all humankind in Africa a long time ago. Upon the discovery of the famous Ishango Bone (approx. 18000–20000 BC) found where we now stand on the border between modern-day Uganda and Congo. There are markings into the bone that appear to be made to work as tally marks. There have been theories that these marks indicate a form of mathematical formations that show the beginning of pattern recognition.

This appears to be one of the earliest relics that depicts recordkeeping. There were more found throughout the region where mankind first developed. The tally marks on these items were used to record possible transactions of some nature. The need to record and store this information has direct links to survival as well. If this was a business transaction tally obviously we were doing it to make sure we received correct compensation or were establishing a marketing strategy. The sequence appeared to also work within the definition of prime numbers.

So was this one of the earliest forms of accounting or did the use of primary numbers lead to creating a system then lead to the development of computer science? Was this also the first algorithm? The desire to solve a problem or to establish a record of an event seems to be at the core of this. There are also beliefs that these markings were meant to record the lunar cycle for possible menstruation recordkeeping (Ishango Bone, n.d.).

Wow! So how did we go from this simple tool for recording events to the systems we use today? In this section we will give a primer to that course of action. Later in Chapter 2 there will be a brief abridged version of that timeline. It will only highlight points from which many things changed, typical watershed moments.

1.3.2 Writing It All Down

One of those watershed moments was obviously the development of the written word however it first came about. The beginnings go back to Lascaux and Ishango but also to the early civilizations and their relics like the petroglyphs, hieroglyphics, to the Sumerians, Phoenicians, Roman, Greek, Charlemagne, and Book of Kells. This path's goal was always to survive, to communicate, and to prosper.

At this point, please be aware that the descriptions and excerpts that follow are meant to just act as a precursor; they will be discussed in greater depth later in the book.

1.3.3 Rock: It All Started Here

1.3.3.1 Petroglyphs

A petroglyph is defined as rock carvings made by breaking through the patina on the rock to expose a different color surface. There are estimates of over 25,000 petroglyph images within a 17-mile boundary point. Over 90% of these images are linked to the Puebloans who lived in the Rio Grande Valley since 500 AD. with a large number of new settlements in later years from 1300 to 1860 AD.

Also according to the National Park Service, petroglyphs, which are widely seen throughout the American Southwest, have a unique meaning. In many cases these symbols went beyond decorative and are believed to reflex the complex customs and cultural belief systems of these past and current regional tribes (Why Were the Petroglyphs Made? – Petroglyph National Monument, U.S. National Park Service, 2016).

1.3.3.2 #Communicating

This next area will give a quick review of the development of the alphabet full versions and portions of which is used today in many countries and cultures. The reason for this is to show the constant goal to connect and to interact between civilizations globally. Today that is the basis of our modern interconnected world. Although this portion may seem of little concern it is the arc from which we jumped into this global connection.

1.3.3.3 Sumerian Alphabet

The early alphabets and mathematical symbols were derived from civilizations like the Sumerians. Recordkeeping was ornamented out of a need to communicate, think of it as their "emojis." Next up chronologically are the Sumerians, however, nothing in these diverse groups is truly sequential; there is always overlap in time due to distance and isolation.

The language and alphabet developed by the Sumerians was then displayed as characters that are listed as the earliest forms of the current Western alphabet. These symbols would later evolve into Cuneiform and later in the 26-character Phoenician alphabet that later again evolved into the current alphabet used primarily in Anglo-Western culture (Kleiner, 2019).

1.3.3.4 Hieroglyphics

A civilization that is heralded as one of the most impressive during the course of recorded time is that of the Egyptians. And from that society came one of the most intricate and expressive forms of communications, hieroglyphics. Given the structure of hieroglyphics, the viewer actually sees the topic of the discussion through the use of illustration. The timeline of hieroglyphics ranges from 3100 BC to the 5th century AD.

Hieroglyphs were originally written on papyrus and later carved into stone on monuments, such as tombs and temple walls. Hieroglyphs consist of over 700 different symbols that represent syllables and sounds (Penn Museum, 2019).

Later more practical forms of hieroglyphics were used. The Egyptians invented hieratic, a cursive form of hieroglyphs that was used for writing with reed brushes, pens on papyri and ostraca. Demotic was a form of writing used at first primarily for formal documents, letters, and tax records. Later Coptic used the Greek alphabet plus a few signs derived from Demotic to form its alphabet.

1.3.3.5 Phoenician Alphabet

Phoenicia lay in the perfect spot, located along the coast of the Eastern Mediterranean. For this reason, Phoenicia was a hub of trade and barter. In order for these diverse groups to trade, a common language was needed (Cartwright, 2016).

As writing became more prevalent and spread throughout the known world, a more unified alphabet was needed for forms of trade and barter, the need to survive was at the core.

So again as one can see, our need for survival was always at the root of our need to communicate. Trade, barter, food, and goods were intertwined. How could we maintain our lives without communicating? As this book goes along, we will see how that need to survive drove our expansion into higher forms of technology by starting with the simplest form – the alphabet.

1.3.3.6 Greek Alphabet and Language

As civilization progressed, diverse groups adopted the restricted alphabet in the hopes of a more communal form of communication. As discussed before, the Phoenician alphabet had a concise, more common format that was accepted and used throughout the region. Again this was for the purpose of common trade and barter. The Semitic alphabet of the Phoenicians was the precursor to the earliest Greek alphabet. Consisting of 22 characters, the Phoenician alphabet had vowel sounds attached to the symbols. This was modified by the Greeks into separate characters for these vowel sounds (Powell, 1996).

1.3.3.7 Latin or Roman Alphabet

The reason for this segue to the Latin/Roman alphabet coincides with an exploration of the more "western" European form of civilization and society. This need to communicate more widely to a diverse group and to an ever-expanding societal structure has now carried across boundaries across mountains, seas, and cultures. Our need to communicate in a common voice is the backbone behind our economies and class structures. As stated before, the focus here is on how our need to communicate created a new economy, society, and individual (Baddeley and Voeste, 2012).

1.3.3.8 Paper: Turning the Page

So as we progress very far through time we enter a period in which our desire to communicate and to record that communication started to join with technology and in particular the technology of print and distribution. In European and Western culture, we attribute a major breakthrough in the world of print and distribution of knowledge to Johannes Gutenberg and the process of movable type and the printing press. However, centuries before Gutenberg's press, Pi Sheng in China had already used movable type pieces to develop a printing system of Chinese lettering.

Pi Sheng (Chinese: 畢昇; 972–1051 AD) is credited with inventing the first form of movable type four centuries before Gutenberg. Being both an alchemist and blacksmith, this form of movable type was created through baked clay and jade.

As many grade schoolers know, Johannes Gutenberg (14th century, Mainz [Germany], February 3, 1468, Mainz), brought movable type to the Western world. His printing press simplified the process and streamlined costs and increased productivity. His first publication was – the Gutenberg Bible.

Gutenberg's press was the first of its kind to use movable type and he is credited with creating the first "information revolution" – the first Incunabula. The use of this new invention quickly spread to multiple regions of both France and Germany. As this new process quickly flourished, the content changed from religious materials to scientific. This led rapidly to the Age of Enlightenment – quite the opposite of today's Digital Incunabula (Palermo, 2014).

The real issue about the invention of the press is its ability to distribute knowledge in an affordable manner. Prior to the printing press, calligraphy was the only means of the transference of knowledge. Scribes created manuscripts through calligraphy and illustrations which took years to finalize and wound up in one final "book." This obviously did not allow for mass distribution. To this day one of the most recognized illuminated manuscripts can only be viewed in one place and through appointments. The Book of Kells is located at Trinity College in Dublin.

The Book of Kells (c. 800 CE) consists of the four gospels of the New Testament. Known for its intricate illuminated manuscripts and its gold leaf pages. Each page was hand-lettered in a uncial format by monks of St. Columba's order of Iona, Scotland. The Book of Kells has an astounding history of having to be hidden from conquering hordes for centuries (Hall, 2022).

Religious materials were always at the forefront of the early days of print and distribution of knowledge. As you can see the Book of Kells was a production of the four gospels. Gutenberg's first and best-known printed book was the Gutenberg Bible. Since the church had financed the work of many artists and artisans, of course, they had great influence over what was produced and distributed. Once the printing process became affordable different opinions were able to be spread, an example is Martin Luther's 95 Theses. This event came to be considered the beginning of the Protestant Reformation (Hanna, 2017).

As civilization progressed the ability to publish logical scientific counterpoints created a great deal of concern and disruption. This disruption was calmed by the focus on classical beauty and serenity of the Renaissance.

Some of the well-known names of the Renaissance are:

> Leonardo da Vinci (1452–1519), Desiderius Erasmus (1466–1536), Rene Descartes (1596–1650), Galileo (1564–1642), Nicolaus Copernicus (1473–1543), Thomas Hobbes (1588–1679), William Shakespeare (1564–1616), and Michelangelo (1475–1564) (Shearer, 2007).

For a full history of this period please refer to *Famous Men of the Renaissance & Reformation* by Robert G. Shearer. This small excerpt is only meant to continue this timeline. Our focus is on the advancement of communication and its involvement with technology.

During this time the works and publications of both Copernicus and Galileo challenged the belief that the earth was the center of all things. Nicolaus Copernicus (1473–1543) was a mathematician, astronomer, and Catholic canon, whose model of the universe placed the Sun rather than Earth at its center. This challenged the belief of the existing authorities and thus led to him being distanced from the church. Later many mathematicians, scientists, and philosophers would use this new theorem to challenge the beliefs that came before. One such person was Galileo Galilei whose most famous contributions to science included astronomy, motion, law of falling bodies, and strength of materials. His discoveries through the use of a telescope led to the Copernican heliocentric system, but his lobbying of that system culminated in a persecution against him (Sanderson, 2010).

This is all related to the fact that as the distribution of information became easier, knowledge spread quicker and to a greater populace that had ever happened before. Individuals were starting to challenge the current political and religious control and its relative hierarchy. As you should be aware, both Copernicus and Galileo were punished in different ways for challenging the powers that be. However, like all bodies in motion their rebuttal created an equal and opposite reaction.

1.3.3.9 Pixels: The Apple Falls from the Tree

During this time one of the greatest scientists and mathematicians was born and had probably the most long-lasting influence on the technology that we use today – Sir Isaac Newton. Newton's discoveries and writings have been called the basis for modern technology and all further mathematics. Newton is known for his invention of many modern practices such as calculus, the study of gravity as a force and later optics (Smith, 2007).

As the power of print and distribution became more commonplace and scientific knowledge spread more easily throughout the civilized world, opinion gave way to fact. Newton's premier work *Principia* was published during this period. *Principia* (1686–1687) is heralded as the foundation for all modern mathematics, physics, calculus, and used to this very day. Newton formulated the theory of universal gravitation around 1665 (Newton, 2016).

Sir Isaac Newton's theories on color – the visible spectrum – influenced the world of communications through his breakdown of the electromagnetic spectrum. In his work *Opticks*, Newton identifies how light works in relation to color. You may question how this relates to technology. Is it the colors used on interactive products? Does it have to do with the design and aesthetics in a UX design? No – the means by which we transfer information is based on this spectrum. Remember lasers carry the information from point A to point B. How they get that information into that beam of light will be discussed later in Chapter 2. This is what Blu-Ray actually means. Red light lasers (light amplification stimulated emission radiation) were first developed in the 1950s by Bell Laboratories.

Albert Einstein praised Newton and his work as the point of origin for all of the work that followed. In the world of communications, the identification of the blue and red light part of the spectrum holds the possibility of the laser and the use of that wavelength to carry information. This later developed into fiber optics.

With such breakthroughs science replaced superstition, proof replaced hearsay and civilization progressed. As stated before, all of this led to the Age of Enlightenment (1685–1815).

1.3.4 STEM All the Way Back Then

This leap forward in individual thinking did not fall to waste. During the period of the 18th–19th centuries, large advancements were made in the world of science, technology, engineering, and mathematics. Yes – they had STEM all the way back then. This period of time and the interest in sciences generated some of the most useful advancements that we still partake in today.

As society strives to go forward our vehicle has always been to speak to each other and with each other and to reach inside that exchange and convey a purpose. We have used whatever means available to communicate and this goes from words to images to sound to technology to possibility.

1.3.5 Photography – Capturing the Light

One such advancement during this time was the further development of the camera obscura. Taken from the ancients, the concept of light drawing an image was first discovered by philosophers such as Aristotle and Lucretius (Herbert, 2021).

Why is this important? Once the modern camera obscura existed the recording of events both formal and informal changed how we experienced life. The advent of photography as a means by which to record everyday life changed how painting and fine art were also determined. Impressionism, it is said, was adopted as a painting style to combat the intrusion of photography as a means of portraiture. From this watershed moment photography was to start a movement toward a much bigger form of communication. The motion picture industry. But before the motion pictures could start the need to showcase imagery with text enhanced the world of communication by illustrating the story in greater form. And today as we know everybody has a tremendous arsenal of images that we increase every day. So how did photography get its start? Through the work of Joseph Niepce in 1827 using a camera obscura method.

> Niepce produced the first lasting record of his work. Using a plate coated with bitumen he recorded an eight-hour exposure… The plate was then washed with a solvent and placed over a box of iodine, producing a plate with light and dark qualities. Niepce named the resulting image a Heliograph.
>
> *(Whitmire, 2016)*

Prior to the use of photography there was printmaking, etching, and lithography. These arts and crafts (printmaking, etching, and lithography) were all done by artisans. Each image is created by hand, drawn by the artist and artisan. These people were trained by the classics. Anatomy was at the core of this discipline.

But now through Niepce's use of light and chemicals – specifically Bitumen – images now could be captured through glass onto a photosensitive material and then reproduced. This process would be progressed through many years of trial and error and by many individuals.

Try to remember all of the imagery you see in an etching in essence has to be done in reverse and backward. In other words, if you want something dark you etch less. If you want something going to the right you etch it going to the left. Everything is reversed and backward.

One of the most well-known and revered artists, printmakers, and etchers was Albrecht Dürer, whose work has long been considered the apex of this artform. Using diverse materials such as woodcuts and copper etched with a sharp stylus, Dürer created stunning works of both human and animal figures enhanced by extensive shading and detail. Please remember this work was achieved in mirror image and all shading was backward – light was dark and dark was light (Wisse, 2002).

Once the black-and-white print process became commonplace, the next move was on to color. Color gave an entirely new depth to the communication process. And embraced Newton's theories of color and how they are perceived by the viewers. This process was known as chromolithography. Chromolithography is basically what it sounds like – chromo – meaning color and lithography a form of mass reproduction of images. Remember prior to this color was only seen in one-off paintings and some woodcuts but now this form of imagery could easily reproduce thousands of copies and used in books (Lambert, 2020).

As printmaking became more popular it was included into the pages with typography to illustrate the meaning and describe the text of the page. By this it is meant that the technical print process overcame the illuminated manuscript. The etcher's plate would be fit into the form next to the typography or on the next page as what was called a "tip-in." Done originally through communication forms as early as hieroglyphics, this became very important in both religious and scientific publications. This evolved from one page at a time from printing press to electrified printing presses that could now produce thousands of pages an hour. Once printing plates could be flexed into cylinders and the more modern print process inclusive of color now the wonders of the world could be brought to life. Now maps, illustrations, experiments, could all be brought to life in real-time color. As photography grew the ability to place etched photographic plates into publications exploded the print, newspaper, magazine, and book industry. Print – especially magazines reigned for decades. Newspapers numbered in the dozens in just one typical large city. Magazines filled the racks of street kiosks. There was a magazine for every interest imaginable.

This would happen again with the development of color television in the 1960s. But before television we need to figure out how to capture motion.

1.3.6 Everything, Everywhere, All at Once – For Sure

One thing to consider is that at this time many things were happening at the same time. Knowledge had spread and was received by many throughout the world. Education – formal education specifically higher education was becoming more commonplace. Knowledge for knowledge's sake now was brought to the masses. Vocational schools such as engineering and technology in general were growing across the country as well. So when the timelines of events seem to overlap, remember there is no direct line from discovery to discovery. But for now let's just look at the accomplishments of the time which were credited to one globally recognized engineer and inventor – Edison.

1.3.7 Lighting Up the World

Heralded as the Wizard of Menlo Park, Edison is credited with many inventions that we still use today. However, after much discussion and discovery there were many other contributors to these events and these breakthroughs. One such man was Nikola Tesla. We will discuss Tesla's involvement in the spread of communications further through Chapter 2. But first, what was the relationship between Edison and photography and who was Thomas Edison?

> Although Thomas Alva Edison [b. 1847] was the most famous American of his time, and remains an international name today, he is mostly remembered only for the gift of universal electric light. In all, this near-deaf genius… patented 1,093 inventions, ….
>
> *(Edison et al. 1998)*

1.3.8 Every Day Photographer

As photography had become part of everyday life many inventors, scientists, and artists were working toward recording civilizations' accomplishments in the most realistic fashion. One of these men was George Eastman. Eastman is responsible for developing a camera for the "everyman." A camera that reduces all the steps toward getting the image into your hands. Eastman focused on inventing a simple camera with enclosed film that could be sent to a film-processing facility and then prints were sent back to the "photographer." Eastman labored at slimming down the photographic process to where anyone could take photographs.

> With a series of innovations, the [Eastman Kodak] company created easy-to-use cameras that made photography widely accessible, established the practice of professional photofinishing, and developed a flexible film that was a critical contribution to the launch of the motion picture industry.
>
> *(About George Eastman, 2019)*

Remember the camera obscura existed in its earliest form as far back as 400 BC. The concept of capturing light into a box and displaying that image was as simple as a pinhole and a wall to display what light and image came in through that pinhole. How that simple concept changed the world is one of the bases of communication and influence to this very day. With Eastman's invention now literally every person could have that control. How we used it fell to our basic ethics and principles. Today due to manipulation techniques common photographs are not allowed as evidence in court cases. This one simple device has recorded our greatest accomplishments and our biggest failures (Lowe, 2019).

Realizing that the photographic process prior to Eastman required the individual to have experience in how to use the chemicals, specific materials, equipment, a very long process, and an extensive set of skills brought about by trial and error, Eastman's invention opened up the world of photography.

From Niepce to Daguerre the process was time-consuming and expensive. Eastman's invention brought photography to the home front. Many of the world's population came to embrace this simple invention known as the Kodak Brownie.

> <u>Kodak #1 camera</u>
>
> By far the most significant event in the history of amateur photography was the introduction of the Kodak #1 camera in 1888. Invented and marketed by George Eastman (1854–1932), ...
>
> To underscore the ease of the Kodak system, Eastman launched an advertising campaign featuring women and children operating the camera, and coined the memorable slogan: "You press the button, we do the rest."
>
> *(Fineman, 2004)*

From advancements like this – bringing the camera into the hands of everyone, Eastman Kodak moved photography from analog to digital. It took almost a century for this to come to fruition but remember computers didn't even exist until the 1950s. So none of these advancements could be used to develop digital cameras until the 1970s and only brought to fruition in the 1990s and then of course into your phone in the 2000s. The first actual digital camera was first developed by Steven Sasson, an Eastman Kodak engineer in 1975 (Trenholm, 2021).

1.3.9 The Beginning of Motion Pictures

In the prior section the term "roll film" was used. This seems like a simple event but in reality "roll film" or flexible celluloid gave the premise for the ability to create motion pictures. With the advancement of flexible photographic film and the light bulb, now these two advancements would come together to create a new form of recording platform. But it would take one person who many considered not a very stable individual to see how this could be used in further discovery and exploration. Seems it does take the "crazy" ones to take the big step. Enter Eadweard Muybridge – you ain't seen crazy until you met the real Crazy Eddie.

> Muybridge also used his camera to explore the potential of motion pictures. He devised techniques to freeze animal and human locomotion, to depict movement as sequences of still images, and to reanimate these in some of the first projected moving pictures.
>
> *(CA, n.d.)*

This quickly mentioned "exploration" of motion pictures does not adequately describe the effect of Muybridge's work on the same effect that was first noticed by Lucretius – the persistence of vision. The beginnings of motion pictures all hinged on a simple bet between a few extremely wealthy individuals. Leland Stanford, the railroad magnate and founder of Stanford University, was the impetus behind the bet that all of a race horse's hooves leave the ground at one time.

Using the "Persistence of Vision," the theory that the human eye interprets motion once an image is seen more than 12 frames a second, Muybridge setup 256 cameras to capture frame by frame the gallop of that race horse to understand if the horses' hooves left the ground at any one point in the gallop. This simple "bet" between two railroad magnates, Stanford and a friend created the entire motion picture industry. Later these images were placed in sequence on a wheel where light was shone through one point as the images twirled in front of the light. Since the images moved quicker than 12 frames a second, there appeared to be motion. There will be more about the growth of motion pictures later in Chapter 2. But for now we will premise the next steps as how Edison (being Edison) completely attributed this to his own personal efforts. The next section explains the growth of entertainment inventions from practical commercial applications to daily personal use (Solnit, 2010).

1.3.10 Tesla and Marconi

Harkening back to Newton's discovery of electromagnetism, scientists became involved in developing devices and uses for these new forms of energy. One of the names that you hear about is Nikola Tesla. Tesla is credited with close to 2,000 patents globally. Tesla is also credited with the discovery and harnessing of wireless electricity and its use to transmit information. Tesla's contribution to Edison's discoveries will be discussed later on in this book. But here is just a quick bit of information on the genius of Nikola Tesla and how this contributed to the development of wireless communications.

> At the height of his career, the pioneering electrical engineer Nikola Tesla became obsessed with an idea. He theorized that electricity could be transmitted wirelessly through the air at long distances – either via a series of strategically positioned towers, or hopping across a system of suspended balloons.
>
> *(Gaura and Brusey, 2021)*

1.3.11 Farnsworth and Sarnoff

During these days of media growth and experimentation, the debatable greatest force in media was to be born – television. As with many of these innovations no one person was the origin point however one such person seems to have brought many forces together at one point in history. Just like an origin story of many of today's popular comic book novels, this story starts in rural America with a simple young farm boy. Enter Philo T. Farnsworth, a 14-year-old Utah farmer's son. At this time Farnsworth was plowing a field of potatoes and noticed how the leaves while in straight rows formed images. From this simple vision Farnsworth deduced the concept of capturing and transmitting an image into pixels (later used as the shortened term for picture images).

As goes further in his story, Farnsworth fought many legal battles with RCA (NBC) and at first won the right to his patent for electronic television but RCA drew out the battle to allow for Farnsworth's patents to expire which was helped by the start of World War II. By the time the war was over Farnsworth's funds and his many patents had expired and he had to concede to RCA (Stashower, 2002).

1.3.12 Enter the Computer Scientists

As much as the artists and image makers provided growth during the 1600s to the mid-1900s, the scientists built the backbone of most of the changes we feel today.

Mathematics has always been at the basis of computer science. From the earliest use of the Ishango Bone to the common abacus dated back to 3000 BC to today's sophisticated algorithms, the principles of math, algebra, calculus, and trigonometry have driven the advancements we enjoy today.

Some very sophisticated analog computers were developed by the Greeks as early as 80 BC. On board an ancient Greek shipwreck discovered in 1901 off the island of Antikythera was a device from that period which consisted of rusted metal pointers and gears now called the "Antikythera mechanism." Once this object was rebuilt it could be used as a mechanism for predicting the motions of the planets and stars. Similar to the previously discussed Ishango Bone, this device was used to predict information from previous knowledge. In essence this is the basis for computer science (Freeth, 2022).

During the Renaissance and into the Age of Enlightenment, the growth of the study and teaching of mathematics and science encouraged our use of machines to aid in our problem-solving needs. Using the work of Newton and others the advancements toward implementing these theories into actual practical objects took hold. Individuals such as Napier, Pascal, Schickhard, Babbage, Lovelace, Jevons, and Hollerith and many others, engineered machines to employ these mathematical theorems into tools such as "counting machines" and the first analog computers.

Ada Augusta Byron, Countess of Lovelace (1815–1852), was the first woman involved in this new "science." Being the only legitimate daughter of Lord Byron, Lovelace had the gravitas to move Babbage's theory forward into a working, practical device (Lew, 2018).

The section "Enter the Computer Scientists" will be brief due to the extensive information needed to properly describe the advancements in the area of computer science. That sounds like an oxymoron but computer science is too big a component concerning the advancement of modern-day communication to not address it in a full chapter. In Chapter 2, we only discuss the lineage of computers and their use in media.

1.3.13 Alan Turing – Beginning Modern Computing

Very few people know the name Alan Turing. In the movie *Jobs* (2015), Steve Jobs (Michael Fassbinder) is being interviewed by a journalist. As they walk through the Moscone Center, posters for the new Apple "Crazy Ones" marketing campaign are hanging about the hallways. They come upon a poster with Alan Turing's face on it. The journalist asks, "Who is that?" Jobs replied, "If you have to ask I am taking him out of the campaign." A true shame.

Well who was Alan Turing? Obviously very few people are familiar with the name. Turing is credited as single-handedly starting the computer revolution in 1939. This fact cannot be overstated. This one person was the catalyst for the advancement of computers taking their place in our modern times.

Alan Turing was a mathematician born in London, England, in 1912. Always fascinated by mathematics and cryptography, Turning would later be utilized by the British government during World War II to aid in breaking the German enigma code. During this time Turing developed the first electronic computer – which he called "Christopher" in honor of a boyhood friend. His later work was to develop the "Imitation Game," which would ultimately determine if something was a computer or a person. This would later be called the "Turing Test" (Hodges, 2013).

These accomplishments and their subsequent events are recognized as the start of the computer revolution. Even after all of the aid and innovation he developed, he was shunned by the same community he helped to save. After his service to his country and the world, Turing was arrested and found guilty of homosexual activities which were outlawed in England at the time. He was given a choice of prison or chemical castration. He chose castration, since he believed this would allow him to work on his computers. Over time he discovered that the chemicals were causing issues with his brain and left him unable to do his work. The story goes that he injected an apple with cyanide and took a bite, subsequently committing suicide (Cowell, 2019).

There are stories that say this is the reason for the first MacIntosh logo – a multicolored apple with a bite missing. Remember Apple was founded in the San Francisco region of California and even in the 1970s (especially) this was home to one of the largest LBGTQ communities in the United States, and Jobs being Jobs who knows? Even the original logo designer disclaims the story – stating, "I'm afraid it didn't have a thing to do with it," he said. "It's a wonderful urban legend" (by Holden Frith, Special to CNN, 2011).

1.3.14 National Science Foundation Is Born

The purpose of the U.S. National Science Foundation is to act as an independent federal agency that supports science and engineering throughout all 50 states and U.S. territories. It was started after World War II when science was recognized for its ultimate power via the atomic bomb and its destruction of Hiroshima and Nagasaki causing the end of the Japanese theater of operations. The U.S. president at the time was Harry Truman, becoming president after the death of Franklin Delano Roosevelt, and it was through his administration that the NSF and through them ARPA (the Advanced Research Projects Agency) was created (Science Foundation, 1994).

One of the earliest and also ongoing developments to occur at ARPA was the Internet which in its earliest form looked nothing as we know it today. The concept of interconnected computers has been around since the 1950s. It was this need for communal knowledge and the decentralization of this information through the network of computers that in essence created the Internet. But it was also born out of fear. ARPANet was built out of the government agency ARPA in response to the fear of the Soviet Union's launch of Sputnik (Weinberger, 2018).

Sputnik was a basketball-sized satellite that was launched on October 4, 1957, by the Soviet Union and struck such fear into the heart of the U.S. government that from that day forward there was only one purpose, to advance the state of U.S. space exploration and the overall future of technology in the country.

When Sputnik was launched it was viewed as the beginning of potential destructive weapons in space. From here with further advancements the Soviet Union could deploy weapons to conquer the United States. To combat that possibility a new "network of computers" would be built. This was called ARPANet and is the precursor to the Internet we know today (Leiner, Kahn, and Postel, 2003).

This structure of interconnected networks gave way to the term "Internet" and yes it all started in the late 1950s to the early 1960s. As we all know the Internet is the backbone that connects all of the computers throughout the world. And the World Wide Web is merely the device that allows us to use this network. We will discuss the advent of both of these innovations and the people who did the work so the rest of humankind could benefit.

1.3.15 Personal Computing

The steps from Turing's original machine to today's computers are a long road. It traverses through room-sized computers that run electronically with gears to the early days of the digital structure that relied upon transistors to the use of the printed circuit. All of those steps will be covered in more detail in Chapter 2 but for now, we will jump to the origins of the personal computer.

For years the goal of a simple stand-alone computer eluded the world of engineering. That is until one company known for duplicating paper got very scared and decided to invest in the future of information storage. If you have even heard the term "Xerox," you would have to be at least 50 years old. The name is not synonymous with computing let alone the personal computer but that is exactly where the personal computer we all use today started. In what has become the epicenter of technology Silicon Valley – Palo Alto. Fear is a great motivator.

> [Xerox] PARC, formerly (1970–2002) Xerox PARC, research company established in 1970 ... in Palo Alto, California, U.S., ... Many innovations in computer design were developed by PARC researchers, including the Alto, the first personal computer; the graphical user interface; the laser printer; and Ethernet, a ubiquitous computer networking technology.
>
> *(Dennis, 2023)*

1.3.15.1 The PARC Team

Here is a short list of all the people and what they invented while working at PARC. And by short I mean very short where hundreds of people were all working toward one goal, to bring about a new digital age: Nina Amenta, Anne Balsamo, John Seely Brown, Bill English, Charles Geschke, Adele Goldberg, Rich Gossweiler, Rebecca Grinter, Ted Kaehler, Alan Kay, Martin Kay, Robert Metcalfe, Eric Schmidt, Larry Tesler, John Warnock, etc.

These people were at the forefront of what we use today. They invented the graphical user interface, documents, folders, the mouse, touch-screen, ethernet, networking in general, laser printing, computer graphics software, artificial intelligence (ChatGPT), and just about everything else you come in contact with on your phone or desktop every day. Oh yeah and co-founded a little company called Adobe. But yet you never think of Xerox and computers together. Why is that? (Dennis, 2023).

1.4 50+ YEARS LATER IT IS STILL ALTO'S WORLD

In 1969 the team at PARC developed the first desktop computer – the Alto. The Alto changed personal computing. And the changes stuck. The Alto was the forerunner of all of our current computer systems as was most of what was invented at PARC. Nothing we have or use today would be available if not for this team and the main product. From the menus and commands to the hard drive to the use of documents to e-mail to the network it all runs on. This overlooked and barely known group of scientists and artists created the brave new world we live in today. Two names come to the forefront but are hardly known – Adele Goldberg and Alan Kay.

Adele Goldberg was one of the first women at this level of management and supervision in computer science and development. Goldberg was responsible for the development of the graphical user interface that we use today – menus, documents, etc. And as the story goes, Jobs and Gates blatantly stole the concept and the engineering but not until Goldberg issued a memo that stated she was completely against giving this away.

Goldberg is a woman computer scientist who has such credentials, that once Jobs and Gates witnessed what her team had developed they would change direction. It needs to be noted that Rear Admiral Grace Hopper and Katherine Johnson were also leading computer scientists prior to Adele Goldberg.

PARC's accomplishments and innovations are too numerous to list and like all good things has now entered its last phase and will be closed in 2023. More about these individuals and PARC in general in Chapter 2.

1.4.1 Pirates of Southern California

With all of this growth occurring many companies were getting started at the same time. Individuals jumping ship and going to competitors was a normal occurrence. Growth was robust and this space was truly on fire. This being said, it is very common to see the same person involved in the breakthroughs at numerous companies. In selecting a location for PARC, Xerox noted that the Northeast was notorious for ship jumping and how this would affect the privacy and security of the sensitive work they would be doing.

Along with ship jumping came concept stealing. Really no other way to put it. Goldberg and her team were one of the first victims of such actions. And who were the assailants? None other than Jobs and Gates. A great film on this is *Pirates of Silicon Valley*. It pretty much details how Apple and Microsoft went from the black-screened, green-lettered, keystroke environments to Goldberg's graphical user interface. This book will go into this in greater depth later but if you wish to understand it better see the film for a more in-depth explanation.

1.4.2 The Individual Takes the Center Stage

The birth of the Internet occurred in stages. Not unlike a long drawn-out pregnancy – there was pain and then there was joy. And now there is pain again – the adolescent years appear to be upon us. As discussed before, the commercial Internet was generated from the military DARPANet (Defense Advanced Research Project Network) which became ARPANet – (Advanced Research Project Network). Its goal was to make sure all of the information stored on computers would survive a cold war missile attack. This again stemmed from the launch of Sputnik.

This concern led to the creation of a network of interconnected computers across the East Coast first, Washington, DC, Boston, New York, etc. then it spread across the country west to the universities along the areas in California. The problem this conquered was with having one straight line of connectivity that the signal would drop and also it made it vulnerable to an enemy strike. By creating a network of computers in diverse regions the information could be stored and backed up at several locations. In essence a web of computer networks that allowed for the redundancy of information. In simpler terms the beginning of our modern-day commercial Internet.

When we say the "Individual Takes the Center Stage" what does that mean? Well simply put, no longer was a team needed for every interaction online. Prior to the commercial Internet and the World Wide Web, in order to access information and other users there had to be a series of steps played out. Time had to be reserved, machines took up entire rooms and they had to be monitored by teams, terminals were kept outside these rooms because in order to run these large computers temperatures had to be kept low. This was an entire process. Once systems like the Alto, Ethernet, TCP-IP, IMPs, SmallTalk, and finally the World Wide Web were developed, the ability to command this amazing amount of power was now relegated to a single keyboard and a connection.

Today we access information in our car while driving to get that latte and snap pictures to post globally before you even pay for the coffee. It is almost absurd to overlook such an accomplishment. And these "user-generated" comments and content is what has driven the web to become a global revenue generator that it is today – a platform that sways economies, society, politics, and individuals. As we go through this book, we will investigate the effects of this new platform in how it relates to the individual's lifestyle, personal interactions, and more importantly mental health. But for now, when did data become worthwhile?

1.4.3 Back to the Back Up: Storage Systems

In contrast to the web's open-access free-sharing mentality, prior methods of storing data were the typical archival systems. Typical of this was magnetic cores. These "cores" were used to store the data and the recorded materials. These cores were very large in size and required high maintenance and strong energy levels. What were these magnetic cores?

> Tiny donuts made of magnetic material strung on wires into an array: the idea revolutionized computer memory. Each donut was a bit, magnetized one way for "zero," and the other way for "one." The wires could both detect and change the magnetization.
>
> *("Magnetic Core Memory – CHM Revolution," n.d.)*

When we speak about magnetic cores or any form of magnetism – which actually does run our world from the core of our Earth to the core of these machines, please remember harnessing electromagnetism was recognized by one Sir Isaac Newton. But what does "dumping the core" mean?

Turning off the energy supply or "dumping the core" was a tragic event. Imagine having to delete the contents of your laptop or phone whenever an issue arose. And to do this with no form of backup. Yeah – it was pretty tragic. These were the steps that had to be overcome. But later this magnetic method and through the use of laser technologies storage processes gave way to the archival disc.

A common solution to this was storing information to be accessed when needed. Say hello to my little friend – the Compact Disc – Read Only Memory – or CD-ROM. A thin sheet of plastic with bits of information embedded into it. This product allowed for greater amounts of recorded content. Now a typical user could actually "burn" a CD to back-up their entire computer. You could now store and retrieve data at speeds and rates never before seen. The CD-ROM format gave way to the DVD – Digital Versatile Disc and the Blu-Ray. Later we will discuss the arc of these discoveries and developments. Archival methods kept it all to ourselves – now we speak to the world – daily.

1.4.4 Commercial Internet

The beginnings of the commercial Internet had its start many years before what we believe. Most people consider the World Wide Web as the actual start of the commercial Internet. But even though Sir Tim Berners Lee developed the first browser, the "World Wide Web" in the early 1990s the Internet came to the common user a full decade before.

> In 1983, when Barry Leiner took over management of the Internet research program at DARPA, he and Clark recognized that the continuing growth of the Internet community demanded a restructuring of the coordination mechanisms.
>
> *(Internet Society, 1997)*

Once the Internet was opened to commercial use, how to use it properly became the question. And to answer that question came the people who specialize in questions – the academics.

1.4.5 The Start of WWW

It is almost impossible to deny the effect the development and overall use of the World Wide Web has had on the economy, society, and individuals. Its origins were in academia and was used in conjunction with HTML (HyperText Markup Language) which was developed by Ted Nelson in the late 1960s. To state again clearly – the web is a portal that works upon the backbone of the Internet structure. It doesn't work separately. No Internet – no web, simple. Originally the web only carried text but Nelson had originated the <img> tag which would later be used by Marc Andreessen in his doctoral work "Mosaic" to display images. But first who is heralded as the father of the World Wide Web – none other than again an academic – Sir Tim Berners-Lee who worked at CERN (The European Organization for Nuclear Research).

Berners-Lee (a senior researcher and currently the 3Com founder's chair at the MIT Computer Science and Artificial Intelligence Laboratory) distributed a proposal for what he called the World Wide Web browser in March 1989. He then set up the first server and browser to run this new protocol utilizing Steve Jobs' NeXT computer running UNIX and what was later to become Apple's new OS.

> To prevent it being accidentally switched off, the computer had a hand-written label in red ink: "This machine is a server. DO NOT POWER IT DOWN!!"
>
> *(CERN, 2019)*

This is being stated quickly and simply at this point but we will go deeper into the web's beginnings and growth later in this book. For now as an overview, we need to jump ahead to what the end game has been for this new platform of communication. And that endgame is the user getting control. This comes about from Content Management Systems whether WordPress-like or social media. Later we will discuss the further development of the web but for continuity of the subject at hand we will focus on social media.

1.4.6 Friend(sters) for Life: Social Media Platforms

It is tough to overstate the effects of social media on the economy, society, and individuals. Today political candidates are elected to the highest office in the land by having a strong, effective, and persuasive social media presence and overall strategy. Businesses would never be able to compete without a following nor a consumer revenue stream. Individuals feel lost and unattached when they do not get the proper response to their postings. This one segment has grown from the web, to drive our lives. But is it all worthwhile? And how did that come to be?

The democratization of the web occurred when the first non-academic, commercial websites were launched back in the early 1990s. Now that commercial use was becoming common every corporation was starting to feel the need to have their own presence on the web. And they were making sure their message was reflective of their mission and was being seen and heard by the right individuals. Originally starting as a vehicle to allow for discussions via like-minded individuals, it has now taken that premise to its furthest logical (maybe illogical) conclusion with the user controlling the content (supposedly).

1.4.7 Message Boards

Message boards were the mainstay of Internet communications. Simple text messages going back and forth between users – mainly academics who wanted to share their research. These structures were primitive and basically required no acumen besides – type – send. Message boards have been around since the start of the Internet.

In today's world the best example of an Internet forum is Reddit. An open forum of discussion for a wide range of topics. However, back at the start the content was mostly true and not ladened with rumor and conjecture.

Once message boards and forums became the norm the user base quickly adapted to the beginnings of social media. In their earliest forms they were nothing more than really a dressed-up message board. Remember now the <img src> tag was being used frequently. And soon video was introduced as well as background patterns and graphics. One of the first popular social media sites was Friendster. Yep – Friendster – but there was one more that started the trend but never became as big and strong – Six Degrees. Here is a quick chronology of the beginnings of social media starting in the late 1990s.

- May 1997: Six Degrees
- March 2002: Friendster
- August 2003: MySpace
- February 2004: Facebook
- July 2006: Twitter [now X]
- July 2010: Instagram
- July 2014: TikTok

Listed above are the big players in the social media space. As we go further, we will discuss their effects on our three areas of concern – economy, society, and the individual. But for now let's just realize the power they are exerting over our main lifestyle needs and choices (Shah, 2016).

1.4.8 User-Generated Content – Double-Edged Sword

Once social media became ingrained in the daily life of the user the next move was to increase the traffic on these social media platforms in order to monetize the experience. But the question was how to do this? How many individuals would these companies need to generate the viewership to validate the price of ad revenue and overall data mining? What engine could be developed to create the never-ending content needed to get the return visits? Enter the world of user-generated content. The never-ending source of new, fresh, relevant content that would then be weaponized to keep the user coming back and to engage 24/7/365.

1.4.9 Making the "Likes" Liked

Now that the pieces are in place, how to get them coming back day after day? Hence the scientist in computer science. As in all equations the first thing is to understand the goal of the equation. What did we want to accomplish? What model do we follow that has delivered the desired result? This goes all the way back to the Ishango Bone. Purpose first. Here is where the science of psychology meets the science of computers and algorithms. What pieces need to go together to make this work? The goal appears to be conditioning – getting the user to choose the desired response. Who did this before and what were the steps?

1.4.10 Enter B.F. Skinner

As a leading psychologist Burrhus Frederic Skinner's work focused on human behavior. His most well-known contribution is the Skinner Box. A psychologist, Skinner studied human behavior at Harvard while earning his doctorate. His works include *The Behavior of Organisms* (1938) and a novel based on his theories *Walden Two* (1948) (Harvard University, 2007).

Skinner is well known for his development of the Skinner Box, a small chamber that is used typically on animals to observe operant conditioning research. It is customizable to conduct different outcomes by simply moving the partitions in the "maze" to change the route to obtain the goal – water or food supply.

In essence, the screen of your device and the social media platform has become the world's Skinner Box. Where did they hide my cheese? How do they get you to "like" and to see what these companies have determined will generate the most viewership? This is what the algorithm is for – what is an algorithm? Well according to the website "How Stuff Works" – "The algorithm is the basic technique used to get the job done."

Basically, an algorithm is part of a computer program that tells the computer the steps necessary to achieve its goal. These algorithms can be customized to do exactly what the programmer wishes the user to do to achieve its goal. Here is just one of a series of simple step by step goal-oriented algorithms written out in plain English.

Let's say that you have a friend arriving at the airport, and your friend needs to get from the airport to your house. Here are four different algorithms that you might give your friend for getting to your home:

> The taxi algorithm:
>
> 1. Go to the taxi stand.
> 2. Get in a taxi.
> 3. Give the driver my address.
>
> *(What Is a Computer Algorithm?, 2001)*

Now change the airport scenario to desire for you to go back to the same page on a social media site. How do they get you there and keep you there? How do they keep you "looking for your cheese."

How many people think they have free choice in dealing with this environment? Well let's explore some facts from the people who actually worked on these sites. One such person is Frances Haugen. In her many interviews since turning whistleblower on Facebook, Haugen details how these "Facebook algorithms" have gone from engineer written to machine learning algorithms. Here is an excerpt from one of those interviews.

> ...because of Facebook's enormous amounts of user data, it can develop models that learned to infer the existence not only of broad categories like "women" and "men," but of very fine-grained categories like "women between 25 and 34 who liked Facebook pages related to yoga," and [target] ads to them.
>
> *(Hao, 2021)*

Now that you can see the spaghetti thrown against the wall, this should help you realize that psychology is behind these choices. And as in all good marketing the need for your attention is prominent. This is built upon something called operant conditioning. Let's go back to good old B.F. to decipher this B.S. to really see the desired outcome here.

> What are the four types of operant conditioning according to Skinner?
>
> - Positive reinforcers: the addition of a reward. – "likes"
> - Negative reinforcers: the removal of a punishment. – "dislikes"
> - Positive punishers: the addition of a punishment. "harsh comments after a post"
> - Negative punishers: the removal of a reward. – "unfollow"
>
> *– (Harvard University, 2007)*

Should this be allowed? Is this a form of addiction? Is this similar to putting something in a food or other form of consumable that causes you to keep consuming? You know, just like cigarettes. Finally, Congress is looking at regulations and hopefully soon there will be a firm set of regulations on the use of social media and the requirements for these new tech behemoths.

In May 2023 in an article in the *New York Times*, the Surgeon General Dr. Vivek Murthy issued a statement.

> There are ample indicators that Social Media can also have a profound risk of harm to the mental health and well-being of children and adolescents.
>
> *(Richtel et al., 2023)*

A portion of this is common sense and has been heard before. There needs to be a digital diet of sorts. Obviously as always, adolescents need structure and a set of rules to follow for all forms of consumption. In some opinions, this is not the fault of the parents nor the children. The reasoning backing this belief will be covered in further detail in Chapter 3 along with what has caused this to happen.

1.4.11 Protecting the Herd

In regard to the defense of these younger users, the issue of Section 230 has come to light. Section 230 is a law Congress passed in 1996 during the very early days of the online community world. It was originally part of the Communications Decency Act. It simply states:

> No provider or user of an interactive computer service shall be treated as the publisher or speaker of any information provided by another information content provider.
>
> *(47 U.S.C. § 230(c)(1))*

One viewpoint is that this law embodies free speech and the First Amendment.

> The First Amendment provides that Congress make no law respecting an establishment of religion or prohibiting its free exercise. It protects freedom of speech, the press, assembly, and the right to petition the Government for a redress of grievances.
>
> *(The White House, n.d.)*

And some people take it further and believe that the beliefs and principles in Section 230 are at the core of our society.

> Section 230 embodies that principle that we should all be responsible for our own actions and statements online, but generally not those of others. The law prevents most civil suits against users or services that are based on what others say.
>
> *(Electronic Frontier Foundation, 2023)*

But what happens when these services are used unscrupulously and they inflict harm on unsuspecting individuals? Does this occur when we bring the issue of purposeful algorithms into account? Does this go back to the cigarette metaphor? If you are free to smoke cigarettes but you are not aware of the ingredients and their directed purpose but the manufacturer is – does this indicate purposeful malice?

Well two parents along with others believe that it does cross the line from free speech to purposeful harm once their own daughter starts going down rabbit holes leading her to anorexia.

> ...the number of families pursuing lawsuits has grown to over 2,000. More than 350 lawsuits are expected to move forward this year against TikTok, Snapchat, YouTube, Roblox and Meta – the parent company to Instagram and Facebook.
>
> *(Alfonsi, 2023)*

And if the effect on the self-esteem of this young age group and mental health wasn't enough to be concerned with, here is more news. Instagram – a part of Meta Platforms (Facebook parent company) – has been found via investigative journalism pressure that its platform is being used as a "buyer"–"seller" pedophilia ring.

> Instagram's popularity and user-friendly interface make it a preferred option for these activities.
>
> *(Cuthbertson, 2023b)*

This issue of the purposeful harm of social media is still very much up for debate given the benefits of online communities overall. At no point should this be seen as a one-sided, doom scroll but we need to be aware that what we have invited into our lives has as much an adverse effect as the benefits of worldwide ease of communication and commerce.

However, understanding that social media kind of slipped through our gateways is a cautionary tale and now that the new, new, thing – Artificial Intelligence is knocking on our door that cautionary tale of the lack of concern for social media's abusive potential is being paid great attention.

But what is Artificial Intelligence? A good friend, Prof. John Meschi asked, "Is Artificial Intelligence the opposite of Natural Stupidity?" Possibly – great point!

1.4.12 Artificial Intelligence – Defining AI

This last section of this chapter is dedicated to the newest of the new – Artificial Intelligence. Talk of Artificial Intelligence is flying around more than strands of cotton candy at a school fair. This is definitely a new time and a new world. But what is it? What does it all mean? A direct definition of Artificial Intelligence or AI's scope is intelligent machines that teach and learn.

> Artificial Intelligence (AI), a term coined by emeritus Stanford Professor John McCarthy in 1955, was defined by him as "the science and engineering of making intelligent machines."
>
> *(Manning, 2020)*

1.4.13 Fear and Loathing in Silicon Valley and of AI

With anything new and having this much power and capability there are usually many caveats and praise. The praise is obvious. Machines that actually do the dangerous and laborious work for humankind. This is a good thing but many times a workforce – even an artificial one has been known to go on strike. To make demands, renegotiate their position and benefits package. You may laugh at that but others aren't laughing so loud.

Many AI experts are concerned about the possibility of our digital overlords turning the tables on us.

> "Today's systems are not anywhere close to posing an existential risk," said Yoshua Bengio, a professor and A.I. researcher at the University of Montreal. "But in one, two, five years? There is too much uncertainty… We are not sure this won't pass some point where things get catastrophic."
>
> *(Metz, 2023)*

Is this the "Revenge of Clippy" the Microsoft AI helper from 25 years ago? We haven't even touched yet upon the benefits and detriments of AI and its use in software such as ChatGPT. That will be reviewed in the chapter on cognitive dissonance.

The notion of "Deep Fakes" and loss of belief systems in our political systems are on the verge of becoming a common day occurrence. How will we evolve to accept these new events and really understand the truth? Will truth become a passing fancy like chivalry, travel agents, and gender roles?

1.4.14 Benefits of AI

In contrast to the fear of AI there have been tremendous advancements in the area of AI. One specifically is in the medical world where the act of parsing important data can be conquered in a 1000th the time and be uploaded and shared with numerous healthcare teams and services (University, 2023).

As we keep discussing the benefits of AI, we will go wider into a broader realm of disciplines such as:

- Art History
- Design
- Environmental Studies
- Government
- History and Social Sciences
- Language Arts
- Science, Technology, and Society
- Studio Art and Computer Science

1.4.15 Combining Hardware and Software

The benefits of AI are amazing especially when it is coupled with some of the massive new hardware/software combinations like augmented reality (AR) and virtual reality (VR). By now most of us are familiar with virtual reality. We see it in gaming, medical, and educational platforms. It comes to use through a headset environment connected to a computer terminal. But what is AR (augmented reality) and how is it being used?

Let's clearly define VR and AR.

> Virtual Reality (VR) is the technology that provides almost real and/or believable experiences in a synthetic or virtual way. Augmented Reality (AR) can be thought of as a variation of VR…. the goal of AR is to augment the real world with synthetic information such as visualizations and audio.
>
> *(Watts, 2023)*

Dr. Ivan Sutherland is one name that needs to be mentioned at this point for his extensive contributions to the field of computer graphics (Sterling, 2009). There will be more info on all of the individuals that built the software that designers use today later in Chapter 3. For now let's jump ahead to discuss how that software and hardware is being used to create a new reality.

1.4.16 The Birth of the Affordable VR Headset – Oculus

One of the first major players in the VR headset game was a company called (ModRetro) – OculusVR. Started as a form of side hustle where the intention was originally not mass market appeal. Here is a brief history of its origin and skyrocket rise to legitimacy. From a kickstarter campaign to a million dollars in funding in three days.

> On August 1st of 2012, Oculus launched their Kickstarter campaign… Within 24 hours, they'd raised $670,000 from 2,750 people. Within three days, they'd broken a million dollars. Yesterday, [March 25, 2014] they were acquired by Facebook for $2 billion.
>
> *(Kumparak, 2014)*

1.4.17 Apple Takes a Major Step

On June 5, 2023, Apple unveiled its new take on AR – Apple Vision Pro. This new product from Apple combines AR with Spatial Computing via a "snow goggle" looking headset that allows for hand control of the interface – without a controller. Here is a part of Apple's formal press release.

> Through a tight integration of hardware and software, we designed a standalone spatial computer in a compact wearable form factor that is the most advanced personal electronics device ever.
>
> *(Apple Inc., 2023)*

The price tag is typical for Apple – huge and really "insanely" huge. $3,499.00 USD. This may not be a stocking stuffer but as always with Apple how will it change the marketplace?

1.4.18 Immersive Lives

Once these new immersive technologies enter the everyday life of most users the question becomes how will it be monetized? Here is a quick list of the possible uses for VR and AR in practical terms.

- Advertising
- Architecture
- Art and design
- Automotive industry
- Charity
- Corporate Learning and development
- Education
- Entertainment
- Events, conferences, and meetings
- Fitness
- Gambling
- Healthcare
- Interior design
- Law enforcement
- Marketing
- News and journalism
- Real estate
- Recreation
- Recruitment

- Retail
- Social
- Sports
- Tourism
- Well-being

Even more than conforming to prior markets the biggest step may be going forward to new worlds and to get there we need to go backward to the concept of a new life – a SecondLife.

1.4.19 SecondLife (or the MetaVerse's Baby Daddy)

Over 30 years ago there was a place where you could be anywhere, anything, or anybody you wanted to be. Created in 1999 SecondLife appears to be the precursor to the MetaVerse, the environment "developed" by Meta (Facebook's parent company). To understand the relationship of the MetaVerse and the beginnings of digital currency it is important to understand the beginnings of SecondLife and its currency the Linden in 1999.

Created by Philip Rosedale, "Second Life" is an immersive environment of role-playing and community building. Similar to what we know as current today vis-à-vis Minecraft, Roblox, Metaverse, NFTs, and cryptocurrency. SecondLife was decades ahead of its time. Many universities and other institutions bought "islands" and other forms of property where they would "build" replicas of their first life worlds for others to enjoy. Unfortunately, with all experiments that do not grab on to the driving force of capitalism, it has dwindled and become overshadowed by the current drive of Return-On-Investment thinking prevalent today.

The Linden was the currency used specifically in SecondLife. You were able to transfer it to USD. Its comparison to today's digital currency stops short of the ability to encrypt it across geographical boundaries and outside of SecondLife. Cryptocurrency encryption model and its use in NFTs will be discussed in the next section (Jamison, 2017).

1.4.20 Crypto and NFTs

We will also be discussing the proper alignment of new economic systems and their use in diverse areas. One of those economic systems is digital currency and its use and abuse. But for now a quick discussion on cryptocurrency and its position with NFTs (non-fungible tokens). Let's look at a textbook definition of cryptocurrency from the National Institute of Science and Technology (Mell and Yaga, 2023).

> In the case of cryptocurrency creation… the publishing node includes a transaction sending the newly created cryptocurrency to one or more blockchain network users. These assets are transferred from one user to another by using digital signatures with asymmetric-key pairs.
>
> *(Editor, n.d.)*

Yep – a mouthful but pretty simple to understand. Basically, it is a digital currency that is created via a blockchain (a blockchain is a distributed database or ledger shared among a computer network's nodes) that is designed to have an almost impenetrable server network. The issue is that this currency valuation fluctuates incredibly. In the last year, cryptocurrencies (such as Bitcoin) have dropped over 90% causing investors to run for the hills. Here is a quick list of some of the top cryptocurrencies available today from Forbes.com with their Market Cap to the right of the name, June 13, 2023 (Tretina, 2021).

1. *Bitcoin (BTC) – $506.7 billion*
2. *Ethereum (ETH) – $209.5 billion*

3. *Tether (USDT) – $83.5 billion*
4. *Binance Coin (BNB) – $37.7 billion*
5. *U.S. Dollar Coin (USDC) – $28.3 billion*

1.4.21 The Crypto Players

The most well-known crypto player is Sam Bankman-Fried. Bankman-Fried has become infamous with the crazed floundering of crypto and its effects on banking. Just recently Bankman-Fried was tried and found guilty on numerous charges most of which were intended to fraud investors. He has appealed and lost and is now serving a 25-year sentence.

The company that he founded was FTX, a cryptocurrency exchange.

> **What is FTX?** The now-bankrupt company was one of the world's largest cryptocurrency exchanges. It enabled customers to trade digital currencies for other digital currencies or traditional money; it also had a native cryptocurrency known as FTT.
>
> *(Yaffe-Bellany and Goldstein, 2023)*

1.4.22 So What Is an NFT?

NFT stands for (non-fungible tokens). Ok – so what does that mean and why is it useful? Here is an explanation of this financial platform described by Sarah Hammer of the Wharton School of Business (What Are Non-Fungible Tokens?, n.d.).

> NFTs are units of data stored on a blockchain. NFTs may be associated with digital files, such as photos, videos, and audio. Because each non-fungible token is uniquely identifiable, NFTs differ from cryptocurrencies, which are fungible. Common use cases for NFTs are art, music, games, and film.

NFTs have been around for a while but they are used in creating provenance for any form of artwork or digital element for sale therefore endorsing that their worth has increased. However, like many items in a digital world, they evolve, fluctuate, and morph, in other words, the value associated with them through cryptocurrencies is even more volatile than any other form of "currency." We will discuss the process of minting an NFT later on in this book.

1.5 CONCLUSION

The goal of this first chapter was to explain the premise of this book and the need for the next chapters. Each of the following chapters will go into further detail about the stages of development in this new time and new world. We are just at the beginning of this journey. The overall goal is to culminate in an understanding of our future abilities and concerns. Everything is there for the taking. We just need to make sure we know the caveats in our drive for knowledge and the use of these new technologies to enhance our lives and to gain yet more knowledge. Our focus should be to extend our ambitions when necessary and to contract with benevolence when needed and to try to always remember the steps back to sense and sensibility.

2 On enterFrame
Message Meets Medium

The who, why, and when of computers and media in design and development.

2.1 ABOUT THIS CHAPTER

The intent of this chapter is to outline in a simple and concise form the events that led up to the world of interaction design, development, and delivery, as we know it today. To truly create and document a history of interaction design and multimedia and its necessities would take an entire encyclopedia. It is not the intention of this book to cover every occurrence or to settle any arguments. Thousands of names have been left out and for efficiency.

There is a theory called the "hundredth monkey." Simply explained, this theory states that: if an island is occupied by 100 monkeys then it is presumable that the number 1 and number 100 monkeys will come up with the same idea at virtually the same time. Now given the fact that most of the individuals written about in this book are all of comparable intelligence then it is conceivable that at sometime someone else generated the concept with which they are herewith credited. It is the linkages between members of the total group that we are concerned about.

Interaction design and multimedia by nature is a group project. It is almost impossible for one person to create a total UX project. A group that consists of experts in their respective fields typically creates it. New media or multimedia, whatever you want to call it took that group and more to create. We can only hope that as we soar to new heights we remember that we can only reach for the heavens because (as the old saying goes) we stand on the shoulders of giants. This chapter is about many of those giants.

2.2 A VERY BRIEF HISTORY OF COMPUTING AND INTERACTIVE MULTIMEDIA

2.2.1 From Opposable Thumbs to Optical Fiber

In this section, there will be correlated moments from the early days of man to the advent of the computer era. These occurrences have been selected not for their overall merit but for their linkage from point to point. Even then, as now, it is all about connectivity. One point influences another. I have tried to edit down these moments to the ones that are directly connected. True, in the fullest meaning of existentialism, all events affect each other but in this case there was only enough time and space to detail those that are glaringly most prevalent.

Communication has never been about technology, it has been about the use of that technology to communicate. Each generation adapts and assimilates to the powers that it is given. Today three-year-olds are dual platform and multitasking before kindergarten. It is no longer which color ribbon you use to edge the book report. Now it is, have you Zoomed directly with the author? Users desire greater levels of informational structure. The ancillary is quickly becoming as important as the primary. But it wasn't always like this, or was it?

DOI: 10.1201/9781003404347-2

2.2.2 Lascaux Starts It All

In 1941, four young boys were out for an adventure when they stumbled upon one of the greatest archeological finds of the 20th century. Once scientists could properly investigate the caves in Lascaux, France, they discovered drawings on the walls that dated back to Paleolithic times. Here is where the groups had gathered together to visualize the next day's hunt. Here they drew, on the walls and ceilings, images of their prey. A practice that still occurs to this very day in conference rooms around the world.

Upon closer investigation the scientists noticed small chip marks around each of the images. The marks were similar in nature and repetition. The scientists determined that they were spear marks. The hunters must have been visualizing the hunt and been for all intention "practicing" how they would slay their prey. Was this the first interactive learning environment? Did this truly help them to realize their hunt? (Lascaux, n.d.).

2.2.3 Tell Me, Show Me, and Engage Me

How we interact with something is part and parcel of how we communicate. In studies done around the world it has been found that individuals learn better when they are truly part of the lesson not merely spectators. Humans appear to make better decisions when they can inspect and experience first. This is a double-edged sword however. If your product or lesson is weak there is no hiding that, once it is brought directly to the customer.

This environment-first approach has been used since man first discovered methods of communication. Stories are best told when they can be acted out. If one image is worth a thousand words how many words are communicated when that image is moving at 30 frames per second?

2.2.4 Truly Silent Movies

The first example of the moving image is via the Column of Trajan. This monument to communication was constructed by Emperor Trajan and is located in the city of Rome, Italy. A structure that rises 100 feet above ground level, it spells out a history of battles, starting from the top of the column, in images that progress around the column. The viewer can "read" the story by walking around the column and viewing each row as it spirals down toward them. This practice of image as message has been used many times in the progression of civilization. One of the most pronounced and adaptive versions of this concept is in the works of the Pharaoh's hieroglyphics and other ancient cultures. The Egyptians, Mayans, Aztecs, Syrians, Phoenicians, and their ilk all included imagery into the manuscript they chose to communicate. This obviously increased once the concept of drawing with ink directly onto papyrus (paper) was adapted into greater use by all cultures. The folding of those papyrus sheets into a form of sequenced writing led to the concept of the book. These early models gave way to more refined book structures. They were called a codex.

From the codex to the monasteries of Charlemagne image and word started to become common. From these hallmarks came the illuminated manuscripts. The *Book of Kells* is heralded as a strong example of this format (Duncan, 2022). Once these manuscripts became illuminated via image, ink, and calligraphy, they truly became monuments to communication. The creation of these books was time-consuming and tedious.

It was around this time that printing would become more automated. True, the concept of a printing press had been around for centuries from the early Egyptians to the Chinese (Pi Sheng) but now the demand was greater and the audience was larger due to the educational system set up by the monastic societies. The only problem was the distribution to the masses (Weisstein, n.d.).

2.2.5 From Inkpad to Mouse Pad

Enter Johannes Gutenberg and movable type. Gutenberg (c. 1393/1406–February 3, 1468) was responsible for the concept of creating a printing press that used separate letter shapes to form the individual blocks of words that could be inked. This was actually a derivation of the Chinese printing methods from over a half millennium before created by Pi Sheng. Gutenberg and the like started using this method to quickly create blocks of singular letter shapes arranged together for printing pages of type. This allowed for more printing done and less work in order to generate more pages and lessen the cost and time. Once words could be "wrapped" around a shape the introduction of images mixed with words evolved. Now different segments could be "inked up" with different colors. Other publishers utilized this method and hence the world of modern publishing met with the common man (Lemelson, 2020).

Although this method of printing was a great improvement it still required the image to be carved into the wood or other substrate. Once lithography was invented the images could be drawn onto stone then later metal plates with oil-based crayons. Another lithographic improvement was to etch these images out of topically covered plates. This then allowed for more intricate images to be used. This process quickly moved toward a photographic method in the 1800s with the invention of pinhole cameras and later on photographic lenses.

Once photography was adapted to create print images it was later discovered that it could be integrated into the production of printing plates. This was achieved by the use of halftone screens or patterns. The use of multiple plates for different colors soon followed leading to chromolithography. Chromolithography basically means utilizing many colors for one image via the lithographic process. Now color images could be reproduced in a cheap, relatively quick manner for distribution around the world.

Where one-color lithography developed the newspaper and book publishing industries, chromolithography and the use of sand-casted typefaces led to the magazine and poster industry. Magazines such as *Harper's Bazaar* and *Colliers* were the forerunners of the modern-day magazine publishing industry. These publications along with the newspaper industry quickly led to the distribution of images and content across the entire United States and of course other continents. The concept of the magazine was quickly accepted into the mainstay of American life. From this structure a young man of missionary parents would build one of the largest magazine empires known to man. His name was Henry Luce and his vision would change the face of the nation. The merger that his company, Time, Inc. would take part in helped propel the new media industry above and beyond and later the premature but ultimate "crash."

2.2.6 Agents of Mass Distraction

Corporations such as JC Penney and Sears saw the need to advertise in these new publications nationwide. In order to address this issue, they developed the concept of employing advertising agents. The job of the agent was to establish an office in a city away from the main corporate office and make contact with the local publications. The agents would then purchase space in the publications for the retail corporations. They would develop the copy and image with a local lithographer, photographer, or an artist. This image would then be integrated into the publication. The agent would receive a commission from the publication for the sale of the space. These advertising agents would develop into agencies. The first were companies such as Volney Palmer, Leo Burnett, and J. Walter Thompson. The agencies would then expand the concept to include other retail or corporate accounts. They would then contract out to local artisans and studios to produce the work. This is where the modern-day workflow structure of the advertising agency to specialty studio started. From this process was born the practice of outsourcing concepts and content. To this day the major advertising agencies in the world still receive a traditional 15% commission from publications for ad space purchases.

The first individual to start the concept of teaming copywriters and designers was Bill Bernbach. Bernbach and David Ogilvy emerged separately in the late 1940s to early 1960s and would become the fathers of modern-day advertising. Bernbach's campaigns for companies such as Volkswagen were legendary. His "think small" concept is still the main thrust of the Volkswagen campaign to this very day. Understanding your audience, keeping the message clear and concise, plus studious selection of a design base has kept advertising on course for the last 150 years. Regarded as a necessary evil, advertising has been embraced by all forms of publications, mediums, and has finally been regarded as the engine for the current social media explosion. Later in this book, we will discuss the thrust of advertising and the shake out from which it has survived and proven to be the structure from which all business works (Richards, 2022).

2.2.7 Hey That Moved

At the same time that chromolithography captured the publishing industry a few individuals were wondering if a horse's hoofs came completely off the ground when it galloped. This simple question (albeit a weighty one) would cause the world to see images that moved. Leland Stanford, a railroad magnate and then governor of California, commissioned Eadweard Muybridge (1830–1904) to create such a study. After years of research and many patents, Muybridge would prove that in fact all four hooves of a horse did leave the ground simultaneously. Muybridge worked for years progressing to all forms of animals and humans in motion. However, he was not satisfied with the still images of this event. He would progress to images processed onto glass that would spin in front of a lamp thus producing, for all purposes, moving images. The only problem proved to be the lack of this carrier being flexible enough to produce the final desired effect. Eastman would later solve this problem of the need for a more flexible substrate. Thomas Edison would later use these results to create the modern motion pictures. It has been noted that Louis Le Prince invented the first actual motion picture projector.

> Louis Aimé Augustin Le Prince (28 August 1841–disappeared 16 September 1890, declared dead 16 September 1897) was a French artist and the inventor of an early motion-picture camera, possibly the first person to shoot a moving picture sequence using a single lens camera and a strip of (paper) film.
>
> *(Fischer, Paul, April 2022)*

However, this new world came about it changed how we communicate for years to come. Imagine the Trajan Column for a mere 5 cents a showing. Edison's "movies" would become part of the American landscape. These parlors would be integrated into most major cities. Edison would also start to produce his own movies. Where Edison's attempts were simple-minded and common, the Brothers Lumiere would attempt to bring the events of common life to the screen. This style of filmmaking would anticipate the works of such film visionaries as Eisenstein and artistic filmmakers such as Brunel and Duchamp.

2.2.8 Horse-Drawn Movies

Four brothers from Pennsylvania decided to sell their father's horse and use the money to invest in this fledgling industry. They purchased a projector, rented a hall, and borrowed chairs from a funeral parlor. Here they would charge 5 cents a showing for whatever "movies" they could get their hands on. Within one year they made enough money to open their own theater and purchase the horse back for their father as a Chanukah gift. The brothers' name was Warner. Within a decade their empire – Warner Brothers – would be generating over $180 million a year. The rest, as they say, is history. Later major studios and independent filmmakers would emerge. The names of D.W. Griffith, Mack Sennett, Charles Chaplin, Mary Pickford, Lillian Gish, and Douglas Fairbanks, Sr. would become part of the lexicon of the world (Thomson, 2017).

2.2.9 The Shout Heard 'Round the World

For years motion pictures would be devoid of sound. In 1927, Al Jolson would step out onto the screen and open his mouth to sing. It left audiences speechless. This was accomplished via a method of integrating audio and video. This method has had many incarnations but most commonly known was "Vitaphone." Strips of audio tracks would be incorporated onto the filmstrip. On these tracks would be sound impressions. These impressions would be synchronized to the image via an audio projection system. The result would be sound synchronized with the motions and lip-syncing of the actor. Many other versions also came about from the major motion picture studios of the time: 20th Century Fox, Paramount, and a reluctant Warner Bros. There are newspaper clippings which have Jack Warner asserting that sound in movies was just a fad. That it would never last. This prophecy would be heard a thousand times over as any and all new technology developed. Once these "talkies" hit the silver screen, the modern motion picture industry would begin (Buhler, Neumeyer, and Deemer, 2010).

2.2.10 Look Ma No Wires

Thirty years before the innovation of sound in the movies came the advent of the wireless radio. Developed separately by both Tesla and Marconi (circa. 1890), wireless audio transmissions revolutionized not just the entertainment field but also all industry. Nikolai Tesla is credited with being the first to invent the wireless format. Tesla, unfortunately, was interested in bigger fish and didn't bring this invention to the populace. Enrico Marconi, however, capitalized on these discoveries and adopted Tesla's theories toward practical applications.

Once manufacturing of the wireless radio became popular enough for home purchase, it became a fixture in the American household. Families would gather around the "talking furniture" on weeknights to hear their favorite radio show. The Great Depression of the 1930s led to the adoption of this device as almost a mechanical fireplace. Entertainers, politicians, and religious leaders would use this new technology to deliver their message around the country and later around the world. RCA, Radio Corporation of America would morph into NBC – National Broadcasting Corporation later after television became a reality.

2.2.11 Sounds about Right

From Edison's wax cylinders to compact discs, sound is probably the most human of electronic interventions. We easily respond to hearing songs or sounds of any kind and most commonly from nature. From the concept of the echo to a fully digital studio, audio appears to provide the quickest most easily recognizable response from most people. They are comfortable with it and seem to be most at home speaking into a microphone. That is until playback.

2.2.12 Furniture That Stares Back at You

The great radio personality Fred Allen is credited with this comment about television, "…who wants a piece of furniture that stares back at you?" Well, similar to Mr. Warner's belief about sound in motion pictures, this statement proved to be as much a fallacy as any before and many others yet to be said later in the 20th century.

In 1921, a young farm boy from Utah named Philo T. Farnsworth had a vision while plowing a potato field. This vision was how to transmit images via the electromagnetic field prevalent in the air. Philo was a mere 14 years old when this occurred. The concept would later be brought to realization in 1927 when he and his team of five would transmit via a camera the first image to a receiving system, a vacuum tube. Although just an image of the smoke from his cigarette, this would be confirmation to his theories and the beginning of a new era that would become a major watershed in the field of mass communication. By the end of 1929, he had moved onto the development of color television. It first debuted in 1940 commercially.

Once these concepts were brought to fruition many major scientific laboratories would come knocking on the door of Mr. Farnsworth's San Francisco-based "Green Street Lab." Radio Corporation of America or RCA would be one of the first to imitate Farnsworth's invention. This led to a court battle and the issuing of the first television license in 1938.

As with its predecessor radio, television became part of the landscape. By the 1939 World's Fair, all the world had seen and touched this new piece of furniture. Now was the time for all the major players to take the initiative. Stars of the silver screen would become stars of the small screen. Vaudeville had been considered the lowbrow entertainment circuit of the 1910s and the roaring 1920s. Radio and television enlisted these performers and brought them to the country in living black and white, albeit on a six-inch screen, housed in a mammoth cabinet that weighed more than the new Philco frost-free refrigerator. Once the new technologies of television and radio became immersed into the American psyche, the next step was to bring it all back home (W Russell Neuman, 2010).

2.2.13 Beta (or VHS)-Blockers and the Last Mile

Television (video) was then and is now the king of mass communication. From the top of the antennas that once stood high above the Twin Towers to the local transmissions from regional broadcast centers, nothing brings the world to users like Philo's magic box. The next logical progression in this delivery system was a methodology that has its basis prior to Tesla's invention, the copper cable.

Cable television and, its distant blood relation, the videocassette recorder (VCR) both originated around the same time. Due to the need for stronger reception in rural landscapes, the use of cable television rose to popular use during the early 1970s. At the same time the VCR became ready for a popular marketplace. Simply connect it to your television via copper wires and you can watch what you want when you want to watch it.

One of the first incarnations of this new technology was actually superior to its more common version. The Sony Corporation first invented the Beta format in 1977. The images were transferred to magnetic tape via an electromagnetic field. These images that have now been captured are displayed stronger than those of VHS. VHS (Video Home System) is also captured through the use of electromagnetic fields. Beta was used as the industry standard in commercial television. Why then, if the image is better and the convenience the same, did this technology not become the standard for home use? Simple, the developer blocked the release of this architecture to third-party manufacturers. Once this occurred other manufacturers commercialized VHS format. This then allowed for VHS to be reproduced for home usage at a much more affordable cost. Remember Tesla and Marconi. It's not about how great the concept is; it is about how many people are using it that leads to commercial success. What once cost $1500.00(US) to purchase is now $69.95(US) at Wal-Mart if you can find one! (The Betamax Case, n.d.).

Cable television on the other hand has increased in price since its inception. Where typically distribution costs diminish when distribution sources increase, cable television seems to go in the opposite direction. Obviously, a lack of competition is the reason for this phenomenon.

Cablevision, one of the largest delivery systems of cable television in the Northeast United States, has steadily increased its rates since going mainstream. This is supply and demand in reverse. The reason for this is simple. Once Cablevision laid the basis for the copper lines they became tollbooths for the end users. The last mile is the name of the concept. The concept of controlling the home marketplace proved to be truly a visionary occurrence.

Today mergers of television networks and streaming portals are as common and controversial as chocolate and peanut butter. Original case in point, America Online (AOL) and Time Warner. At first these mergers appeared to be a strange idea, but they became tasteful and enjoyable to the user (but yet this then caused the biggest failure in modern business history).

Where else could Esther Dyson's statement "Content is king" be truly developed and tested. The problem that comes about, however, is will there be 557 channels with nothing on? Ultimately it will be the viewer who decides. The concept of interactivity needs to be redefined starting from the parameters of a new construct. This needs to emerge from one of multiple-choice to one of truly interacting with the medium. Passivity has to go away. This, however, is not just up to the viewer. The content creators need to redefine their paradigm. For this new media to truly flourish and return an investment it must provide solutions not merely new problems. How to do this has yet to be figured out (Brockman, 1996).

2.2.14 OK Computer – Say Hello to My Little Friend

This section will not try to completely pursue the history of computing, which in itself is a multi-volume set of books. What will be done is an outline of some of the watershed events that led up to where we are now. The turning point in all this mass communicating obviously is the advent of the computer. As much as television has had an impact on society at large the computer has made an impact on first a select few, then slowly, a compounded impact on the entire world. There is no doubt the way this device has intruded into and in most cases benefited our daily life. The word "communicating" and the concept of the personal computer now have intertwined as no two concepts have ever done before. How they got this way is a book within itself.

2.2.15 From the Ishango Bone to ENIAC

The Ishango bone was mentioned briefly in Chapter 1 but here we go a bit more into the effects of this tally making. For two millennia man has been working toward the use of combining tools and determining our purpose here on earth. As we have become more adept our tools have also become more advanced. The ancients used the calculation of numbers to create our monuments and monoliths. From the use of simple counting machines like the Ishango bone type of tally stick and the abacus to the calculations of Napier and Pascal, we have found a symbiotic relationship between mathematics and the arts. The concept of the Golden Mean in the time of the Greeks tried to explain to us how that equation works. As our use of mathematics, algebra, calculus, and geometry progressed it empowered us. Showing us how to ask bigger and more in-depth questions. From the study of the planets to the study of ourselves the correlation between our measurements and our environment, between our habitats and our vehicles has been developed into our architecture and later into our psychological studies. To aid us in these questions we have developed counting machines. One element consistently repeats during the development of these machines – electromagnetism and how it was harnessed and understood.

Electromagnetism is inherent in the world around us. It is how sound waves are transmitted as well as the structure we use to distribute any form of vibration. Throughout this book there will be references to this natural phenomenon and how it has aided communications, every aspiration, and dream.

As early as 1900, Herman Hollerith introduced the automatic card feed in his electromechanical information machine to process census data. The first analog computer was developed by Vannevar Bush in 1925 to solve differential equations. The British team first used the Colossus computer in 1942 as a code breaker against the German's enigma machine; Alan Turing headed the team. This is a name you need to remember (Forbes, 2019).

In 1946 ENIAC (Electronic Numerical Integrator and Computer) was unveiled at the University of Pennsylvania. This machine designed by J. Presper Eckert and John Mauchly was the first electronic digital version of a modern-day computer. Developed jointly by the military and the University of Pennsylvania, its job was to study ballistic computations. In that same year a computer laboratory was founded at Manchester University, England, by a grant from the Royal Society. The age of the computer has become a worldwide focus (Levy, 2013).

Now that these machines were operational, the next step was to create a form of memory storage system. Again an electromagnetic system was used. In 1947, F.C. Williams began a working model of his memory system. Later that year the ENIAC was converted from a counting machine into an elementary stored-program computer. By 1948 magnetic drum memory was introduced as a data storage device for computers. Past this point the advances came fast and furious. By 1949 computers were running programs from storage systems. Specific code or programs are being developed to add multiple functions to one system. Later in 1949 computers were constructed at Massachusetts Institute of Technology to be based on real-time computations. From here on MIT would become synonymous with the development of computer engineering.

By 1952 symbolic notation was being used to program computers. This is very familiar in regards to the C++, DHTML, or Java-like languages of today. Subsequent programming languages followed such as Pascal, Fortran, and COBOL. At the same time a growing company, International Business Machines, appoints Thomas Watson, Jr. as president. By 1952 IBM would develop the IBM 701 – the Defense Calculator. This was followed in 1953 by the IBM650, the first mass-produced computer. Originally it was anticipated to introduce just 50 items into the marketplace, a number that changed to over 1,000 by year's end. Later in 1957 IBM delivered the new 709 computers. Now things would really change.

In 1957, ARPA (Advanced Research Projects Agency) was developed by the military to ensure technological superiority for the United States. A necessity brought about supposedly by the launch of the Russian satellite – Sputnik. This would become the basis for the development of the Internet. ARPAnet (later the Internet) was developed by a group of academics and military personnel from within this country's foremost technology-based universities. Once ARPAnet is up and running the next step is to focus on delivery of this computer-formulated information. As this became more focused on National Security the name was changed to DARPANet (Defense Advanced Research Project Agency). This will be discussed later in the section concerning all forms of delivery systems ("ARPA Is Born," 2022).

2.2.16 Enhancements Take Center Stage

In the late 1950s and early 1960s things started to really heat up. Integrated circuits and transistors were becoming substitutes for the old concepts of vacuum tubes. Landmark discoveries are being made across the world in the development of storage and controller devices. Texas Instruments introduced the first silicon transistor in 1954. By 1959 Robert Noyce of Texas Instruments gave a public demonstration of an integrated circuit. By the end of that same year developers have created over 200 programming languages and computers are now using paper tape of these programs to read and store information. This standard of paper tape would remain the mainstay for computer language well into the 1960s and 1970s.

Now the floppy diskette and the daisy wheel printer would be on the scene. Information could now be transported and printed relatively easily. Xerox, a developer of office machines, would establish the Palo Alto Research Center (PARC) for computer research at Stanford University. This center would become the scene of one of the grandest thefts of computer innovation for all time.

Later in 1970s sand was melted and developed into silicon for use as the base for computer chips. Intel engineers developed the Intel 4004 microprocessor dubbing it a "computer on a chip." Its processing power was 60,000 instructions per second as opposed to the standard 1,000 instructions possible by prior means. By 1971 the regional area where these developments were occurring was called "Silicon Valley USA" in an industry publication. Seymour Cray, father of the supercomputer, started Cray Research and debuted the Cray I, the world's first successful commercial supercomputer in 1972. By this time word about these new-fangled devices had gotten around. Young people were getting interested. One of them was a fellow from Southern California named Steve Wozniak. Wozniak and a friend would later make some pretty strong headway into the world of personal computers but first we need to discuss one of the most fertile grounds for computer technology and probably the least known by today's standards.

2.2.17 PARC – the Great Hidden World

Xerox's Palo Alto Research Center (PARC) was truly the epicenter of all modern-day computer technology. Understanding that the world may be going toward a paperless society, Xerox created PARC in an effort to understand and conquer this new world of communications. Now some 60 years later it is closing down and being deployed to SRI International a non-profit scientific research group (Robinson, 2023).

Some of the names that we will speak about later from PARC are Alan Kay, Dan Ingalls, Adele Goldberg, Larry Tesler, and Robert Metcalfe. These individuals alone represent the impetus behind helping to develop such work as the Alto, Object Oriented Programming, GUIs, Ethernet, Smalltalk, WYSIWIG and basically Apple Computer's Operating System. One of the most influential developments at Xerox PARC is the concept of What You See Is What You Get applications – WYSIWYG, as it is called, has had great impact on all future computer operating systems.

Another groundbreaking system developed by the researchers at Xerox's Palo Alto Research Center was a computer called the Alto. The Alto was the world's first desktop computer that used the graphical user interface, the mouse and Ethernet in 1970. The computer was truly the world's first desktop computer. Douglas Engelbart developed the mouse, Alan Kay developed the GUI system, and Robert Metcalf developed Ethernet. Metcalf uses the term from pre-Einstein and Maxwell theories of the way "ether" supposedly carried sound waves in the air. Metcalf describes Ethernet as a modified version of the "Alohanet" created by Norman Abramson in the late 1960s at the University of Hawaii. (Around the same time, work begins at a Stanford University laboratory headed by Vinton Cerf on Transmission Control Protocol.) Together these components start the structure of what would later become the World Wide Web but first let's stay with the desktop (Robinson, 2023).

2.2.18 Alto – the World's First Desktop Computer and the Revolution It Caused

Alan Kay and his team at PARC had been working on something a bit more adventurous: the first notebook computer – the Dynabook. Kay's concept was to create a computer that would be no bigger than a traditional textbook. Since technology could not at that time produce such a product the PARC team set their sights a little lower. Not that much lower though! In 1972 PARC started work on the first personal computer, the Alto.

The Alto was named after its place of origin, the Palo Alto Research Center. Having been developed by a team of engineers and scientists, the purpose behind the Alto was for PARC to create a computer small enough to fit comfortably in an office environment. And that is exactly what they did. Consisting of four major parts: the graphics display, the graphics mouse, a keyboard, and disk storage/processor the Alto did everything it set out to do and more. The strongest component of the Alto was the nature of the software it used to operate, but more about that in the section below.

Along with the Alto came many other new innovations such as SLOT. Gary Starkweather, a physicist who specialized in lasers and optics developed SLOT. SLOT stands for Scanning Laser Output Terminal. SLOT's main purpose was to be able to print out any image that was displayed on a computer terminal by taking the digital information and reproducing it onto a photosensitive drum of a copier-like device. But that was only one side of the equation. Another individual – the newly hired Ron Rider, was solving the other side of this equation. Rider developed the front end of this protocol by allowing the printer to read the images generated by the computer. This device was called a research character generator. This simple connection created for the Xerox Corporation one of their most successful line of products of all time – the laser printer.

This was all well and good but how do you get the information to these devices when they are not hardwired? Enter Robert Metcalfe, who is responsible for developing Ethernet. Ethernet was developed as a response to the time-sharing problems being faced when using the current international network system developed by ARPA. The name of this system was ARPANET, which through many changes and reworkings has become the Internet.

Metcalfe created a system that would transfer packets of information. These packets contained all the necessary bits of information needed to transmit the main portion of the packet to the actual file. To get into how this actually works would take no less than a chapter and has taken many books by many authors to explain. For now just realize that once Ethernet was finally developed it allowed for transmission of information over great distances and locally for local area network printing. Metcalfe has since gone on from PARC to create the company 3Com. 3Com would become a leader in all forms of digital information distribution.

Another development to come out of PARC at the time was the invention of the GUI. Basically a GUI is the screen display of information used to allow users to interact with a computer program. Some of the developers of this concept were Larry Tesler, Tim Mott, and Adele Goldberg. The first step in developing this new method of using a computer (actually Dr. I.V. Sutherland had given the basis for it with his Sketchpad) was the software Bravo. Bravo was a simple text editor (prior to WordStar) that worked in a similar fashion to most other DOS-based word processing system text only and minimal quality to the display. First engineered by Butler Lampson, the text editor was turned over to Charles Simonyi, a recent Stanford PhD graduate to fully develop. Simonyi enhanced the text editor by learning how to include such items as diverse typeface selections, diverse margin settings, and page numbers. These new text-editing enhancements allowed the software to send the page as it appeared to the printer. This was the birth of *WhatYouSeeIsWhatYouGet – WYSIWIG–* obviously a milestone in computer graphics software development. The next step was getting the computer to be easier to use.

After witnessing how typical end users (read not engineers) used a computer, PARC researchers Tesler and Mott created a program called Gypsy. After consulting with these typical users the team determined that the way to create a better working environment for the user was to mimic the method of traditional paperwork. If information needed to be rearranged typically we copy and paste it from section to section. This is exactly what they did.

Tesler's main breakthrough visually was in developing the use of icons and menus. Since the Alto computer was already using a mouse and storage systems, Tesler employed the method of drop-down menus and clickable icons. The idea did not take hold immediately with the corporate higher ups but once a test group was conducted it took about one hour for even the most skeptical to join the ranks of the converted. Tesler's invention of the graphical user interface along with Simonyi's new updated version of a text editor started the page composition revolution that led to the development of such programs as PageMaker, QuarkXPress, and InDesign. Tesler went on to work directly with Apple Computer while Simonyi later joined Microsoft where he was influential in the creation of such breakthroughs and mainstays as Microsoft Word.

The next step to be taken was to develop extensive graphical display programs. As usual PARC took that first step. Enter Alan Kay and his team of Adele Goldberg and Dan Ingalls et al. In the late 1960s typical computer programming languages made distinctions between data and the process to make that data work. The more data and the more instructions required added to the complexity of the program and subsequently the time to work out these requests made by the end user. While working at PARC Alan Kay was developing a program to teach children the process and art of computer programming. Kay realized that the structure currently in place to develop programs was very laborious and decided to figure out how to minimize that effort as much as possible. Kay's idea was to compress the needed function and procedures into discrete modules of programming. These he called objects. Objects are, very simply put, combinations of data and procedures that are put together into these "objects." The objects can now be used as modules to which "messages" can be sent to create functions within a program, in other words sort of like macros or actions in much of the applications we use today. A simple concept but not a minor development! The system would be referred to as object-oriented programming (OOP).

The language that this system, developed by Kay and his team, came to be called is Smalltalk. From this new language emerged the ability to manipulate images on a screen with ease and flexibility. Smalltalk was released in its first functional version in 1972. By 1979 an event occurred, which is nothing short of crossing the Rubicon, Steve Jobs and Bill Gates visited PARC and for some reason (some say there was a $1 million stock option) were given access to most of the work done by its development teams. Dr. Adele Goldberg was the lead on this part of development. The story goes that Dr. Goldberg objected to handing over the work to Jobs and Gates and only did so after a memo was issued. Goldberg, as noted before needs to be acknowledged along with Katherine Johnson, and Dr. Grace Hopper as the leading pioneering woman in computer science.

From that visit the use of icons, folders, drop-down menus, and object-oriented programming appeared in almost all of Apple and Microsoft's future product developments. The basis behind object-oriented programming not only impacted the development of these two software and hardware giants but became implemented throughout the computer programming world of the late 1980s and early 1990s. OOP – object oriented programming, has been credited with being the basis behind many powerful breakthroughs in the development of computer programming languages such as C++ and Java. We will talk about that some more in the sections below. PARC continued until 2023 but the advancements it made in the 1970s truly broke open the world of personal computing. It should be noted that PARC was formally closed in 2023 (Robinson, 2023).

2.2.19 Enter the Small Time Heroes

Around 1975 one of the first computer kits appeared on the cover of Popular Electronics. It is called the Altair 8800. This created the beginning of affordable desktop computing. Another young man and his friend take notice of this development. He would become the world's foremost college dropout. His name is William H. (Bill) Gates III; his friend's name is Paul Allen. This partnership would work together with the manufacturers of the Altair and in 1975 develop the first microcomputer operating system generated from the language BASIC. The company they would later come to establish is called Microsoft, first headquartered in Albuquerque, New Mexico. From this partnership would come a company that would forever change the way the world works? But as that was happening two other young fellows were thinking a bit differently.

Another two young men that come to mind during this time were Steve Jobs and Steve Wozniak. Being the quieter of the two, Wozniak appears to have had first an engineering mind and secondly an entrepreneur's spirit. In 1972, he developed a "blue box" tone generator and made free phone calls from his dorm in UC Berkeley. This spirit would serve him well in years to come. On the other hand, Jobs was a dynamic personality with great vision and a charisma that was superior to most. In 1976, Steve Wozniak met up with his friend Steve Jobs (both were working at Atari) and together the two started one of the world's most innovative computer companies, Apple Computer.

Wozniak and Jobs continued working in a ying-yang form of partnership. Wozniak the engineer, and Jobs the full-fledged entrepreneur. These two appeared to be different sides of the same coin. Both men were prophetic visionaries but yet possessed different visions. Jobs' concept of world domination is similar to that of Gates, but Jobs it would appear wished to do it through some form of spiritual transformation of man and machine. Gates is just focused on the goal. These two companies, Microsoft and Apple Computer, would eventually control the world of desktop computing. But first, big things start to shake (Isaacson, 2011).

2.2.20 OOPS There It Is

Now everything is starting to heat up exponentially. Xerox PARC through the work of Alan Kay develops Smalltalk the beginning of Object-Oriented Programming Systems. It is later used to create specific applications, like AppleTalk by Apple computer. The first PC clone is developed. PostScript, a page-description language, is written by Adobe systems (another group from PARC) to work with these new GUI-based computers. The race is a foot. From this period in the early 1980s the computer has now become part of the American landscape. *Time* magazine names it the "Man of the Year."

Around the same time as these smaller, more personal machines, Seymour Cray develops the Cray-X-MP and starts the concepts of parallel processing. This new computer is three times faster than his first super computer, the Cray-1. Even though Cray's machines were for industry use and truly mammoth in physical size, the years 1983 and especially 1984 became point zero for the acceptance of the personal computer into the home market (Perry, 1985).

2.2.21 Apple Hurls the First Stone

At halftime for the 1984 Super Bowl, Apple Computer would make a statement concerning the personal computer that to this day still is heralded as a breakthrough in not just computing or advertising but a statement that would reach to the core of consumerism and the American psyche. In a television commercial, Apple depicts the old world mentality of computing by showing an Orwellian-based society. Large groups of people are all walking together in line working toward the safety and security of the mainframe. Everyone dressed in the same drab clothing. All headed toward a great hall to hear the voice of the leader, which is projected on a large screen. From the back of the hall a woman dressed as an athlete runs up and throws an Olympian hammer at the screen to break the hold that it has on the populace. The last thing the viewer remembers is the Macintosh logo. As an old advertising man I drool every time I remember seeing it or read about it. It is brilliant and timeless advertising – Steve Jobs at his best.

Programs such as MacPaint and MacWrite, developed for Apple Computer to work with their new line of Macintosh computers, are debuted. Soon after, programs such as PageMaker by Houghton-Mifflin (and later Aldus and later Adobe), and a vector-based drawing program named Illustrator from Adobe appear on the scene. Such began the world of electronic publishing.

2.2.22 1986 the Year of Cottage Cheese

In 1986 a new world began, a world of communication that has opened the doors for everyone to become a "publisher." This is good and bad. Just as today it is taken for granted, "if it's published on the Internet it must be true" speaks the same as "I saw it on the TV." Back then in the beginning of this "Cottage Industry" as long as it came out of the printer it looked "professional," therefore, it must be correct. These early programs disregarded the need for proper kerning and tracking of type. It took for granted not how a word appeared on the page, but just for the fact that it was there. How the letters aligned or how the type was used was of almost no concern. Soon the art of typography was reduced to a "ransom note" appearance. Base alignment of type disappeared. Widows and orphans would ransack the page. Images would be printed that were pure mud. Dot patterns would be used in halftones that looked as if the person pictured had developed acne at the age of 45. It was horrible and beautiful at the same time. The freedom it placed back into the hands of the everyday person was worth the suffering that it caused. As with any new invention it had to find its way into the right hands. This took time and patience. But many still believe that there was a great loss.

2.2.23 Oh Delivery Boy!

Society has always looked to develop tools that alleviate the steps necessary to complete a task. From using ropes and pulleys to move the stones for the pyramids to a remote control for the television we have used technology to allow for more leisure time in our already sedentary lifestyles. It would appear that we get bigger as our tools get smaller. Someday, and dare say soon, our chubby little fingers won't be able to punch the necessary keys or turn on the infinitely tinier devices. Our jobs develop as fast as the machines. Surely there is a breaking point coming soon, but not necessarily, as we know them to be.

As we stand in the early stages of interactive television to take over our homes we need to know how we got there in the first place. In thinking about this a short story by Kurt Vonnegut called "The

Eupho Box" comes to mind. In this piece a town's people get this new device called a "Eupho Box" and then get frozen in its grasp. (Vonnegut, 2014) This basically sums up many fears of new technologies. Information is gathering up at tremendously large levels like a levee. Previously we had filters to edit this flow into an orderly fashion. Individuals would determine what was important and what wasn't. Not necessarily the best form of civilization but one that has worked so far. Now we wait for the new delivery boy.

2.2.24 All Things Large and Small or "Do the Math"

Everything we use today has come about from some form of mathematical equation. From the shoes we walk in to the chairs we sit in. Somewhere at some time someone had to figure out the measurements and the engineering necessary to make this product. The same is true for all our delivery systems. To better understand this let's start with the concept of hertz. The term "hertz" comes from its discoverer: German physicist Heinrich Hertz, who made important contributions to the study of electromagnetism. In computer hardware development, as in the production of all electronic devices, the typical unit of measure is the hertz. A hertz is a unit of measure for one beat per second. In other words how frequently something moves. Now if something moves at 1,000 beats per second that is called a kilohertz. A million beats per second and that is a megahertz. A billion beats per second and that is a gigahertz.

Different radio frequencies are measured in their specific beats per second or hertz. Your favorite radio station is at 89.1 megahertz. This is obviously much more complicated than it sounds. The concept here is what we commonly refer to as a broadcast level. Every radio station broadcasts at a different frequency. The difference between computer's use of frequency and a radio's is that computers use a consistent megahertz level to categorize its speed where a radio uses it for a specification (Buchwald, 2011).

2.2.25 Maxwell's Slivered Spectrum

As always it is all in the delivery. Maxwell's spectrum is the mother load of this concept. The band of frequencies that we can broadcast on is infinite. James Clerk Maxwell first discovered this in 1856. He determined from previous research, formatted from Faraday's theories, that there is an electromagnetic field. This field can be divided up into frequencies. These frequencies allow for the broadcast selections. There must be a form or entity that exists from which to hitch our proverbial wagon. Originally thought to be some invisible vapor-like ether. Albert Einstein concreted the concepts of electromagnetism later, in 1917.

So obviously a structure to transport information on would not be the problem, the problem lies in the speed and vehicle on which it is possible to deliver this information. But before we focused on harnessing the delivery structure we needed to focus on the process by which we would transport this delivery. To understand this we first must decide on what moves the fastest. It was decided that light has the fastest speed ratio of anything we can understand and deal with. It was from this understanding that the maser (a precursor to the laser) was first created. This device transmitted a focused beam of light toward a target. It is the ability for the light to generate enough energy in order to move at a focus target at its natural uninhibited speed. The hard part is directing it (Sutter, 2021).

2.2.26 Maser and Lasers and Photons Oh My!

Charles H. Townes was the first scientist to develop a working maser, the precursor to a laser. Maser stands for microwave amplification through the stimulated emission of radiation. Townes achieved this in 1955 while working at Columbia University and Bell Labs. Simply put, a maser bounces microwaves back and forth from mirror to mirror with one mirror being slightly transparent.

Then when they are accelerated enough, the light projects through the transparent area, a circle, in a strong focused ray.

A laser is similar in technique but works with a hard crystal – a ruby in most cases. Laser stands for light amplification through the stimulated emission of radiation. Either way the end result is similar, it is a collective ray that in later use will carry information. Think about the possibilities. If the fastest speed we know is that of light and glass has no inhibitors like copper then how fast could this travel? This is where the laser meets the fiber optic network. But yet today we are upset if the Wi-Fi doesn't automatically connect.

2.2.27 Next Stage Booster

Currently we are at the end of a transitional stage from analog to digital. The modem was one of the first devices created to translate this problem. MODEM stands for Modulator / Demodulator. Simply put it converts one form of signal to another. This is the noise you used to hear when going online with your 28.8 modem. It has become as ingrained into our minds as that of a doorbell. Speed is the champion and the enemy at the same time.

Latency is the main issue in the distribution of information. Copper wires and telephone lines deliver information at different speeds. The fastest means of delivery currently in popular use is optical fiber. Optical fiber is basically glass (yet more sand in the mix). Since it is glass, it is the perfect conductor for light. Because our goal is to move things along at the speed of light this looks good.

AT&T's laboratory in New Jersey first developed optical fibers in the late 1950s generating solutions from the efforts of Robert D. Maurer at Corning Laboratories and a combination of discoveries starting with Will Hicks and Elias Snitzer. Due to the efforts of individuals like this it is now possible for each strand of wound glass to be reduced to an extremely fine thinness. Each fiber of that strand has the capacity of carrying over 25,000 gigahertz of information. Each strand contains a pant load of fibers. You get the picture (Hecht, 2004).

The only obstacle in the way of this delivery is the actual switch. The switch's job is to first accept information from the network cable and then direct the information to the desired end user. If the switches are analog (as most were) the speed diminishes to the potential of the switch. Once fully optical switches were finalized and deployed the last miles seemed like inches. Now that the machines were ready, the chips were in, the programs coded, the need was established, and the next stage was a delivery structure.

2.2.28 Gentlemen and Ladies Start Your (Search) Engines

ARPA was the precursor to DARPA and then changed back to ARPA (then in 1996 changed back to DARPA confused enough?) ARPAnet was the precursor of the Internet. The Advanced Research Projects Agency (ARPA) was created in 1958 by the military and academia. Its first duty was to devise a communications system that would connect all military and academic centers across the country. This need became quickly apparent once the Soviets launched the first of its Sputnik satellites. In 1960, Paul Baran, while working at the Rand (Research and Development) Corporation, created the packet-switching principle. This principle would allow for information to be sent in small packets across the emerging network. In the early 1960s this new ARPAnetwork was functional although in quite the embryonic stage. Academics from leading universities were developing systems to operate within this new framework. One of the first was Fernando Corbato from the Massachusetts Institute of Technology.

In 1962 he developed CTSS, the time-sharing system that would create a way for multiple users to share computer time. Baran continued his research into packet switching and by the end of 1962 he had fully developed this network model sending system that allowed for chunks of information to be sent and then reassembled at its final destination.

Commercial applications for these new network concepts were also being developed. American Airlines and IBM completed its Sabre project thus allowing travel agents to make airline reservations from anywhere with a terminal. By 1965 ARPA was sponsoring continued research into a cooperative network of time-sharing computer terminals by obviously taking advantage of Corborato's earlier work and Baran's communication efforts. With these accomplishments the value is not lost on others. Theoreticians and the like were already coining such terms as hypertext to mean any text that is linked to images, audio, or other text not in close proximity. Academics were also publishing papers on the value of this new network and how it will affect future endeavors in all disciplines.

The military and academia are truly working together at this point. By 1969 four universities were selected to become hosts for this new network. The military was also outsourcing development of hardware and software from private corporations to meet the demands of this new form of communication. New devices such as packet switches (information message processors) are being developed by a company in Cambridge, Massachusetts. The company selected for this chore is Bolt, Beranek and Newman. They would come to be known simply as BB&N. These were the new young Turks of this fast-exploding era of communications. These were the true pioneers of this breakthrough venue for doing just about everything (Pelkey, 2021).

2.2.29 No Matter Where You Go, There You Are

In elementary school we were taught about the advent of flight just a mere 100 years before. How individuals like the Wright brothers dared to envision this new form of travel. Many often wondered what it was like to be alive during an era when everything changed. Now we know what that feels like.

In 1971 BB&N was sending the first network e-mail message. By 1972 the same group created an e-mail program capable of sending personal messages thus allowing ARPAnet to enter the first commercial stage of its development. That same year Telnet allowed users to log on from wherever they had a terminal that was connected. By 1973 a new step was taken when file transfer protocol (ftp) was developed. This allowed for long distance transferring of files from one computer to another. That technology is currently referred to as "ftp-ing." Now no longer would this new network be constrained to the United States, now England, France, and Norway had connections.

By 1974 Robert Metcalfe started the concept that would develop into "Ethernet." The memo is called "Ether Acquisition" and describes how Ethernet would work. The basis for which is the packet-switching protocol. Information is sent with certain coding that allows it first to be segmented and then to be gathered back up at the end of the route. The information is buffered en route. Metcalfe is part of Xerox PARC (Palo Alto Research Center) and it is here where he develops the first local area network (LAN). There will be a lot more about PARC later in this chapter. PARC is the epicenter for most of the computer technology generated in this country during the late 1950s to the 1980s.

The reason why this is so important is because this now implements the concepts of transfer of information over an inexpensive efficient structure. The basis is that it runs over lines similar to telephone lines and requires a simple interface card plus the software. There was no mammoth infrastructure to develop just simply point a to point b. Now introduce a router or hub (basically a directional unit) and you can break this out to a vast amount of end users. Ethernet became an extremely successful means by which to transmit information over relatively short distances and is still used to this day. After LANs became successful the next step was the WAN or the Wide Area Network. This utilized boosters to help maintain the integrity of the signal from router to router and hub to hub. By no means did LANs or WANs take the place of ARPAnet (Internet). The structure was still growing and being more and more widely accepted by academia and the government. The first commercial e-mail service, OnTyme, was launched in 1976. It quickly finds a very small and limited market.

This is due to the obviously small (in comparison to today's standard) install base. This phenomenon occurs every time new technology is introduced.

2.2.30 Building a Better Mousetrap – Optics Broaden the Net

Give a user a fish and he'll input for one day. Show them how to fish and they'll slow down your connection to a crawl. Getting everybody into the same library is a great idea from which to exchange ideas but think about the bottleneck at the checkout. Enter fiber optics.

In 1977, companies started developing protocols to work with fiber optic cable. Prior to this, networks were running over conventional telephone lines. As discussed before, the amount of information distribution possible over this new structure was geometrically greater than simple copper lines. Latency would be a greater issue. Fiber optics was first developed in theory in the late 1950s. An amalgam of scientists fed each other, some intentionally some unintentionally, with information.

2.2.31 Time to Come Out of the Closet

As stated before, latency of the substrate is the issue. The glass fibers used in the fiber optics worked well but the natural impurities in the glass slowed down the signal. Once the impurities were removed it was now possible for light to move as far as one kilometer with a significant readable percentage of the signal left in tack. Remember this is moving at the speed of light, the fastest speed known to man. With these impurities removed information passes quicker and with greater abundance than ever before. Latency, however, has continued to be the problem, not in the fiber optic cables but in the switches that route the signals to end users. Recently this problem has been overcome as well. Now optical switches are in place. Companies like IBM, HP, Sysco, and start-ups like Cassion are producing optical switches for use in fiber optic networks. These switches allow the signal to not be compromised. Today consumers are starting to reap the benefits of this technology with cable modems, digital television, speedier networks, and faster teleconferencing. People like Will Hicks, Paul E. Green, Jr., Robert D. Maurer, and Elias Snitzer were the pioneers of this new telecommunications technology. Elias Snitzer is continuing his work in the field of fiber optics at the Rutgers Optical Center in New Jersey.

2.2.32 Let's Get Ready to Stumble

By the late 1970s companies such as Compuserve and The Source were created to act as public information services. That same year Usenet, a public use network, was set up to allow individuals to exchange information and concepts via messages to user groups and newsgroups. More and more individuals are starting to use this new infrastructure. The word is spreading. Now the competition of the capitalistic society has its framework in place. Make the better mousetrap, a very appropriate way of looking at the issue. How do we get the viewers? Remember OnTyme? It had the great idea first, similar to Tesla and Marconi, Muybridge and Edison, Gates and Jobs. The idea is not so much the question as is "what do we do with it?" For years the question has been heard at conferences around the country "How do you make money on the web?" The answer has always been the same, "make a better mousetrap." Market your products and ideas to the right consumer at the right cost. It's as simple as Orchard Street gone worldwide.

By the late 1970s different delivery methods were being developed. Digital videodisks are starting to appear. Cellular telephones are being developed as far away as Japan and as close to home as Chicago. This new networking thing is great. It is just slow as all get out and its connections are always filled. Organizations like Because It's Time Network (BITNet) and the Computer Science Network (Csnet) are starting up. We know what the message is, now how do we get it out as fast and as cheaply as possible.

2.2.33 Back to the Future Part 138

Now that the early days were coming to an end it became necessary to start to utilize this new technology for what it truly could become. The early 1980s were the launching pad for all this mishmash of concepts, actualities, and hopes. Now we had to figure out what the hell to do with all these wires, switches, circuits, motherboards, cables, and the like. And to think we actually accomplished it with a 73-year-old in the White House.

In 1982 commercial e-mail service began with users circling between 25 cities. The word "Internet" is relegated to this network of connected terminals. Voila! It all begins. Later this same year Bob Kahn and Vint Cerf developed TCP/IP (Transmission Control Protocol/Internet Protocol) and it became an Internet standard of operation. TCP/IP is a set of connectionless transport protocols that are still used to support the Internet's services. This new standard will control, check, correct, and track any transmission and its errors. Now we have a standard, something the boys are still fighting over every day. In 1983, the acceptance of this standard heralds the "creation" of this new distribution network – the Internet ("Dr. Vinton Cerf and Dr. Robert Kahn," n.d.).

By 1984 users spawned the need for the Internet to get organized. There are now so many sites of information that it becomes necessary to designate domain names. Therefore, a system has to be devised, hence the DNS (Domain Name System). Around this same time the subculture of society is becoming aware. Acting as it always has, like the end neuron of the sociological endocrine system; writers such as William Gibson create work like "Neuromancer" and in it coin the term "cyberspace." Artists, writers, and performers alike are soon to follow.

2.2.34 The Web Starts to Be Spun

As was stated earlier in this chapter the Internet started as ARPANET, a project funded and backed by the government via its Advanced Research Projects Agency. The goal being to send information to many replicated systems instantly – or close to it – for reasons of security. The one thing that the Internet could not do easily was send and receives images.

In 1980 a gentleman named Tim Berners-Lee was working at CERN – (Centre Européen pour la Recherche Nucleaire). Berners-Lee wrote a paper called *The Enquire System: A Short Description* (Tim Berners I Lemelson, n.d.-a). In it he outlined how to invent something that would enable anyone to format information and have it stored through a new system (over Internet lines) where it would be able to be retrieved at any time from any node connected to the network. After conceiving of this protocol Berners-Lee had to develop a simple easy-to-use language through which information could be posted and retrieved. Berners-Lee utilized Ted Nelson's concept of HyperText to create the basis for Hyper Text Markup Language. In 1984 the Internet and TCP/IP debuted at CERN and by 1989 Berners-Lee outlined the advantages to a hypertext-based system and garnered funding to make it come about. Later Berners-Lee worked on developing a browser that would enable users to simply create, edit, and view documents already created in HTML. Once this was accomplished the next step was to create a method of viewing this information and to have servers capable of storing it and retrieving the information. This was accomplished through the CERN and it subsequently housed the first web server and a copy of the browser.

The first browser developed by Berners-Lee to view these pages done in HTML was called World Wide Web due to the fact that if you wished to view anything contained in this format this was the only way you could see it. The next stage was to utilize the World Wide Web to distribute greater values of information. Now that the protocols and procedures for working on the World Wide Web were available to all, new products were being introduced. One such product was the result of a graduate student named Marc Andreessen.

While working with the National Center for Supercomputing Applications–NCSA, at Illinois in 1993, Andreessen and a team of graduate students debuted their first version of what they called "Mosaic for X." This was a point and click browser based on the graphical user interface system.

The beauty of the system was that Andreessen had implemented the <img> tag created by Ted Nelson and used by Berners-Lee in his browser; this led to the ability to display images across the World Wide Web. Later that year the group released its first version of Mosaic for Windows and Macintosh operating systems. It was distributed as a freeware product.

Later in 1993 Andreessen left Illinois to work in California at a small start-up software company. He later met the founding father of computer-based three-dimensional modeling and animation Jim Clark. Clark was much older than Andreessen and had spent many years in academia and subsequently many years in the new technology haven of Silicon Valley. Clark's huge funding from his SGI dealings allowed him to start the company "Mosaic Communications Corp." In 1994 that name changed to Netscape and then the World Wide Web revolution started.

2.2.35 Making the Revolution Readable

Once the World Wide Web allowed for graphics to be used the transition from print designer to web designer was coming fast and furious. Many have attempted to make the transition but none so successfully as Roger Black. Black's history in print is astonishing. He is responsible for – as he says it – "designing more publications than most people have ever read." His resume reads like a who's who of print design. Black has been art director or designer of *Rolling Stone*, the *New York Times* magazine, *Newsweek*, *McCall's*, *The New Republic*, *Fast Company*, *Reader's Digest*, *Foreign Affairs*, *Advertising Age*, *Esquire*, etc. (Heller and Ballance, 2001).

But Black has also been heralded as the one designer who truly made the transition to web design without ever compromising quality and design ethics. One of his more famous web incarnations was the website @Home. After that Black was one of the original founders of the Interactive Bureau in 1994, which then merged with circle.com. Among his major online clients were MSNBC, USA Today, BarnesandNoble.com, and the Discovery Channel Online just to name a few. In Chapter 4 on design there is more on the scope of Black's (and other designers) work.

2.2.36 The Icon's Icon

The one person responsible for that nasty "bomb" when your Macintosh went crazy and many other computer icons for clients such as Microsoft is Susan Kare. Kare's computer interface career has now spanned almost 40 years. Kare started her career in 1983 with Apple Computer via her friendship with Andy Hertzfeld, an Apple programmer. She developed for Apple during the years 1983–1986. Her job was as an icon and font designer for the new computer the Apple Macintosh. Kare client list reads like who's who of the computer industry: Apple Computer, Inc., Electronic Arts, Handspring, IBM Corporation, Infoseek, Microsoft Corporation, Netscape Communications, Palm, Inc., SGI, Inc., Xerox Corporation, etc.

Kare (maybe inadvertently) has created a new form of visual communication, the computer icon but she has in the past reflected on the origins of these icons as part of the mosaic system originated by early civilizations. Originally a fine artist, Kare sees the relationship between her new forms of artwork and all levels of communication. Susan Kare is truly someone whose work has become part of the world's lexicon.

2.2.37 John Maeda – Creating a New Aesthetic

In the motherland of all new computer-engineering accomplishments sat a young man who built a brand new kingdom. No it is not Arthur and Guinevere but even bigger and dreamier; it is the Aesthetics and Computation Group at the Massachusetts Institute of Technology. In the halls of this institution sits the MIT Media Lab. Founded by Dr. Nicholas Negroponte and others in the late 1970s the Media Lab currently boasts as home to the Aesthetics and Computation Group. The ACG is now called the Physical Language Workshop and was headed up by Dr. John Maeda. Dr. Maeda was also the 16th president of the Rhode Island School of Design (2008) and now the vice president of Design and Artificial Intelligence at Microsoft.

In 1999 *Esquire* magazine listed Maeda as one of the 21 people to watch in the new millennium, quite an honor for such a young man. Dr. Maeda's credentials read like a dream résumé covering everything from the Sunday *New York Times* magazine cover to digital creations for corporate clients. He appears to be the end goal and the ever-changing morphing creative, all in one. According to the MIT website, Dr. Maeda's résumé showcases numerous awards and achievements.

> John Maeda heads the Lab's Physical Language Workshop. A recipient of some of the world's most prestigious design awards, Maeda has most recently been honored with the National Design Award and Japan's Mainichi Design Prize.
>
> *(John Maeda, 10/5/2023, n.d.)*

Dr. Maeda even has an Absolut ad to his credit – done completely in code. His work appears to be simple but yet once the geometry is studied and you realize that he is creating all of this from code you start to get the idea. The theory that could be presented about his work would go as follows: if what you see is pleasant to the traditional eye and it is completely digital (read code-based) then it would appear that the artist (?) has bridged the gap between the true cerebral and the true physical.

On the personal section of his website Dr. Maeda states his quote from an article in the *New York Times Science Times* written by Claudia Dreifus on November 11, 2003:

> Amidst all the attention given to the sciences as to how they can lead to the cure of all diseases and daily problems of mankind, I believe that the biggest breakthrough will be the realization that the arts, which are conventionally considered 'useless,' will be recognized as the whole reason why we ever try to live longer or live more prosperously. The arts are the science of enjoying life.
>
> *(NYT 2003, n.d.)*

The great part about this entire quote is that in it Maeda finally describes the relationship between science and art. This appears to be the essence of the entire effort by the Internet Art movement. How can we combine the beauty and logic of science, mathematics, and art? Even Duchamp and Poincaré both felt it necessary to explore. The real question may be why is it necessary? What concerns mankind so much that we need to see a merging of all thoughts and how or why do we need to keep it separate? In the next section we will discuss the growth of directing information to the home?

2.2.38 You've Got a Trail

America Online has become part of the American landscape from its supposed acquisition of an American empire – Time Warner – to the removal of its name from that marquee. The rise and fall of America Online (AOL) is truly the story of free enterprise and entrepreneurial spirit. Coming out of an amalgam of media giants and garage company work ethics, AOL was spawned from the belief that individuals would be willing to pay for specific services as long as they were delivered to them. This purchase pattern helped to build the company from its early lineage of CompuServe and The Source.

In 1978 Bill Von Meister founded The Source. The purpose behind this company was to send diverse forms of information and entertainment directly to people in their homes via telephone lines. This was the very first online service directed at the average consumer. Its business model was the impetus for AOL.

Future concepts generated by Von Meister include the delivery of music to home subscribers. This idea was conceived in 1980 but was quickly brought to failure due to the fact that the music industry felt it would destroy their retail sales. The next stop for Von Meister and company would be the computer game industry. In 1982 Von Meister debuted the GameLine Master Module for the Atari at the Consumer Electronics Show. The only problem with the product was that it actually didn't exist yet.

But then again this never stopped a true entrepreneur. One of the original companies that invested in Control Video Corporation (CVC) the manufacturer of the GameLine system was Daniel Case, brother of Stephen McDonnell Case – future CEO of AOL. In 1983 Steve Case viewed the work of Von Meister at an industry convention and became fascinated with the concept of delivering information to the home computer user. Case, himself, was a subscriber to The Source since its inception.

By 1985 CVC changed its name to Quantum Computer Services (to help shed its reputation of failure) and started to work with Commodore Computers. Commodore requested that Quantum get its business in order and license the rights to PlayNet's software. Quantum's next moves were to continue with the concept of online gaming and delivery of content to subscribers. Quantum Link or Q-Link debuted at 6 pm on November 1, 1985, to a mere 24 customers. But nevertheless, the basis for AOL was born. Q-Link focused on developing with Apple Computer to provide a base for home linking of consumer's computers. Apple Computer was one of the biggest promoters of the new service in the early 1980s. But given the extremely keen minds at Apple, they cut ties with the new company by 1989. To make matters worse Q-Link's service (then called AppleLink) requested and got $2.5 million from Apple Computer to relinquish the use of the Apple name back to Apple Computer. In 1991 the DOS version of AOL was released and now the service becomes totally cross platform. By the end of 1991 Quantum became America Online.

2.2.39 Opening the FloodGates

By the beginning of 1992 Microsoft started to get the idea that this new Internet thing might just take off. Bill Gates and Paul Allen, a partner in Microsoft with Gates, attempt to buy AOL. This of course produces a tremendous amount of interest in AOL as well as all online services. AOL refuses Gates' attempts to purchase the company but strikes up a deal to develop an online service for Microsoft. The question is – was this all Gates ever really wanted?

By the end of 1992 AOL licensed software to Apple to help them create Apple's eWorld. At this same time Mosaic/Netscape was born and allowed for AOL to link to graphics. By the middle of 1993 (by this time the CD-ROM has been in everyone's computer since the late 1980s), AOL started its barrage of CD-ROM promotional mailings.

Now Bill Gates tastes blood in the water and starts to focus on his next meal. However, the beginnings of Microsoft's web endeavors were not very auspicious. The entire venture started out with one computer tied to the Internet and tucked beneath a desk at the end of a long hallway (or so goes the story on Microsoft's website). Since these early days in 1994 Microsoft's venture into the web has been focused down to its embedding of its Internet Explorer browser into the main structure and culture of its operating system. Gates' struggles with the U.S. Department of Justice have become legendary. But it wasn't just all "shock and awe," there were simple starts to understand.

Originally microsoft.com was an offshoot of the company's quarterly newsletter. In this newsletter were software update information and other bits of content. The publication was sent around on CD-ROM written in SGML. Soon the information could easily be transcribed from SGML to HTML (Hyper Text Markup Language) and repurposed for the website. The only problem was that the information was being requested on not a quarterly basis but because of the web's tremendous reach it was being requested on a daily basis. So this meant that update implied daily update not quarterly. From these early requests it became apparent to the management at Microsoft that the web was here to stay and to grow. But what was going to make the web a true piece of the landscape was usability for all and for all a good night and an early morning rise. And what better way to rise than with the sun.

2.2.40 SUN Also Rises

To tell the story about Sun Microsystems is to write 100 books. Sun is the epitome of the start-up that goes incredibly well. Originally started in 1982 by Andreas Bechtolsheim, William Joy, Vinod Khosla, and Scott McNealy, Sun Microsystems has become the standard for networking computers in the world. Now in its 40th year Sun has given birth to such standards as Java, Solaris – operating system, NFS-network file sharing standard, UNIX(R) System V, and an entire cast of revolutionary computer networking standards. Truly a groundbreaking corporation and influence not just on the computer marketplace but on the entire world of commerce and ecommerce.

Announcing its Initial Public Offering in 1986, Sun sets the standard for ridiculously successful IPOs. The difference for Sun is that its IPO is realistic and probably by today's standards a steal and undervalued. This influx of money allowed Sun to open offices around the world. But what it may be known best for is a little thing called Java. Java is a completely universal software platform designed from the bottom up at Sun Microsystems. The uniqueness of this platform is that it allows developers to write a program once to run on any computer. Truly allowing for information to be shared freely from platform to platform a true breakthrough in the proprietary world of computer lingo and language. Please realize that Java and JavaScript are not related at all. JavaScript was simply a name given to a language in order to be associated with a successful venture or so the story goes.

Sun's influence on the web can also never be overvalued. Its Sparc/Solaris workstations allowed for incredibly fast Internet access and relays. Sun's workstations and overall technology revolutionize the way we use the web for the delivery, retrieval, and storage of information like no other company has done before a complete and utter slam-dunk.

To cover this topic is impossible in a short section but one should visit http://www.sun.com daily to read the new innovations that this company is developing. They have changed the world more so than Apple or Microsoft and they have done it without the overt controversy and manipulation – they simply make good products.

One of the original founders Dr. William Joy wrote an article for *Wired* magazine entitled, "Why the Future Doesn't Need Us." This was such a groundbreaking work that it is up there with Negroponte's "Being Digital." A must read for all involved in this cyber world. But what really happened to all the hyperbole and the rush to fame and fortune? In this next section, we'll discuss some issues and propose some hypotheses – like the burst of the bubble and why it came back. How and why did the Internet bubble burst? How could something that progressed at such a rate suddenly come to a halt and why has there been nothing more than a stabilization? Who was the first company to start the slide and what kind of thinking created this movement toward stagnation? Back in the heyday everything was going at a rate that was truly unbelievable. Why and how did this hype come about and when did it start.

The start of the Internet hype or more accurately the hype of the World Wide Web came about approximately in 1993. Suddenly everyone was hearing about this new Gold Rush. The flow of information riding telephone lines from your home into the bank was going to be the economic wave of the future. Many doubted its ability to survive and in 2000 the dot com bubble burst. Many believed it to be hibernation. And now it has come back with a vengeance.

> In 2022, the reported total value of retail e-commerce sales in the United States amounted to over one trillion U.S. dollars – impressive, but the figure pales compared to the total annual retail trade value of seven trillion U.S. dollars!
>
> *(Coppola, 2024)*

These statistics are true today but why did it slow down?

2.2.41 Copper, Fiber Everywhere and Not a Reason Why

What happened to all the momentum? Some industry experts believe that one of the major reasons for the slowdown in the growth of Internet momentum came from the user base just not desiring the connection speed that the producers or suppliers expected them to want. Simply put companies like WorldCom, Lucent, and Global Crossings expected users to not be able to live without high-speed Internet connection. They leveraged a lot of money on these desires and some – most – were caught short when the market need did not consume the supply. Many suppliers were caught with a glut of copper and fiber connections. Some even went to jail over IPOs not reaching goals. Sounds strange but true. However, they were both right and wrong.

2.2.42 Right Product but the Wrong Generation: The Start of Gen-i

Another main reason for the slow movement in the growth of the Internet has been the fact that the generation that it was intended for is now only just materializing. Generation-i is only in its infancy. It is only now appearing as the main consumers of this new medium. Gen-i is based on all forms of interactivity, Internet access, inter-connectivity, isolationism, immediacy, intensity, intelligence, information, inoculation, infusion, ingenuity, and informality, infectious, etc. They are ready, willing, and able to digest at a tremendously increased speed. The only question is about comprehension. Many are 30 years old and don't remember a world without the Internet or barely remember dial-up access. This is the generation that this new medium was intended for. They understand how to properly consume and digest it. We will have to wait and let the two merge together to see the real worth and benefit of this Gen-i.

The shift was slow to come and it may have not met some company's proposed business plans but it is coming. The companies who have successfully made the journey were the bridging companies. These are the ones who have one leg in the last millennium while spanning the digital divide into this new millennium. These are the companies that have the consumables of this new future. These are the ones who are dealing in Negropontes bits while still being able to charge consumers for atoms. They are the ones who have waited for the new currency of this brave new world to congeal. But what kind of a new world will it be for the future of online services bits or atoms? A student recently said in a discussion that we all feared an Orwellian future but we never noticed that Huxley had it right. It wasn't about being watched – although we are, it was about making sure we consumed – which we do. But we are also being watched closely.

2.2.43 Cut Off His Food Supply

In the early days we saw the introduction of broadband come about like the droplets from a gerbil's water bottle. Slowly everyone is getting a taste as we run around on the wheel providing the fuel and the desire to get more. We stay online longer to download and have generated phone bills to subsidize the development of faster connections. The bad news is the technology for faster connections has been around since the late 1980s so why the delay? Simple, if you were in business to provide a service that is basically inexhaustible why would you give all you have instantly to a market without seeing just how much they would pay over a longer time? That sounds like a contradiction but take it apart to really see it. Free to Freemium.

You have the ability to provide fast Internet access to all consumers, but at what price? Now you can have two business models to follow the slow raise or the fast raise. The slow raise is the one being done by the cable companies and the fast raise is the one the oil companies use. The slow raise looks like this: remember your spouse – you don't necessarily notice the weight gain if you are too close but once you step back you may notice it. So if you are looking to raise your rates you do it slowly so your consumer gets adjusted to it and no one screams too loud. The mistake is to gain 15 pounds overnight. Everyone would scream. This is the approach the oil company uses. Just hit them hard and let them scream – the cars still need the gas.

So why were the cable companies so slow to provide the service? Simple if they went too fast we might figure out that we don't need the cable to make the connection. We might have gone to cellular or satellite. The competition will get the jump and the consumer will have the initiative. Well if the technology exists does it just have to become affordable? Personally I don't buy that answer but nevertheless here we are. Cable bills have gone up 300% in the last ten years while service is almost exactly the same. So is there an answer? Not really we just sit and hold onto our phones waiting for the connection to come through. But the real question is what will be delivered and will it change our lives? How fast will we get it delivered and how will the content adapt to our needs? If you don't believe that it has become a necessity just try turning your e-mail off for one day. Try letting your kids do a homework assignment without surfing the web for the information. Try not touching your smartphone for just one day and see what happens – you'll look like Ray Milland from *The Lost Weekend*. Don't know the reference? Well you're in luck just Google it.

2.2.44 Section Summary

First and foremost it is necessary to state that thousands of names have been omitted from this section. This was not a purposeful act. There simply is not enough room for them all. The intention of this chapter was to simply whet the reader's appetite to find out more information. Pioneers have generally been forgotten throughout time. This is the shame of innovation. It is the cost and chance of infamy. Some will make the headlines and some will make the work happen. A Machiavellian attitude wouldn't hurt in cases like these. The smart ones knew the score and figured out the end before it happened. Never forget the difference between Tesla and Marconi. One was a genius and one was ingenious.

Technology is forever changing both itself and the generations that it affects. This is something that can never be stopped. From the days of Ned Ludd to the current days of Google and Meta, society has been on a collision course with its own desires and subsequent needs. Where we wind up is partly due to our own course and our own steering mechanisms. We build them, we need them, and we own them. There is almost no going back but there is also no forward – there is only today. A great quote from Einstein paraphrased for this occasion – "time is only necessary so that everything doesn't happen at once."

2.3 INFORMATION ARCHITECTURE (IA) – FROM DaVinci TO KRUG

2.3.1 "IA is Like Building a Bridge That You Hope Someone Will Need to Use..."

Information Architecture or IA as it is known in the field of UX design is a separate and integral part of the process. Simply put in 1999 by Shil Kimen, "information architecture is a combination of organizing a site's content into categories and creating an interface to support those categories." Why is this important? This is the problem-solving moment. All the technology is of no worth if it is not producing the desired result.

The process of Information Architecture initiates the entire production cycle from the original sit-down meeting with clients to the final production stage. At the start of a UX design project there would be a sequence of initial meetings commonly referred to as a listen and learn. A site map or flowchart and production schedule is devised. From this initial meeting the technical and design teams would start their workflow. From this point the technical team writes up a series of technical specifications. These specifications are then brought to a physical stage by the visual design team and later by the coding team. Both portions of the project now start to mesh and the project then comes to a first stage of usability and functionality testing. This process is repeated until a final product is readied for launch. Overseeing this entire process is the information architect or product designer. The qualities of an information architect are at best diverse and at worst schizophrenic. They must know a lot about every part of the process and should be able to almost fill in where necessary being part visionary, part designer, part programmer, part salesperson and all production manager. This section is focused on how IA came to be and who were some of the philosophers in the original development of science.

2.3.2 DaVinci the Hacker

Some scholars believe that Leonardo daVinci was one of the original information architects. DaVinci's method of study and achievement appears to have always been requisite of a plan and a follow-up process. Truly called the first "Renaissance Man," daVinci's models and projects spanned many disciplines from medicine to painting to architecture. Known as the Architect of Florence, daVinci was responsible for the planning of the city and later as a premier originator of the defense of the city. Although these capabilities are attributed to Leonardo, due to his inability to read Latin he considered himself to be an "unlettered" man. Amazing in comparison to today's students, who think that because they received an 800 or over on the SATs they automatically become a member of Mensa.

Nevertheless, some consider daVinci to be one of the original "hackers." In her article "DaVinci: The Pith Behind the Man" (Wired, Jan 29, 2003), Michelle Delio states, "That compulsion to tinker has led many modern hackers to claim Leonardo retroactively as one of their own." Giorgio Vasari wrote of Leonardo, in his classic reference work *Lives of Artists*, "In Arithmetic, during the few months that he studied it, he made such progress that he frequently confounded his master by continually raising doubts and difficulties." The ability to delve into the mechanics and structure of devices or ideas is what it takes to be a true information architecture.

Other accounts of daVinci's place in IA history are backed-up by the media design company Thread. On their site in the Thread, articles section "Design: The secret to good information and so much more," *Feb. 17, 2003*, D. Knemeyer writes, "At Thread, we follow a Design process based on that adopted by Leonardo daVinci, one of the greatest innovators in human history. His own personal Design process included four steps: Collect, Create, Relate and Donate. Here is our four-step process, loosely based on that matter.

1. Discover – Learn the environment. Define the objectives.
2. Plan – Determine the end product. Structure how the end product will be created.
3. Produce – Create the end product.
4. Manage – Deliver the end product. Ensure that the end product successfully accomplishes its objectives, both now and in the future."

With this being stated you will start to see that good planning is as appropriate now as it was in the 15th century. There are many others who believe in this structure as well. Some of these philosophers are Edward Tufte, Nathan Shedroff, Jakob Nielsen, and Steve Krug.

The reason for selecting these philosophers of UX design is that they are almost diametrically opposed in their philosophies. Tufte believes that the proper use of graphics and any other form of information architecture or distribution can make all the difference. Nielsen believes that graphics are almost not necessary, where Krug asks that the designer should present in a manner that makes the user say "Don't Make Me Think." Now all of these individuals are correct to some degree but let's present their respective histories and part of the base philosophies (Krug, 2006).

2.3.3 Tufte – Visual Design

Edward Tufte is professor emeritus at Yale University, where he taught courses in statistical evidence, information design, and interface design. Professor Tufte has written books and numerous articles on the display of information. Tufte has also won to date 40 awards for content and design. Tufte's main premise is that he believes in the use of graphics to portray an idea. He also believes that an idea can be misrepresented if the graphics are used improperly. These concepts are brought to light in his famous trilogy of books – *The Visual Display of Quantitative Information* (1983), *Envisioning Information* (1990), and *Visual Explanations* (2002).

Tufte believes in the use of graphics to display an idea – this differs from the beliefs of Nielsen. Tufte believes that graphics are the key to good information and reiterates this by relating to E.B. White's statement, "No one can write decently who is distrustful of the reader's intelligence, or whose attitude is patronizing." Obviously, Tufte believes in the opposite of Krug and Nielsen in so much that he wants the reader or user to think. This trilogy also contains many examples of strong graphic visualizations such as ancient Chinese maps to charts of entire galaxies. Tufte believes in a common-sense approach to site organization and to the basic principles for data design. Similar to the beliefs of Nielsen and Krug are his beliefs that the designer should remove unnecessary ink or content. Tufte believes that the job of an information architect is to defend the simple usability of a site against the bureaucratic superiors and their desire to make simple – complex (Tufte, 2012). His structure is so similar to what we discussed before that in a review of his books Tufte has been called "the Leonardo daVinci of quantitative information" by the *New York Times*.

As stated by Krug, "the visual display of quantitative information isn't a very sexy phrase" (Krug, 2006) and it might have been true if it were not for the world of computers and their ability to display such information. But if it wasn't for the fact that the world is truly immersed in the use of these machines – whose main purpose is to display quantitative information – Tufte may have passed unnoticed from popular view. But now his theories and work are pointed directly at the heart of what is now the largest revolution in societal matters since the industrial revolution. Today, to truly communicate in an era where according to Esther Dyson, "Content is king," we will need to understand better how to create, digest, and rework the numbers and data that is being delivered to us on a minute-by-minute basis.

2.3.4 The Unified Field Theory of Design

The Vienna-based International Institute for Information Design was formed by Nathan Shedroff, chief creative officer, Vivid Studios in his paper "Information Interaction Design: A Unified Field Theory of Design." Throughout this paper Shedroff refers to a process not unlike daVinci's model. Shedroff strikes together the concentric circle idea of information design, interaction design, sensorial design all of which is being held together by the needs and manipulation of content. These circles are what he believes to be the structural components of information interaction design.

Information Design, according to Shedroff, is the means by which a designer addresses the organization and presentation of data – its transformation into valuable, meaningful information. This sounds similar to Tufte's beliefs but seems to go further toward actually putting it all together in a more practical method.

Interaction design is essentially the art of story creation or basic storytelling. It would appear that Shedroff means that the medium we select has a great deal to do with how we tell the story. This is true no matter what revolutionary delivery system has occurred and what effect it has had on the distribution of information.

Sensorial design is defined as the "employment of all techniques with which we communicate to others through our senses." He does understand that this entails a huge series of skill sets – audio, video, graphic design, and writing.

The theories that may be of greater interest are those of Richard Saul Wurman (who Shedroff attributes in this paper). Wurman expresses in his book *Information Anxiety* that "data is fairly worthless to most of us" in a raw sense. "To have informational value, it must be organized, transformed, and presented in a way that gives it meaning." This appears to be where an information architect goes to work. Curiously, Wurman does construct a framework similar to that of daVinci's. Data (research creation and gathering) goes to Information (presentation and organization) goes to Knowledge (conversation and storytelling) finalizes in Wisdom (contemplation, evaluation, interpretation, and retrospection). This appears to be a grander version of the Thread Studios structure (stated by Knemeyer) modeled after that of daVinci. Great minds think alike!

These theories seem to make perfect sense and seem to be more so involved with the current diet of the multimedia world. However, they are linked to a full multimedia experience – film, video, and performance more so than the limited world of the Internet or CD-ROM.

2.3.5 The Nielsen Rating

Jakob Nielsen has nothing to do with the television rating system. But he does have a lot to do with the rating of usable website structures. Jakob Nielsen, PhD, is a user advocate and principal of the Nielsen Norman Group, which he co-founded with Dr. Donald A. Norman. Up until 1998 Nielsen was a Sun Microsystems' distinguished engineer. He has written widely on usability and has published his semiweekly alert box column since 1995.

Having coined the term "usability," Nielsen still holds court even today with his theories on web usability and overall content design. The main difference between what we have discussed and Nielsen's theories are his original dislike for graphics. On his site http://www.useit.com, Nielsen states several reasons "why this site has no graphics." Nielsen states that download capabilities is one of the reasons for not supporting graphics. Since not all users have DSL or cable modems web pages need to be 3 kilobytes or smaller. Users, he believes, will leave the site if download exceeds 10 seconds. This corresponds to 30 kilobytes per second modem speed. But the part that appears to be the actual truth is that he states, "I am not a visual designer, so my graphics look crummy anyway." Plus since useit.com is his personal site he does not want to spend money on a designer. So now we kind of see why he may not like graphics. This is directly opposed to Tufte's beliefs that graphics tell a greater story than text alone although Nielsen does, however, like to have a large photo of himself (also available is a high-resolution version) on the site. These graphics he approves of. His process is simple and does suppose the initial client meeting. In an interview dated 2001, Nielsen spells out his process for site design.

- Simplify both text and graphics
- Use plain language
- Adopt existing conventions
- Be specific
- Reveal content by examples not hype
- (working with momentary fads to display information)
- Show benefits
- Carefully copy edit

But he does go into greater detail to explain to designers how to improve their sites' IA and its usability. Nielsen is a great advocate of basic usability studies and he does also favor test groups. He actually coined the term. He also believes that users should be involved throughout the entire design process. As methodology to improve sites, Nielsen states the following criteria.

1. Determine goals and write user tasks.
2. Determine the user profile.
3. Conduct the session.
4. Evaluate the data.

There's a lot more you can do. Even after all the basics are accomplished Nielsen believes that this is just the start. Once the first round of testing is over he prefers or recommends the design team go through other rounds to be sure the launch is perfect. Nielsen believes that it is almost impossible to recover from a bad launch.

Nielsen comes to the IA forum with great credentials. However, he still appears to prefer staying distant from the term "graphical user interface." This is where the next individual seems to combine the issue of text and graphics.

2.3.6 Krug – Don't Make Me Think

Steve Krug is the author of the famous web usability book *Don't Make Me Think*. Mr. Krug also practices what he preaches in so much as he has been commissioned by numerous well-known companies such Apple, AOL, Netscape, BarnesandNoble.com, Circle.com to either evaluate or redesign their commercial or corporate websites. In it Krug's main message appears to be that users do not want to think. They simply want to find what they are looking for quickly and easily. This is true but at what cost?

Krug's main credo is that web pages should be self-explanatory. Nothing should be cryptic. If a button is supposed to be a button then make it look like one. Now this may be considered by some to be design but if "form follows function" then the button design is as critical to the IA as the site map. For if users cannot work the site then the IA is irrelevant. He states, "when I'm looking at a page that makes me think, all the thought balloons over my head have question marks in them." This point is correct but at what level are estimating our users to be at? Surely this estimation has changed with the web starting its tenth year of common popularity. Also what about the factor of marketing and uniqueness to the product image? (Krug, 2006).

Krug also believes in setting up your site as if to be a marketplace where aisles are readily available. This seems to be the most agreeable part of his theory. Enter the store, find the right department or aisle, make the purchase and leave. This is a straightforward flowchart. In and out with a goal accomplished. This is great for the user but what about the marketer? Doesn't he/she want you to stay awhile and take a look around? What about putting the candy by the register so that the children can whine and complain? When you are designing a site or more correctly when you are creating the architecture for a site, aren't you planning for the most "bang for the buck" philosophy?

Krug also refers to Nielsen when he speaks about "search dominant" and "link dominant" users. These are the individuals who want to come to the site and quickly input their search requirements or click on a link to make the appropriate item appear. Good for the user, bad for the marketer. Krug also makes a strong distinction between focus groups and usability testing – two strong components of the IA structure. Focus groups are typically 35 people who are reacting to designs or ideas that are shown to them. One person at a time typically does a usability test in order for the one person to figure out what they are looking at or whether they can accomplish the specific task at hand.

Now Krug is more so a design critic than an information architect. But what you need to understand is that one works hand in hand with the other. If the site map or overall plan is confusing the design will ultimately be as confusing as well. Not the specific graphics but the relevance from page to page not even mentioning the navigational system design or typography. An item's dominance on a page is a method of directing the user. Krug understands this very well and is truly focused on a common-sense approach to design and usability (Krug, 2006).

2.3.7 Section Summary

In this section on information architecture, the four individuals discussed presented either related or unique viewpoints to the nature and essence of the field of IA. Some viewpoints were practical, some philosophical, and some appeared to be just personal beliefs. After reviewing their philosophies it would seem that all in all the field of IA is basically in its infancy even after ten years of popular web usage. The one thing that becomes apparent, however, is that usability, functionality, visual design, marketing, and programming are all intertwined. It's how you put them together and for what purpose that either makes the project a success or a failure.

How we use new media is still to be determined. The advent of AI, VR, AR, avatars, holograms, human computer interaction, nanotechnology to name a few will change any viewpoint that we have today. Although it may be true that usability and functionality are the keys, what will enable a site or app to truly stand out and garner the market share necessary to continue growth and expansion?

Another question is what will enable the message to be personal if we adhere to the doctrines of these or other usability gurus? How will each message stand out from the clatter of all the rest? Although it is true that form follows function, what is the function of the particular project in question? Is it a promotional, entertainment venue, informational distribution medium, or an educational environment? Obviously, these environments are all different – each deserving of a unique and richly formatted design. If form truly followed function then wouldn't we all drive the same cars, wear the same clothes, and look exactly alike? Is uniqueness a necessity or is that meant only for fine art?

Although it is true that finding what you want is at the base of all interactivity – who says what you want? – just the user? If that's the case then where do we find progression? Or maybe it will stay the way it has always been – maybe the marketer is telling you what's available and we'll just kind of pick what is there? The only way for this new media world to progress may be by getting dragged forward by innovators and originators.

But what was our first entry into this world of information – basically words. The first websites were text based with no imagery, hard to believe today but originally the online experience was to read!

2.4 CONTENT SELECTION AND CREATION

2.4.1 From Bush to Berners-Lee

2.4.1.1 SGML for All

Well before the future of the personal computer had all but been finalized, the first step for all of these computations was practical use. One of the first industries to take charge was the world of computerized typesetting. Typesetting is the art and science of imaging letter shapes onto a page for print purposes. In other words, how do we turn typewritten or text editing copy into advertisements, books, magazines, etc.? Well since most word processing software has its own version of text markup built into it a standard had to be developed that would allow the coded text to be readable by other computer systems. This was done as a solution to the problem of different computerized typesetting systems using different markup language in order to change text from regular face to bold face, etc. Back in 1967 at a meeting of the Canadian Government Printing Office, William Tunnicliffe presented a talk on the separation of information content of documents from their format. This was the start of a system that was later called Standard Generalized Markup Language – SGML. At the same time Stanley Rice, a New York book designer, devised a set of parameterized "editorial structure" tags.

From this start the Graphic Communication Association initiated-meetings to continue to implement and develop this concept. From this process was born the GCA General Coding committee or Gencode. Taking off from this standard a group from IBM-devised Generalized Markup Language. This GML was based on the generic coding concepts of Tunnicliffe and Rice. The next goal for this standardization was to make it possible for any markup to be interpreted by either a human or a program.

Once these standards had been agreed upon, the next step was to devise a protocol. Since typesetters had used a variety of coding for font and style changes, the working consortium was set up and coding formats were established by the late 1970s. This format is in use to this very day but from it grew the style of tags used in developing HTML – HyperText Markup Language (Florida Cooperative Extension Service/ Institute of Food and Agricultural Sciences, 1992). But before this could be used properly other programs had to be developed (Longman, 2000).

2.4.1.2 C++ as an Average

Since the start of computer programming and even before the formulations and techniques put forth by the work of Pascal, Boole, Hollerith, and Babbage, the need to establish a set of techniques that would allow for less work when creating computer programs became apparent. But along with our previous discussion of the work of Alan Kay and all of the members of PARC, another name and another innovation needs to come into play.

In the late 1970s a computer scientist Bjarne Stroustrup started to work with the computer language of C to develop a way to give support to developing further work with object-oriented programming. Stroustrup's developments came to fruition as the language of C++. In his writings Stroustrup defines C++ as a "language supporting object-oriented programming by making abstraction techniques affordable and manageable for mainstream projects." Put a lot simpler what Stroustrup did was make it possible for developers to utilize OOP technology without having to reinvent the wheel every time they wished to start a new project.

Stroustrups' work has been heralded as one of the strongest contributions to the world of object-oriented programming in the 20th century. In 1995, *BYTE* magazine described him as one of "the 20 most influential people in the computer industry's last twenty years." Stroustrup had already been called one of "America's twelve top young scientists" at the age of 40 by *Fortune* magazine in 1990. With accolades such as these you would think his name would be as common as Gates or Jobs.

C++ is such a prominent computer language that literally thousands of programs, some of which we use every day, are based on its capabilities. On his website Dr. Stroustrup has put together a very brief list of just some of the applications that use C++ in their development. The list is staggering: Adobe systems' Illustrator, Photoshop, Acrobat, InDesign, GoLive, etc. Alias|Wavefront's Maya, Apple's OS X use C++ in some aspects, Microsoft XP, NT, Windows 95, 98, Me, Nokia's Mobile Communications radio-station/internet bridges, not to mention almost the entire gaming world. Macromedia's DirectorMX and FlashMX, two of the world's leading authoring tools for multimedia, were both developed using a C++ structure.

With all of this integration and economizing going on, what purpose would all this serve? What were all these programs and applications actually invented for? How could we just get the computer to create documents that could be replicated and distributed? Because after all wasn't that what started this entire rodeo to begin with?

2.4.1.3 PostScript – Illustrator and Photoshop

Object-oriented programming opened up an entire world for the graphics industry. As mentioned above programs like Adobe Illustrator and Adobe Photoshop broke open the world of computer graphics like no other program before. After leaving their positions at Xerox PARC John Warnock and Charles Geschke started Adobe Systems in 1982. Since Xerox did not see any future for Warnock and Geschke's new language – PostScript, the two moved to Mountain View, California, near the Adobe Creek hence the company's name.

The first products actually sold by Adobe systems were its digital fonts. The PostScript language doesn't count because it was licensed and not sold. But soon after came Adobe Illustrator. Illustrator is a vector-based drawing program that allows for the flexibility and finesse that Sutherland may have intended. Working with Apple Computer Adobe released this new drawing program as Adobe Illustrator 1.0 and quickly 1.1. An extremely powerful and reliable program for creating technical drawings and line artwork Illustrator has reigned supreme for almost 35 years. The true heart of this program is its use of encapsulated postscript format, which allowed for a high level of printing finals with a very small file size. Without its ability to encapsulate information it could have rendered file sizes even larger than Photoshop's raster format.

The next major breakthrough was Adobe Photoshop. Photoshop was the vision of two young men who were the sons of an academic and photo enthusiast in Ann Arbor, Michigan. Their names were John and Thomas Knoll. Their father Glenn Knoll was an amateur photographer and a computer hobbyist. Thomas became interested in the Apple Macintosh Plus and while writing his doctoral thesis he noticed he couldn't get quality grayscale graphics on the screen. So like any other ingenious twenty-something year old he decided to write a subroutine program to fix that program. This continued on into writing other subroutines and other programs. The story continues that his brother John who was home vacationing for his job at Industrial Light and Magic (George Lucas's company) noticed what his brother was doing. With both of these young men merging their talents and abilities, the Knoll brothers devised a raster image-based computer program called "Display." Raster is a method different from vector whereas all the pixels are used to display the final image giving it the ability to show a greater range of tonal values. The file size is much larger than vector but so was the image quality. Thomas' programming skills increased and John asked if he could make the final image files savable in other formats for use in other programs? These early efforts led to other versions, the most professional of which was called "ImagePro." Once this was stable John started to shop the program around by his company's offices in Silicon Valley. Many companies turned him down but finally they hit upon the offices of Adobe Systems. In September 1988, John was fortunate to be granted a demo at Adobe. The rest as they say is history.

Now that the world has the capability of creating both vector-based and raster-based images and they could be printed to a color laser printer, what other avenue of distribution should be conquered next? Although it was a good thing to be able to print out an image through a network next door, what about around the world in a nanosecond? (Pfiffner, 2003).

2.4.1.4 WWW and HTML

It is hard to think of HTML – HyperText Markup Language without thinking of the World Wide Web and it is hard to think about the web without thinking of the Internet and to discuss one it is necessary to talk about the others. The Internet started as ARPANET, a project funded and backed by the government via its Advanced Research Projects Agency. The goal of the project was to develop a network capable of replicating computer information or telecommunications over large distances. This development was possible due to the advancements made in computer technology – the ENIAC and EDVAC during and after World War II. The Internet was primarily used for sending text and file sharing. At its origin these were extremely slow moving and laborious tasks. But the one thing that the Internet could not do easily was send and receive images. In 1980, all that started to change.

2.4.1.5 Mosaic

Now that Berners-Lee and others had made the protocols and procedures for working on the World Wide Web available to all, new products were being introduced. One such product was the result of graduate work by a young man from the University of Illinois. His name is Marc Andreesen. As a graduate student he was fortunate to be working with the National Center for Supercomputing Applications – NCSA, at Illinois. In 1993, he and a team of graduate students debuted their first version of what they called "Mosaic for X." This was a point and click browser based on the graphical user interface system. The beauty of the system was that Andreesen had implemented the <img> tag created by Berners-Lee into his browser. This led to the ability to display images across the World Wide Web. Later that year the group released its first version of Mosaic for Windows and Macintosh operating systems. It was distributed as a freeware product.

Later that same year Andreessen left Illinois to work in California at a small start-up software company. He later met the founding father of computer-based three-dimensional modeling and animation Jim Clark. Clark was much older than Andreessen and had spent many years in academia and subsequently many years in the new technology haven of Silicon Valley. Clark is responsible for devising the geometry engines behind the Silicon Graphics Industry products used in movie making and special effects – the SGI machine series. At the time before meeting Andreessen, Clark had been working on what he called a "computer appliance." This idea was spawned from Time – Warner's venture into the ITV – interactive television marketplace. Once that didn't pan out and with huge funding from his SGI dealings, Clark decided to start the company "Mosaic Communications Corp." In 1994 that name changed to Netscape and from there the revolution started (Payment, 2007).

2.4.2 The Gangs All Here

From this point on the pace truly got fast and furious. Soon enhanced technologies were being developed for this new method of image distribution. Apple had already developed its QuickTime and later QTVR (QuickTime Virtual reality) technologies for digital movie creation, soon after both Windows and other third-party companies were quickly developing video and audio formats for distribution across the web. Companies such as Macromind, now by this time called Macromedia, had devised ways of using its flagship authoring program Director as a web development tool. Using its new Shockwave format, Macromedia quickly took hold of the two-dimensional animation and interactive file format delivery system for the web community. Macromedia has stayed in first place by following the Director version of Shockwave with its Macromedia Flash. Flash had basically blown the doors off of the web delivery system. With its ability to carry Mp3 quality audio and now by licensing it, the swf format to developers of three-dimensional graphics systems Macromedia had become the industry standard for truly enhanced web technologies when it came to deploying graphics and fully interactive websites. YouTube would not exist if it wasn't for the Flash's video compression format. Couple this with Macromedia's entire suite of HTML and other web-based technologies – Contribute, DreamweaverMX, Fireworks, Freehand, Fusion, etc. and the possibilities became endless seen today in the future World Wide Web enhancements and development.

But even greater enhancements are being made on the backend side. Languages and protocols like ASP (Active Server Pages), DHTML (Dynamic HyperText MarkUp Language), XML, PERL, CGI, JAVA, Python, Rails, etc. are making Berners-Lee vision of how the Internet should be used a reality.

2.4.3 Developing Imagery

Previously discussing the development of PhotoShop what needed to be mentioned was that the Knoll brothers developed their program Display by taking apart a simple flatbed scanner and reworking the software to modify the imagery generated by the scanner. Similar to Farnsworth's ability to transform physical imagery into pixels for transmission to a television, this now allowed everyone to take analog to digital with the click of a mouse.

2.4.4 Flatbeds Don't Suck

Flatbed scanners are great for most art that is flat (hence the name) or what is called reflective images. These are photographs or two-dimensional drawings. These are appropriate for flatbed scanners. The problem with a flatbed scanner – no matter what input dpi the scanner is capable of lacks one major component. They can't fight gravity. Whenever you place an image down on a scanner the problem that comes about is the fact that everything in this world works against gravity. What gravity does is cause an image or piece of artwork to curve ever so slightly. This slight curve can cause light to reflect off of the surface on an angle. This causes the image to be diffused or slightly blurred. This impedes the quality of the image. No matter if you stack an entire set of encyclopedias on the scanner's lid, the image will still slightly curve away from the bed. This is even worse with slides. No matter how you strip the slide out of the carrier, the image will still curve a bit. The only way to get this right is with a slide scanner or a drum scanner.

2.4.5 You Can't Beat the Drum

A drum scanner is typically used whenever you have a slide format for artwork. The image is produced as a slide – typically a four-by-five inch kind. These images are high in quality and very rich in color content. A slide is prepared for a drum-scan by using something called Newton or anti-gravity oil. The oil is applied to the slide and then the slide is applied to the curve of the drum (really a glass cylinder). The drum is then mounted onto the lathe of the scanner where it rotates at an incredibly high speed. A laser scans the image and relays the information to the hard drive. The resolution for drum scanners is somewhat near the 5,000 dpi level. This produces a very crisp image and an extremely large file. Typically this is process for print media but it is always great when you get an image like this for use in a multimedia project. When setting the preferences for any scan you want to get as much content in as possible. This means scanning an image at least at a setting of 300 dpi. You need to do this because you want to get as much tonal value as possible.

2.4.6 Finishing It Off

Once you have scanned the image it is ready for input into the project. This couldn't be further from the truth. You need to be sure that the image will display correctly on screen. Unlike a print project where you are looking for mid-tones with a multimedia project you are looking for strong robust color. Don't over-saturate to where the colors don't look realistic but remember a monitor drains away color. You are working with the RGB color scale. Here is where you are considering additive color. This means that when all colors are added up – red, green, and blue, the resulting color is white. This is a very shorthand approach to these color theories but one that is easily remembered. (To truly understand color on a computer, there are a number of books that you can get.) Once you scanned, cropped, and color-corrected your images, you are now ready to save it for use in your project (Kieran, 1994).

2.4.7 Saving for the Future

There are many formats available for scanning and saving images. The most vanilla of all is the joint Photographers Expert Group format or JPEG .jpg. This format allows for input into almost any multimedia project. You can also use a .gif format. It is a bit old school but can give you the best results for flat type-based navigational images. But at what setting should you compress the final images? It all depends on the quality of the final image. If the image is going to be large and is mission critical then exporting a .jpg to a "12" compression – highest quality is the best to use. If it is small and inconsequential, a .gif or .jpg will work. Today the new winner is the .png format. This allows for clear backgrounds around imagery. PNG stands for Portable Network Graphics which gives you a smaller file size but great quality. Remember the higher you set the compression number the larger the file. As you go through the process you can determine what setting works best for your project.

2.4.8 Summary Section

Even though chocolate is believed to be the most requested flavor, the reality is, taken from market surveys, that good old-fashioned boring vanilla is the favorite. This should teach us all a lesson. Sometimes exotic flavors aren't what the public desires. But we need to know better to help them along. Although we want to push the market along, we need to be careful in doing so. Many are the executives who constructed their entire presentation on a proprietary compression system. Don't get me wrong; it was an industry standard just not currently enjoying a huge market share. Come the day of the presentation to over 150 of his peers and the file wouldn't playback via the projector to the screen. It played great on their laptop, we could all see it there but just the rest of the world could see it. True the compression ratio and quality was excellent on the laptop but unless he wanted to walk it around the world it just wouldn't do.

Well what did we learn from this? Be sure of your content and of your deliverables. Be sure that the format you are getting from either your client, vendors, or team members meets your needs and fits the overall plan. Don't necessarily push the market by yourself; it is too big of a job. As always it is impossible to list in a limited number of pages all the advancements that computer technology has brought to the world of communications. Many have been left out or overlooked. The ones listed in this section are only done so to create a linkage from brilliant ideas to workable products. What should be noted about this section are the ability and immeasurable foresight, absolute brilliance, and complexity of thought put forth by these young men and women. For they have truly allowed us to enter this brave new world with the tools and skill sets necessary to garner the fruits of their labor. Hopefully, these efforts have started to answer the question put forth by Vannevar Bush "Of what lasting benefit has been man's use of science and of the new instruments which his research brought into our existence?"

In the previous section we discussed the need for a strong information architecture. But there is a more philosophical question that comes up. Why is it so important to know your audience? Should you play down to them or up to the next level? There are basically two schools of thought on this? First is the obvious one. You want to reach as many people as possible so you make a product that conforms to the current industry standard. If the world is using Mp3 you construct it with Mp3. Remember the "pioneers are the ones with the arrows in their backs" attributed to Douglas Engelbart.

But how and when do you try to move the market place along? Also how do you know when to take the chance on newer versions of players or compression formats so not to look like an old hat when you hit the market? Remember this also impacts how your vendors are finalizing their parts of the projects. Power perceived is power achieved. The questions are how do you make sure to stay on top and how has it been done before?

2.5 DESIGNING FOR THE GRID

2.5.1 From Bauhaus to the New Bauhaus

To understand the design of these new media projects we must first somewhat understand the history of graphic design, specifically the history of the use of the grid structure in design and directly editorial design. The reason for studying editorial design is that most new media – the web particularly – has evolved from an editorial base. The designers chosen for this section do not represent the end-all group and are not meant to segregate any others. They simply are the ones that have directly influenced this new emerging area of design.

All functional design has its roots in the two schools, the Bauhaus and de Stijl. These schools (or movements) first brought to light the concept of substance over surface, or as it is known within the design movement "form follows function." Hence the precession from strong architectural theory to strong information architecture – define the purpose for this site or project. But to argue this a bit further, what determines function? Is it the intended use of the object or visual or is it the appreciation of the object or visual? Or is it a combination of the two? In studying design it would be best to have a source or reference from which you can determine the validity of a design. To simply say, "you like the way it looks" may not be enough. What if the object being designed is meant to be fundamental in securing your safety or vital signs? The function then may be premier over the form. In Section 2.2 on information architecture, some theories and philosophies were discussed with the names ranging from Leonardo daVinci to Jakob Nielsen to Steve Krug. All in all what was determined in that section was that no matter the project's scope, size, scale, or importance, a plan is needed and must be followed. In other words, what is the function of the form? From that understanding the following schools, theories, and individuals have been selected to outline the history of new media design.

The first concern with designing for new media is how different it is from conventional print design. Typically design was passive – just a viewer, now this new design is interactive and heavily dependent on the user – not a viewer. Now designers need to anticipate how a user will act. How they will respond to a design's flow or structure. Will it work successfully for all concerned – client and end user. New problems but old solutions – "form follows function" (Lupton and Miller, 1993).

2.5.2 UX Design

UX design addresses this problem directly by becoming basically the print design of this century and new medium. It is like print design but it now has a group of stakeholders inputting their needs and requests. The designer now has to not just simply communicate one message but make sure it gets through to diverse users – not viewers – on multiple levels. UX designers have to anticipate the choices and needs of a diverse yet focused group. This requires an entirely new process.

David J. Kelley is the founder of Ideo, one of the world's leading design houses. He is also a professor of mechanical design at Stanford University. Mechanical design doesn't seem a normal fit to typical visual design but today we use more of the "mechanical" aspects of the design to produce a successful or useful experience for the app or website. Kelley's approach is called D-School and his theories are encapsulated in the title "Design Thinking." Here is his main belief about the integration of mechanical design and UX design (David Kelley | Stanford University School of Engineering, n.d.).

> Design Thinking is based on building empathy for user needs, developing solutions with iterative prototyping, and inspiring ideas for the future through storytelling.

This new area of design heralds back to the early days of modern design. In the beginning days of print design, the goal was mainly to fit the content on the page. Early calligraphic pages were mechanically positioned to fit. The goal was to replicate each letter perfectly to match the previous letter. Structure was key. As centuries progressed, as type and page design became industrialized, typefaces were used to fit the area more so than for uniformity. But as our technology progressed we looked to organize the design in favor of the message. In the 1940s–1950s the design look headed toward the uniformity of the grid. This grid format was built into early web design as an important specification to optimize loading time for images especially. Andreessen's original browser Mosaic was developed with the goal of showing images. The name itself speaks volumes as to how important is the concept of the grid.

2.5.3 Why the Grid?

Why is the grid such an important element to this world of interaction design? First, the grid is the basis for all proportion and cohesiveness to visual design. It helps to maintain an order or flow to a design or a page. To some this may seem to be opinion (that will be discussed in the Why portion of this section) but in reality it is a tried-and-true method of design that is addressed through this section. Second, what is the relationship between the grid format and digital communication? As you might know all computer screens are based on an x and y-axes. The word "pixel" comes from the words "picture" and "element" and they are the coordinates where information is directed. When images are loaded, they are measured out first in the upper left-hand corner of the screen. They are then displayed where they are finally meant to be seen. This is all dependent on the pixel setup. Pixels are redrawn on screen in the same manner starting from the upper left-hand corner and working across the screen, each position being marked by a coordinate on the x and y-axes. Since web pages are displayed on a computer screen, they are drawn the same way. Everything having to do with displaying information on a computer screen works off of this system. The question now comes about as to who started all this and how is it used today?

2.5.4 Descartes – I Think Therefore I Grid?

Who first introduced the theory of the grid to mathematics? What is it? And what purpose does it serve? The individual who first introduced the Cartesian grid to the field of mathematics was philosopher Rene Descartes (1596–1650) in the 1600s (Descartes is most famous for his theory "I think therefore I am" – Cogito, ergo sum). Descartes was a sort of prodigy in many disciplines. The myth or story goes, one day he was looking at a fly on the wall and wondered, "How do I know where that fly is?" Subsequently, he decided to map out the position of the insect by using an x and y-axes. This is the story that is told but who knows the truth? The issue is that this grid-based formula has permeated all the way through architecture to design to furniture making to computer programming.

The grid formula is considered to be the figure of pure modernism–possibly replacing the proportional classic measurements of the "Golden Mean." One of the original members of the Bauhaus school of design, the architect LeCorbusier (Charles Edouard Jeanneret, 1866–1965) referred to its geometry in the foreword to his 1924 book *The City of Tomorrow and Its Planning*.

Geometry is the means, created by ourselves, whereby we perceive the external world and express the world within us. Geometry is the foundation. It is also the material basis on which we build those symbols, which represent to us perfection and the divine. It brings with it the noble joys of mathematics. Machinery is the result of geometry. The age in which we live is therefore essentially a geometrical one; all its ideas are so orientated in the direction of geometry. Modern art and thought – after a century of analysis – are now seeking beyond what is merely accidental; geometry leads them to mathematical forms, a more and more generalized attitude.

In the Bauhaus school, the interest in this modern form (grid structure) pervaded the other disciplines of study: from art to architecture, to graphic design. In this next section we will discuss the basis of the two most renowned schools of visual theory, art and design.

2.5.5 Bauhaus and De Stijl

Started in Weimar, Germany, in 1919 by Walter Gropius and later moved to Dessau, Germany, in 1933, the Bauhaus school came to epitomize the concept of functional design. Coming from the two German words for "build and house" – the Bauhaus' main purpose at the start was to find a means to construct objects for use by individuals. Gropius was soon joined by many of the world's finest architects, painters, engineers, sculptors, and craftsmen. From here the credo of "form follows function" was first spoken to the world. The Bauhaus was responsible for many of the world's advancements into the modern era of design and living.

But the actual creation of what we have come to consider the basis for all graphic design was a movement called de Stijl. The de Stijl or "the Style" was started in 1917 and lasted until 1931. The reason for both schools (Bauhaus and de Stijl) suspending studies in the 1930s was the obvious growth of Nazism in Germany. Nevertheless, what was accomplished by both schools is still relevant and in practice to this very day. But de Stijl was more than an art movement; it became a way of life. Members of the movement were also responsible for maintaining a monthly periodical – *The Style*. Best known worldwide for its painters – primarily Mondrian and van Doesburg, the main focus of the movement seemed to be the use of total abstraction. Mondrian's paintings caught the attention of the world by his use of simple geometric shapes – primarily squares and rectangles – and the use of primary colors.

De Stijl was responsible for the use of the "grid" in many other forms of art and design. Based directly off of the Cartesian grid from the 1600s. The first use in graphic design of such a format was through the movement's magazine *de Stijl*, originated in 1917, by Theo van Doesburg. The magazine acted as a vehicle for views shared by many of the de Stijl artists but was primarily used by van Doesburg to push his ideas and beliefs about abstraction. Nevertheless, the magazine became a force in the development of this art movement. Vilmos Huszar, another member of de Stijl designed the magazine covers. Critics of the time have described the magazine as "austere and unpretentious to a point of plainness" (Lupton and Miller, 1993).

The layout of the magazine changed over time going from a portrait structure to a newer more friendly landscape proportion (web-based?). Other changes reflected the use of simple austere – but yet well proportioned – sans-serif typefaces. After van Doesburg's death in April 1930, the movement slowed to a halt by 1931. Whether either movement is still realistically active after that point is irrelevant. The message and direction that both the Bauhaus and de Stijl movement left behind is still going strong to this very day. Many historians believe that de Stijl "was the formulation of a visual language that transcends time and geography" (Friedman 227).

2.5.6 Swiss Design and the Grid

The Swiss Design movement was initiated in the early 1950s and was actually generated from both Switzerland and Germany. Originally called the Swiss Design movement, it has also been referred to as the International Typographic Style (ITS). The object of this movement was to promote proportion and clarity and for the designer to create a timeless perfection of form. It would appear that this movement's purpose was directed toward a "form following a function."

Some believe that the origin of this movement can be traced directly back to the designer Ernst Keller (1891–1968). At the age of 27 Keller started teaching at the Zurich Kunstgewerbeschule (School of Applied Art). His course load included advertising layout and the development of a course in design and typography. Opposite to the beliefs of the current ITS, Keller believed that the solution to a design problem came from the content of the problem not from a preset formula. (This is contrary to the beliefs of many of the information architects discussed in Chapter 2.) But, nevertheless, Keller's roots (as well as the ITS) can be traced back directly to the Bauhaus, de Stijl, and the new typography of the early 20th century.

Two of the Swiss designers who studied at the Bauhaus were Theo Ballmer and Max Bill. Both men are considered to be direct links to that movement's beliefs and theories starting in the new constructivist methods and later emerging as the design style of post-World War II. Ballmer, who studied for a brief time at the Bauhaus, later made use of de Stijl's concepts in graphic design. These concepts included the use of an arithmetic grid system of horizontal and vertical alignments. Bill also was influenced by many members of the Bauhaus movement and later moved to Zurich where he developed his own style and methodology.

2.5.7 Josef Muller-Brockman

Although many designers were starting to embrace the use of formalized structure in graphic design one person sought out an absolute and universal graphic expression. Brockman's desire to communicate on a universal level could be viewed as a precursor to the standardization needs of the Internet. Starting in the 1930s his career spanned several art and design movements from the Bauhaus to de Stijl to the Swiss Werkbund to the International Typographic Style. His body of work, from typographical design to corporate designs to logos to exhibition designs to his most famous photographic posters is as valid and vital today as they were when he created them in the 1950s.

Brockman's resounding legacy may be in how his book *Grid Systems in Graphic Design* published in 1981 is still being used to this day in classrooms around the world. In this book Brockman explained the basic characteristics and usage of the grid. The book is heralded as a prevailing history and manual for use of the Cartesian grid concept in the world of graphic design. Brockman's teachings carried through to America via the migration of many talented European designers.

2.5.8 Use of the Grid in Editorial Design

An area of design that is probably the greatest proponent of the grid system is the publishing industry. The grid system used in the development of an editorial page is based on the vertical columns and horizontal images. These elements are designed to be in proportion to each other. Certain publications are based off of a three, four, five, and six-column format. The integration of art and text has caused the need for a grid ratio to be developed. A grid ratio in editorial design is a mathematical relationship between two or more grid measurements. The ratio is used to determine the size and placement of page elements. The theory behind this ratio is meant to keep the reader in the "flow" of the story and not allow for the individual's eye to jump from area to area. A rule of thumb is that the pica (unit of measure in printing 6 picas to one inch) width of a column should never be larger than two times the point size (unit of measure in typography – 72 points to one inch). The belief is that the human eye cannot read across successfully past that dimension.

According to Shawn McKinney in his essay "Grid: Parallel Views," visual hierarchy "Grids help designers position elements and maintain proportional relationships." Grid modules, and/or special zones can be designated for particular kinds of content, while flow-lines divide pages (or page spreads) into major zones. These elements help create a clear, unified, visual hierarchy that often improves accessibility and readability – these ideas adapted primarily from "the typographic grid" (Typographic Design [2nd ed.], Carter et al., 2018).

In some cases, a modular relationship can be established without using a standard grid. This does allow for greater freedom and can compromise the quality of proportion and overall relationships between spatial elements. In this next section, we will discuss how these style formats and theories came to the United States and what influence they had on the future of graphic design.

2.5.9 Paul Rand and American Usage

During this immigration of European designers in the 1940s American graphic design truly began its next phase of expansion that would last well into the 1990s. Of all the American designers of the mid-20th century, Paul Rand (1914–1996) may be known as the initiator of modern American design. Rand's main focus appeared to be the desire to communicate content over form through the use of eh rid system. Rand's main definition of design was the integration of form and function for the most effective communication of the concept. Truly substance over surface! The Art Directors Club of New York inducted Rand into their Hall of Fame in 1972. On their website www.adcny.org they describe Rand,

> He explored the formal vocabulary of the European avant garde art movements and developed an unique and distinctly American graphic style which was characterized by simplicity, wit and a rational approach to problem solving... Rand was a major force in editorial design.

Rand was educated in New York at Pratt Institute from 1929 to 1932, Parsons School of Design from 1932 to 1933, and the Art Student's League from 1933 to 1934. From 1935 to 1941 Rand was the art director of *Esquire* and *Apparel Arts* (later GQ: *Gentlemen's Quarterly*) from 1935 to 1943. He was also responsible for having designed covers for the Directions cultural journal between 1938 and 1945 and as a design consultant developed identity systems for major corporations such as IBM and Westinghouse. Later in 1988 he surprised the world again with designing the NeXT logo for Steve Jobs' start-up company.

Rand was also an academic teaching from 1956 to 1969 and beginning again in 1974 at Yale University in New Haven, Connecticut. He later resigned from his teaching position at Yale, some think over the disregard for the grid system and too great use of the post-modernist design in the classroom. This is only a theory.

To this very day Paul Rand and his writings (especially his 1946 book *Thoughts on Design*) are considered to be one of the greatest single forces behind American graphic design. Rand was thought to be responsible for influencing a new generation of American designers.

There were many other innovators of early editorial design that were present when the advent of technology in print production allowed the magazine industry to come of age in the mid-1940s. At that time, however, only a few of these publications were considered to be designed well. Magazines like *Fortune*, *Vogue*, and *Harper's Bazaar* led the field as far as visual and page design were concerned by their respective design directors such as Alexander Liberman, Alexey Brodovitch, Will Burtin, and Leo Lionni. At the time, women were rarely seen in such supervisory positions. One in particular is credited with helping to break the convention of current editorial design. Her name was Cipe Pineles. Ms. Pineles would commission large pieces of fine art and illustration to be used throughout an editorial layout, thus breaking the mold set forth by the illustrious grid system. Due to these efforts Ms. Pineles became the first woman admitted to membership in the male-dominated New York Art Directors Club.

In the next section we will deal with certain revered editorial designers whose roots were firmly entrenched in the use of the grid system (Meggs and Purvis, 2016).

2.5.10 Creative DNA?

A major name in the world of editorial design is Fred Woodward. Woodward's body of work has helped redefine the world of editorial design. During his tenure as art director of *Rolling Stone*, the magazine garnered more awards than any other magazine in the United States.

At the age of 43 his induction into the Art Directors Hall of Fame is the youngest to date. Born and raised in Noxapater, Mississippi, Woodward attended Mississippi State and later Memphis State Universities. Switching from major to major until finally taking two graphic arts courses late in his academic career, he suddenly landed his first job at a design studio working on *Memphis* magazine. Woodward's career then started to move to landing positions at several magazines: Dallas' *D* magazine, *Westward*, the Sunday supplement of the *Dallas Times-Herald*, and *Texas Monthly*. In 1987, Woodward settled at *Rolling Stone*. And during his time he redefined the image of the magazine while still maintaining the heritage of the grid system. For such a hand-on-the-pulse magazine to stay true to the grid structure is testimony to the impact and stability of this system. Woodward was later appointed creative director of Wenner Media, overseeing all publishing projects, including books and new media. "Fred is truly a visionary in the field of magazine art direction," says Jann Wenner. "His creative DNA has become an integral part of the *Rolling Stone* spirit."

2.5.11 Black on the Grid?

Roger Black has come to be the prime example of a successful transition from print designer to web designer. Growing up in West Texas in the 1950s and then subsequently attending the University of Chicago in the 1960s, Black was involved in underground publications. From there he had some experience with an ad agency in Texas and then it was out to Los Angeles and then to *Rolling Stone*. He became one of the primary creators of *Rolling Stone* via its art direction from 1967 until 1977. During that time the magazine became synonymous with the youth movement in the United States and possibly the world. Having come out of the 1960s design movement, it is logical to assume that Black was trained on the use of the grid system but what is more important is how he has transferred this logic and proportion to the web (http://www.newyorkmetro.com/nymetro/news/media/columns/medialife/1588/).

In the article "Meet Roger Black," This Media Life, Michael Wolfe states, "He (Black) pioneered the use of computers in design, cut the best deals, and made himself synonymous with the modern magazine. And he doesn't break a sweat doing it."

With just one look at his enormous resume of editorial design accomplishments such as *Rolling Stone*, the *New York Times* magazine, *Newsweek*, *McCall's*, *The New Republic*, *Fast Company*, *Reader's Digest*, *Foreign Affairs*, *Advertising Age*, *Esquire*, and now the *National Enquirer* (to name just a few) you come to realize that Black's statement, "I've designed more magazines than you'll ever read" is true.

With such an enormous amount of credibility why would Black get involved in something that has such limitations as this world of new media design? The answer to that question is in Black's lineage of using the computer in his design work. As soon as the desktop world was raised to a professional level, Black was involved. In the mid-1980s, the Macintosh became the standard for electronic publishing with the use of programs such as Adobe Photoshop, Illustrator, and QuarkXPress – now InDesign. With this change in the production of graphic design the nature of the job was changed forever. And Roger Black was the first to retrain. He quickly became the expert in the field of computer-aided page composition.

This transition was made complete in the early 1990s. At this time the web had finally gotten into full swing. Black jumped in with both feet with one of his first assignments @Home. @Home would later go public netting Black a part of a tremendous IPO (Initial Public Offering). From here he founded the Interactive Bureau in 1994 and then merged with circle.com, a public company controlled by advertising agency aggregator Snyder Communications. Among his major online clients were MSNBC, USA Today, BarnesandNoble.com, and the Discovery Channel Online to name a few. Black had co-founded Danilo Black, Inc. with designer Eduardo Danilo. This new company is focused on offering a multitude of services across all platforms from broadcast to print and Internet. He continues to write books and to speak at conferences around the world. One such book is his 1997 *Web Sites That Work*. In it Black outlines some of his rules for good web design. Although the rules are old hat by now they still deserve repeating.

Rule 1
Large type works better.
Rule 2
Color the page correctly.
Rule 3
The faster the better.
Rule 4
It is all about content.
Rule 5
Smaller pages work best.
Rule 6
Follow the plan.

It is true that most of what Black stated in 1997 has come to be the norm for a good number of websites. In an article by Jeremy Schlosberg, "Roger Black on the Art of Web Design," Black addresses some issues on web design. When asked about the enormous content on site design and development Black states,

> The medium may be new but human nature is still human nature. ...Human beings are the same. Evolution isn't going to change human beings because the Internet came along. Not that quickly anyway. We're not going to mutate before we get to our next deadline.
>
> *(Schlossberg, 2001)*

Black continues to make tremendous contributions to the world of design but here he couldn't have been more wrong. We know today that the tail is wagging the dog. Our digital overlords are leading us around and there are steps being taken to correct that – hopefully, in time before Artificial Intelligence lands firmly to redirect everything.

2.5.12 Hillman Curtis Making the Invisible Visible

Roger Black's point – concerning our ability to absorb information – may be true but what about the Internet transforming how much information we can absorb? What about how much information the next generation can absorb? This question may need to be answered by designers closer to the mode of thinking of this new Generation i (Gen-i)*.

Hillman Curtis is the principal and chief creative officer of hillmancurtis.com, inc. an all-medium digital design firm in New York City. He is well known for his successful solutions to some of the world's largest corporations: Adobe systems Incorporated, *Born Magazine*, British Airways, Lycos, MTV, Macromedia, and Sun Microsystems. Curtis is also a prolific author of numerous web design books (most recently MTIV: Process, Inspiration, and Practice for the New Media Designer) and a frequent lecturer at both national and international conferences. Curtis' approach to web design and development is a direct link back to even Leonardo's approach to all forms of development. On his site he discussed his approach to client relations and development in his "approach" section.

2.5.13 Hillman Curtis Approach – Find the Theme

We believe that great design comes from defining a singular theme and supporting it consistently and clearly. The theme is the heart of a design. It is the emotional center that when communicated will resonate deeply and lead to a successful project.

- Simple Questions Lead to Simple Answers
- Communicate It Clearly
- Honor Consistency
- Eat the Audience

Hillman Curtis may just have it right and you can prove that if you read his client list. But what approach does he make to his design form in general? His portfolio would seem to reflect a great concern for the safety of the grid. Small areas of image and type are typically the standard look for his sites. The difference, however, between Curtis' work and the other designers spoken about here is his great and extremely strong grasp of Macromedia Flash. Mr. Curtis is a recognized Flash guru spawning some half a dozen Flash design manuals (as was mentioned at the start of this section). Most of these books are heralded as the bible for interactive-animation web design. His sites are beautiful in their form and truly interactive in their function – albeit they are similar from site to site.

When asked about the future of web design in an interview, Curtis responds, "I have no idea…seriously…I learned a long time ago not to answer these questions because you're always wrong…that said, I can't help but say interactive video." As far as sources of inspiration he relates directly back to film, usually films from the late 1970s or early 1980s, *Three Days of the Condor*, or *Network*. But when it comes to designers, he goes back a bit as well reflecting on the work of Neville Brody.

As far as judging which media has the greatest scope for a designer, Curtis, after deliberation, goes quickly to the web. He states,

> For me it's definitely the web… simply because it can handle anything and is a true free global broadcast medium, for media it's probably Flash again because of its ability to do almost anything, though Final Cut is my application of choice these days.

Although he obviously enjoys the media, he did have one complaint about Macromedia Flash, "I would have made it come with an ActionScript programmer robot…that also made coffee." As Curtis reflects, sometimes the design and the function do conflict. But personally, I feel that the designer usually wins in the end. Now enter ChatGPT?

2.5.14 Going Off the Grid

Many designers have taken steps to work off the grid. Many revolutionary experiments in design have failed but yet have brought the design world new energy and new life. How far off the grid do we go and with what purpose? In an old interview with Dace A. Campbell, Assoc, AIA – then from the Human Interface Technology Lab at University of Washington (November 1995), Allison Stamides (then of Massachusetts Institute of Technology) starts to challenge why designers need the grid. Some of Ms. Stamides' comments reflect how impossible it maybe to completely escape this Cartesian concept of space. But she does refer to the ability to skew the Cartesian – de Stijl aesthetic. The reason to bring this point up now in this section is many designers put their stamp or trademark into their work by diverging from the accepted norm.

Has the perpetual message been lost in the medium or is the medium creating the form of his message? To answer this properly it may be necessary to reflect back on one of the past century's greatest sociological thinkers – the person who first stated that – the medium was the message – Marshall McLuhan and the impact his theories have had on the work of web designers?

His seminal work "The Medium Is the Message" believed that the medium will shape the user/viewer into what works best for the medium. Television will make us sedentary consumers with little choice. Now the smartphone has made us into worse consumers. Not only sedentary but actually static. Our heads arched over, walking into each other while we endlessly swipe with our burgeoning thumbs (McLuhan, 1964).

Whichever way we choose to answer this problem the solution will probably are unique and extremely personal. The question that arises from all of this theory and preponderance is how does a designer design? How do they choose a format? When does content completely drive the form? Does the user/viewer come into play before the design or after? Should we choose to direct the user/viewer the way we see fit or the way they choose to use it? Should we choose to conform to conventional design theories – the grid – when we cannot conquer the problem? Which method is the correct methodology to perform the task? Who's driving this bus? Answering this is at best fluid today. Each project is unique and has its own set of solutions. No longer is there one answer for interaction design. Every client has specific needs and using the tools and processes of Information Architecture and Interaction Design is the hallmark of a great UX designer.

2.5.15 Section Summary

Have we truly answered anything in this section about the formula to designing for new media? Probably not but we have exposed several theories that you now have to select from and that was the point. The designers selected for this section do not represent the end-all for design on the web or for any medium. They simply were selected for their landmark changes to a direct lineage between order and chaos.

There are many other names that could have been mentioned such as John Maeda, Thomas Mueller, Clement Mok, Matthew Butterick, Naomi Enami, Mach Arom, Julie Beeler, Second Story Interactive, Drue Miller, Blank + Cables, Nike's creative services department, Industrie Brand Partners, Hello Design, Periscope, Move Design, MFP Interactive/Mass Transmit, Fallon Interactive, Axis 360, and Firstborn Multimedia. Hundreds of others were left out not for their insignificance but for the fact that space and time has its limitations.

All in all, interface design, multimedia design, new media design all have their caveats each a bit different but all revolving around the concept of "form follows function." The new question we look to answer is: "what is the function and what form will you choose to present it within?" One area that was not covered at all in this design section is "color and its effect on new media" for a simple reason – it would have taken an entire book to discuss. Color is beyond subjective; it is everything to how we view and interpret. It is psychology, art and design plus sociology, anthropology, and ergonomics all rolled into one area, to discuss it in just a section of a chapter is at best ridiculous.

2.6 CREATING ANIMATIONS

2.6.1 From Thaumotropes to VR

Animation has always fascinated mankind. From the beginning of time the concept of motion has represented a living being. Whether it was to visualize the hunt, study the wrestling capabilities of Egyptians or watch the recent Pixar extravaganza, the idea of witnessing movement from something that was out of the ordinary has amazed mankind. The philosopher Lucretius in 65 BC studied the concept of persistence of vision, the ability to trick the eye to see motion from inanimate objects. As early as Leonardo da Vinci, man has been fascinated with the mechanizations of the human body. Now with the advent of a new means of information distribution, animation is taking a firm foothold in the world of multimedia development. As often it is with a new medium the past gets quickly lost. In the case of animation very few outside of the professionals have heard the names of Winsor McCay, Raoul Barre, Chuck Jones, Max Fleischer, Hugh Harman, Rudy Ising, Ube Werk, Tex

Avery, or Bob Clampett. In this section we will explore the beginnings of animation from the early days of two-sided spinning disks to the multi-speed renderings of Silicon Graphics stations. Many names will be lost and many names will have come to play by the time you finish reading this chapter. For that is the nature of creation today. By the time *Final Fantasy* or *Toy Story* hit the big screen, the production techniques employed to make them were old hat. What should never be lost however is the theory behind animation.

At the start of this industry, the study of human mechanics was of the foremost importance. From the early days of Winsor McCay, whom many consider to be the father of modern-day animation, the replication of this skill was meant to be an art form. However, this came to be lost, as does all art when it is brought to the masses. McCay is credited with the quote, "Animation should be an art... what you fellows have done with it is making it into a trade... not an art but a trade...bad luck." Today it would seem with all the technology, even the simple two-dimensional styles of Macromedia Flash and Director, being thrown at making the screen the art component would seem to be the least important aspect. But one thing is helping to keep the art in this 20th-century invention. That would be the study of the aesthetic of motion and form. This section is meant to be a very brief history of the advent of two-dimensional animation and how it applies to this new form of art and regrettably trade (Solomon, 1994).

2.6.2 Two Sides to the Issue

Before we go forward with this section the first question is "How does animation work?" Simply put it relies upon the ability of the physical world to fool the human eye. Just like film animation needs to be portrayed at a speed greater than 12 frames per second. This is later to be called persistence of vision. Who first used this phenomenon for animation and how was it implemented?

Although there are many who have grabbed on to the origin of this phenomenon, the credit in many circles goes to an Englishmen named Peter Mark Roget. In December 1824, Roget offered up the first use of this phenomena described as "persistence of vision," which means that the human eye could distinguish between objects that are replicated only below 12 frames per second. Roget's description to the Royal Society of London was that objects appear to be moving if there is an interruption between the displays of each image (Roget, 1825). Now what is most fascinating about this discovery or invention is how mechanical devices were necessary to portray the concept. Obviously, this would later come to be an important ingredient of all animation techniques.

Many others have been credited with contributing to this phenomenon. One of these is a Frenchman named Joseph Antoine Ferdinand Plateau (1801–1883). In 1824 Mssr. Plateau invents the phenakistoscope ("a spindle viewer"). This viewer is made up of a large disc that had a series of drawings located around the circumference. This is attached via a spindle to another disc made with metal slots in it. When both rotate the drawings can be viewed through the slots in the metal disc, as they appear to move. Again replicating Roget's discovery of interrupting viewing and bringing about motion. Later that same year the Daedalum, or Wheel of the Devil, was invented by William Horner in England. The device is made up of a strip of drawings on paper. The paper is then placed inside of a drum that rotates. The upper edge of the drum has slots through which the drawings are viewed. The device later resurged in popularity when it was marketed as the Wheel of Life, or the zoetrope by William F. Lincoln in America. (A side note to this section: Francis Ford Coppola of "Godfather" series fame decided to call his own production company "Zoetrope" studios.)

In 1877 Emile Reynaud invented the praxinoscope, which used mirrors to reflect the zoetrope images so that larger audiences could view them. This later led to combining this device with a projector. This would lead to him demonstrating a device called a theater optique that projected animated images (painted on long strips of celluloid) onto a flat surface or screen. Obviously, this was the precursor to the filmstrip. This also allowed for a story to be told not just a sequence of moving images. Once this concept was established storytelling could now be used as the message within this medium. In 1892 Reynaud opens his theatre Optique at the Musee Grevin. The Musee Grevin lasted from 1892 to 1900. Some of the films he projected were *The Clown and His Dogs* and *Poor Pierrot*.

Once this technology coupled with the technology of Thomas A. Edison, the first practical means of making animation came into the play. Around the same time, approximately 1906, Stuart Blackton created a short film entitled *Humorous Phases of Funny Faces*. This early experimentation would set the groundwork for a new age of entertainment, storytelling, and art to emerge from the shadows of the two-dimensional world of print and traditional storytelling.

2.6.3 Three Early Stars – Fantasmagorie, Gertie, and Felix

Emile Cohl was born in France in 1857. Cohl is considered by many to be the originator of the first animated film *Fantasmagorie*, 1908. Cohl first worked as a newspaper cartoonist but later dove head first into animation. In his career he is credited with making over 250 animated films. Cohl's work followed the philosophy of the Incoherents. This was an actual group that believed the world of aesthetics would open up to an artist or anyone if they believed in insanity, hallucinations, nightmares, and dreams. The Incoherents believed that this was the source for all inspiration.

Two other early animators were Winsor McCay and Otto Messmer. Winsor McCay was born in 1871. At first, like most later to be animators, McCay worked as a newspaper cartoonist. His best-known comic strip *Little Nemo* ran from 1905 to 1911. But both of his strips *Little Nemo* and *Dreams of a Rarebit Fiend* never truly gained popularity, but had a strong cult following due to his expressive graphic style. Most of his work for both *Little Nemo* and *Dreams of a Rarebit Fiend* were based on fantasy art styling.

In 1912 McCay created the animated short *How a Mosquito Operates* and later in 1914 he introduced the kinetic capabilities of animation with his milestone film *Gertie the Dinosaur*. This simple story was fabricated in line drawing that still to this day displays a true fascination with the study of motion graphics. *Gertie* was quite the hit and helped propel McCay into a limited starring role. McCay's main accomplishment more so than the actual animation was the fact that he had created a character that had personality and lifelike movements. The animation contained such strong detail that it was only matched by the Disney Studios commitment to character development. "Gertie" is held by film historians as the cartoon character created solely for film, and the first to display a "unique, realistic 'personality'."

McCay went on to create many other short animated films utilizing the same painstaking process of drawing every frame by hand. After popularity subsided, McCay would travel the country in a vaudeville review showing his animated shorts to the audience while he "interacted" with the "stars" through hand movements and other short skits.

Otto Messmer was born in 1892. He attended a parochial high school where he was encouraged to hone his drawing skills. He later attended the Thomas School of Art in New York City. Around 1912 Messmer's brother took him to witness one of McCay's vaudeville acts. It was during that year that he first saw the short *How a Mosquito Operates* and later in 1914 *Gertie the Dinosaur*. These combinations of live action and animation encouraged Messmer to begin contributing his own work to the newspaper syndicates.

Messmer interviewed at Universal Studios as a set painter with the hopes of further expanding his set of skills.

Mr. Jack Cohn (then of Universal) witnessed some of Messmer's comics and signed him to make a test film on December 21, 1915. Cohn, it seemed like other studios, was engaged in attempts to expand the studio's animation output.

This is where one of the all-time cartoon character legends is born. Messmer returned home to start his assignment for Cohn and Universal. Unaware of the advancements made in animation technique such as drawing on celluloid for ease of replication or the peg system for registration, Messmer painstakingly drew the thousands of frames needed by hand and eye. The character he created was called Motor Mat, a young fearless automobilist. Although the work was never released, Messmer (through Cohn) was introduced to Pat Sullivan and Henry Mayer. Messmer wound up working with Mayer on the animated film *Travels with Teddy* – based on the adventures of Teddy Roosevelt. This project seemed to give Messmer the confidence he needed. Once Sullivan had seen the film, he started to employ Messmer on some projects, giving him more creative responsibility. This led to a long working relationship. During this time many requests would come in for animated shorts, one such request came in from Earl Hurd at Paramount Studios. The request was for a news travelog cartoon compilation. Sullivan's studios were so busy that they instructed Messmer to do the work outside of the studio. Messmer did so entitling the short *Feline Follies* which for simplicity contained an all-black cat who was the cartoon's hero – Felix, the year was 1919.

Obviously, this character went on to become one of animations' most well-known cartoon characters – similar to Mickey Mouse. But the reason for mentioning Messmer's story is more so to explain what happened later. Even though Messmer created this work outside of Sullivan's studios, Sullivan still declared title to the animation. Sullivan negotiated all the financial deals for Messmer's character and for many years billed himself as creator. Having received numerous contracts from Margaret J. Winkler (the first female producer and distributor of animated films), Sullivan enjoyed making many Felix shorts from 1919 to 1931. Sullivan succumbed to alcoholism in 1932 but never declared that Messmer had created the character that the world had come to enjoy – Felix the Cat. This allowed for Sullivan to supposedly enjoy millions of dollars in royalties where Messmer supposedly received only a salary. The story is told that only later during a resurrection of popularity in the 1960s that a television producer Joe Oriolo truly gave credit to Otto Messmer. John Canemaker, an animation historian, is also credited with finding Messmer and telling his story. Messmer died in 1983 long enough to relish the fame and recognition that he truly deserved. Felix's appeal continues to this very day.

2.6.4 New York Center of Animation

The concept of animation studios was just starting in the early 20th century and the center of this pioneering effort in the United States was New York City. One of these pioneers was a gentleman named Raoul Barre. Barre would open his studio in New York City in 1913. Credited with inventing the peg system of registration, Barre would certainly be heralded as a true pioneer of animation. Within a short time, more and more animation studios would start to emerge in the New York City area. One of those was the John Bray Studio. Considered by some historians as not the most successful but yet definitely the most influential. John Bray is credited with creating the first original cartoon character series in 1914. The character was entitled Col. Heeza Liar. An appropriate name for work that came out of most of these studios given the fact that the royalties usually went to the studio owner and not the artist or originator of the characters.

The greatest contribution to come out of John Bray Studio truly was in the area of technology. It is thought that the Bray Studio is responsible for marketing and developing the concept of cell technology. An employee of the studio, Earl Hurd is credited with inventing this technique of painting a background and then on clear celluloid drawing the character movements. Since Hurd, like Messmer, was a studio employee, he allowed studio owner Bray to charge royalties for use of this process at other animation studios of the day. The company's name was Bray-Hurd and it licensed this technology until the 1930s. This process saved tremendous hours of work by other animators and when coupled with the peg system of Raoul Barre, we are left with basically all the necessary physical tools to create animation to this very day.

2.6.5 Out of the Inkwell

Another very important name to come out of the New York School of animation was an Austrian-born gentleman named Max Fleischer. Fleischer, like many other famous animators, would get his start by working for the John Bray Studio. Here he would work with his brother David to produce the first Invisible Ink Films featuring KoKo the Clown. The character's name would be shortened to KoKo later for copyright reasons. Fleischer, who had a fascination with mechanics and along with his brothers David and Joseph, would go on to invent the process for tracing live action footage to animation. This process became known as rotoscoping. In his first film *Experiment No. 1* in 1915, Fleischer exhibited this new technique by having his brother David pose as the clown Koko. Felischer and his brother David patented the process around 1915. Along with rotoscoping Fleischer is credited with being the first to combine cartoon animation and sound using the DeForest Phonofilm process. By the time sound had come along, Fleischer had considerable experience dating back to 1924 and his Song Car-tunes. In this short audiences are encouraged to sing along with the bouncing ball. A technique utilized for years after. (Obviously Disney's *Steamboat Willie* has become known as the first cartoon to use sound due to its immense popularity at the time.)

In 1929 Max and David created Fleischer Studios and started producing cartoons for Paramount Pictures. In 1930 Fleischer Studios had its first major hit with the animated series Betty Boop. The first in the series was entitled Dizzy Dishes. Boop was probably the first female animated star and also the first to deal with sexual overtones. The design for Betty Boop was fashioned after singer Helen Kane who was known for the expression "Boop-oop-adoop."

Fleischer's KoKo series employed another technique that has lived to this very day. That is the use of combining animated characters with live action footage. Fleischer in his film *The Challenge* would be depicted as actually jumping into the film with Koko to assume his own identity as the creator. This harkens back to the work of Cohl and Ferdinand Zecca as well as the well-known John Bray. This technique would be employed in other animated and live action films of the 1950s, 1960s and later, such as *An American in Paris*, *Tron*, and *Roger Rabbit*.

Fleischer is also credited with experimenting with Claymation, a form of animation done with clay to develop the frame-by-frame movements in his work Modelling and Clay Town. Fleischer also developed a process that utilized photo collage depicted in the piece Koko's Earth Control.

The development of characters and storylines continued into the 1930s with the introduction of *Popeye: The Sailor Man*. This character embodied a long narrative form and personified the struggle between good and evil with the characters of Popeye and his arch nemesis Bluto along with the beautiful Olive Oyl. Fleischer through his alliance with Paramount Pictures was able to bring to light feature productions such as Popeye the Sailor Meets Sinbad the Sailor, Popeye Meets Ali Baba and His 40 Thieves, and Gulliver's Travels. Later Fleischer was responsible for developing the style of the first superhero when he produced a detailed series of animations for the original Superman cartoons. The series was displayed for the Warner Bros. staff of the very popular, pioneering, and unique series *Batman: The Animated Series*.

Fleischer and his brothers split up in 1942 when David went off to head Columbia's Screen Gems cartoon unit and Max Fleischer himself ironically finished out his career by working for a Bray colleague named Jam Handy producing the screen version of Rudolph the Red Nosed Reindeer and later for Bray studios producing training films for the Navy. Fleischer's style of animation lives on today.

These technologies were such milestones in animation that most forms of computer-generated software today – in a sense – employ the same processes of layering, compositing, rotoscoping, and tweening (a form of generating the in-between stages of drawings). This coupled with the development of full story lines has been the bedrock within which all present-day animation has found its roots.

2.6.6 They're All Together Looney

The magic that is Bugs Bunny, Daffy Duck, and Porky Pig would never have been possible if not for the genius of a few good people. Starting with the creative source and director Charles (Chuck) Martin Jones and continuing through to such likes as Friz Freleng and Tex Avery Warner Bros. Animation studios would create such classic series as Looney Tunes and Merrie Melodies.

It may have been their outrageous styles or their unusual names but the artists and directors of Warner Bros. Cartoons have managed to rise above the name of the studio and be recognized for their own unique brilliance and capabilities. Whatever it was the name Looney Tunes or Merrie Melodies mention of either one of these will always get a smile in anyone old enough to remember the classic slapstick storylines and amazing characters that these people brought to life starting with its original movies. Beginning in 1930 Looney Tunes originated with two artists from the Disney studios, Hugh Harman and Rudy Ising plus another veteran Leon Schlesinger. (Disney Studios will be discussed in the following paragraphs.) Similar to the Disney method, Warner Bros. continued with the stage-by-stage development concept of cartoons. Going from original storyboards to original drawings to pencil tests to inking to background to overlaying celluloid to final coloring to final filming to audio voice-overs. This method appeared ironclad, even though by the 1950s many different units were creating cartoons featuring the same characters: Bugs Bunny, Daffy Duck, Porky Pig etc. These classic works of two-dimensional animations were also well directed under the guidance of names like Tex Avery and Bob Clampett. The method has stayed with almost all studios to this day.

Once the 1940s and 1950s came about, television had made its way into the American home. Here was the perfect venue for a slew of new cartoons and a rehashing of the classics. Two names that stay in the forefront of television cartoon animation are Bill Hanna and Joseph BarbaraHanna-Barbara. Both of these gentlemen came from the MGM studios and were well schooled in the ways of Avery and Clampett. The formula of Hanna-Barbera dominated the television cartoon marketplace well into the late 1970s and early 1980s. Most of these cartoons have attempted to make the leap from small screen animation to full screen live action. A curious transition given the start of animation was to make cartoon characters look like real people now we are trying to make real people look like cartoon characters.

This dominance of the television marketplace has given the (late 1980s) world the likes of Spielberg's "Tiny Toon Adventures" (a prequel to the "Looney Toons") crew and also a spin off from Disney of "Ducktales." In the early 1990s, the new look was to work toward a dark and ominous styling of cartoons like "Spawn" and "Batman: The Animated Series." This new look dealt directly with a comic book styling combined with real motion picture like camera movements. This was no doubt influenced by the use of three-dimensional animation in feature length films by Disney, Pixar, DreamWorks SKG, and other smaller studios (Solomon, 1994).

2.6.7 The Mouse That Roared and Hasn't Stopped

Another city that can be heralded as the start of animation is strangely enough Kansas City, Missouri. Why here? Well it is where a little-known gentleman had his start – his name was Walter Elias Disney. In 1920 Disney was hired by the Kansas City Film Ad Company to develop a series of cartoon advertisements. Even then the commercial arm of the art world contributed to the development of a worldwide phenomenon. Disney is credited – similar to Henry Ford – not with inventing the cartoon but bringing it to the world in a popular fashion. He employed the technology established by Barre and Bray-Hurd to a level of tremendous efficiency. Two of Disney's original animators Vladimir Tytla and Freddie Moore are compared to the likes of the Renaissance masters for their creative brilliance and expertise.

In 1923 Disney started a move that continues to this very day. He headed west to settle in the movie-making capital of the world Hollywood. Disney was not alone in this move however. With him went Hugh Harman and Rudy Ising. These gentlemen would later head such animation powerhouses as Warner Bros. and MGM. This move to the West Coast – which already had become the center for filmmaking – would become a major watershed for the animation business.

Some of Disney's main technical accomplishments are the focus on the true movement of the human form (similar to McCay's *Gertie*), and the use of sound and color. His famous introduction of Mickey Mouse in the 1928 Steamboat Willie – one of the first cartoons to use sound – came about at just the right time. Other animation studios, throughout the country and in New York, thought sound in animation was just a costly fad. But Disney was a visionary and he knew that the two together created a relationship that was both invaluable and ultimately inseparable.

As was the case with the other studios, it was an animator who would help to develop one of the more groundbreaking stylistic changes credited with the studio's name. Freddie Moore, who had started with Disney in the 1920s, is recognized as leading the movement toward creating a more realistic style of motion in character animation. Originally Mickey Mouse moved in what was called a "rubber hose" fashion meaning less pivot points in the limbs and body. But Moore, through Disney's urging, reemphasized the more realistic style as early as 1935.

Disney's studio would continue such groundbreaking work as *Snow White and the Seven Dwarfs 1937*, *Pinocchio and Fantasia 1940*, *Lady* and *The Tramp 1955*, and *The Jungle Book*. But it wasn't until the 1960s that the Wonderful World of Color would come into our homes every Sunday night. Walt Disney's legacy is truly phenomenal; 81 feature films, first popular cartoon character, pioneered the use of music and sound in animation, theme parks, scholarships, and a strong alliance with education.

Before his death Disney was instrumental in helping to establish the California Institute of the Arts. The school was established in 1961 as a combination of two existing schools – Los Angeles Conservatory of Music and Chouinard Art Institute. Disney's vision continues to this very day via a strong alliance with Pixar studios, John Lasseter (LucasFilm), and Ed Catmull (LucasFilm and a major name in computer-generated animation). Once the computer entered the world of animation, from the early days of Sketchpad by Ivan Sutherland to the heady days of Mary Ellen Butte and John Whitney, Sr., everything changed for good and obviously beyond. The convergence of traditional studio skills with new technologies has brought about a tremendous change in how we watch movies as much as how they are made. This all came about in the late 1980s with movies like *Who Framed Roger Rabbit?* and *The Little Mermaid* series. Since this section is dedicated to two-dimensional animation, the discussion of Pixar and any other discussion of three-dimensional animation will stop for now (Baxter, 2023).

A name that must be reiterated, before we progress to computer-generated animation, is that of Mary Ellen Butte. Butte's work established pioneering efforts in motion graphics. In the early 1930s she was busy creating abstract animated films. A little-known figure in the world of film animation, Butte's career spanned from 1934 to 1959. Having grown up in Texas, she continued her studies in Texas and later in Philadelphia. She later studied stage lighting at Yale and focused on creating a "Color organ" that combined the worlds of music and light. During this time, she became familiar with the work of Theremin and Wilfred. Their device the Clauvilux was able to project streams of swirling colors. Later Butte would see the work of Joseph Schillinger. Together they would produce work reminiscent of Kandinsky's paintings. The work was groundbreaking and probably impossible for one animator to accomplish if it had not been for Schillinger's method of reducing all music to mathematical computations and formulae.

Universal Studios released Butte's work, *Dada*, to the public in 1936. In 1939 she began to work in color. By 1940 Butte had collaborated with artist Norman McLaren whose method of directly drawing on film to synchronized music would appear in films such as *Spook Sport*, *Tarantella*, *Color Rhapsodie*, and *Polka Graph*. It would appear, however, that Butte was putting the carriage before the horse. Many critics thought that she was enamored with the technique and gave, at times, little attention to the actual storyline. However, it should be noted that Butte's pioneering efforts have had residual effects.

In 1954 Butte started to experiment with the use of oscilloscope patterns to generate the main "actors" in her new work. Butte's claim to be the first to combine "science and art" is evident in her next films *Abstronic* and *Mood Contrasts*. Butte coupled these "actors" with unique and colorful backgrounds making her work take on a surrealistic style. Butte also made a leap to the small screen with two works that were displayed on television. The first was entitled *Imagination* for the Steve Allen show in 1958 and later in 1959 *New Sensations in Sound* for RCA. For all of her efforts, Mary Ellen Butte needs to be recognized as a pioneer in the art of film animation (Sito, 2015).

2.6.8 To the Very Small Screen

The question that comes about from all this information is: What effect has all this had on new media use of these traditional skills and direction? How can it be transcribed from the large screen to the small screen to the very small screen? Is it the same process? Step by step – frame-by-frame – tweening – rendering – voice-overs dropped in? Well the answer is Yes!

The movement of elements on the computer screen got its start from one gentleman who is regarded as the originator of computer-based motion graphics. His name was mentioned before but due to his contribution he needs to get better recognition. His name is John Whitney, Sr.

Whitney is considered by many to be the "father of computer graphics." His strong use of combining computer technology and an artistic sense have created a legacy of visual movement and musical integration that has affected the world of the two- and three-dimensional animation since the late 1970s. While in Paris during the late 1930s, Whitney studied the work of Schoenberg's Twelve Tone techniques with Rene Liebowitz while working on animated abstract designs. Whitney's first contributions to this new world of motion came about in the late 1940s with his work *Five Abstract Film Exercises*. This piece won him first prize at the First International Experimental Film Competition in Belgium (1949). From this early combination of motion and music Whitney became intrigued with how music produced sine waves and the form these waves generated. How this occurred was not so much a fascination of Whitney's (although later it would in essence become the basis for his work) as how these images transverse time. "Time has become visual," stated Whitney. Whitney's fascination with images was also apparent at home as well as in his work having married an abstract painter, Jacqueline Helen Blum. Whitney and Blum had three sons: John, Jr., Michael, and Mark.

Whitney's fascination became concrete in 1952 with the founding of Motion Graphics, Inc. The company was founded primarily to produce commercial motion pictures, but later produced animated television shows and films. By the late 1950s Whitney started to develop mechanical analog systems to produce his work. These systems were not yet computerized but this would come about shortly with the use of analog computer systems made from surplus anti-aircraft parts. In 1961 he produced Catalog 1961, which acted as a portfolio to display the industry capabilities of his new motion techniques. In 1966 he created permutations, which was created entirely from a computer system that utilized programming generated from the likes of GRAF and FORTRAN. The creation was completely done in black and white adding color through the use of an optical printer.

Throughout the 1970s Whitney would continue his creation of computer-generated motion graphics through numerous support structures such as commercial companies and the National Endowment for the Arts and Information International. In 1975 his landmark work Arabesque premiered. Arabesque is a combination of sound and movement that launched an entire industry. In his book *Digital Harmony* (1980), Whitney explained his theories of motion graphics and the complement of music and visual art. Whitney stated that "Painting and song share the earliest relationships in cave societies." In the late 1980s Whitney and Jerry Reed joined forces to create a series of computer graphics joined with music composition. These inventions included an instrument that generated a combination of "tonal action with graphic action." This Whitney concluded was capable primarily because of computer technologies. Whitney continued to create new pioneering efforts in motion graphics until his death in 1996 (Sito, 2015).

2.6.9 I Really Want to Direct

As all of this new motion and computer integration progressed, other areas of technology started to feed into the mix. Evans and Sutherland were busy creating software that allowed for users to create images directly on the screen of a computer. Sketchpad (Sutherland's PhD thesis at the Massachusetts Institute of Technology) was one of these first incarnations. But from all this came an even greater contribution to the field of two-dimensional computer-generated animation.

In 1982 John Warnock and Dr. Charles Geschke co-founded Adobe Systems. For nearly 30 years Warnock has been an innovator in the field of computer graphics software. Starting at the University of Utah in the 1970s, Warnock later became a principal scientist at the Xerox Palo Alto Research Center (PARC). Warnock had previously been part of development teams at Evans and Sutherland (Sketchpad), IBM, and the University at Utah. The main contribution of Warnock to the field of computer graphics software (I personally feel) is the creation of algorithms that generated the vector format of computer-generated imagery. The uniqueness of vector is that it allows (unlike Raster imagery) the ability for the background of a graphic to be transparent. This then minimizes the file size and allows for imagery to appear to be transparent to other objects "underneath" the selected object. What this ultimately leads to is Adobe Illustrator, Dimensions, GoLive, and Live Motion to name just a few.

From this one innovation many others are spawned. But before that is possible a simple solution is brought to the forefront: the ability to allow simple, mere mortals to control the order, interaction, and direction of information visually to and from a computer screen. One of the first to do this is a program that is little known to today's computer users – HyperCard. In 1986 Bill Atkinson created HyperCard. Released in 1987 by Apple, HyperCard has become almost a legend in interactive development. With a strong and dedicated group of followers HyperCard still remains as one of the world's largest interactive authoring tools. Simple to use and relatively inexpensive, HyperCard was first released free to end users. Due to its quick and extensive popularity, Apple decided to start charging $250 per license for the software. In the mid-1990s HyperCard was adopted by Apple's Claris software division and subsequently brought back to Apple in 1997 and almost completely ignored. Apple still sells the software now for $99.

To any game authoring developer or web designer, the name Macromedia conjured up both disgust and unbridled excitement. Started in 1984 by Mark Pierce, Jamie Fenton, and Marc Cantor, Macromedia – originally Macromind – has become the industry standard for interactive game, two-dimensional animation and web development.

Whether the product is DreamweaverMX, Freehand, FireworksMX, DirectorMX, or FlashMX you are sure to have visited http://www.macromedia.com once in your career. The original product generated by the team was called VideoWorks. In 1987 VideoWorksII was released with full color capabilities. Later VideoWorks evolved into VideoWorks Interactive, which was based on the programming language BASIC. Although never released to the general public, many software companies used the product to develop tutorials and tours of their own software products. Macromedia for Mac-based users created one of the first CD-ROMs in 1987.

Later in 1988 VideoWorks Interactive Pro morphed into MacroMind Director 1.0 via the programming genius of John (JT) Thompson and Erik Neumann. VideoWorks/Director had been rumored to be used to create the computer graphic displays for StarTrek: The Next Generation. Once this adoption into the world of commercial motion pictures had been made multimedia was just about to launch to the general public. With the advent of Director 3.1, the program crossed platforms with a Windows-based player for its final projects. In the fall of 1993, MacroMind released Director 3.13 and MacroMind became Macromedia. Now the new release is supporting the use of Apple's new format QuickTime. Now videos could be played directly in the window of Director, along with the audio and Raster image motion that it always possessed. Once the three visual capabilities were rolled into one interactive program true multimedia was coming of age for the typical developer.

In 1993 Director 4.0 was released and quickly became the state of the art for the interactive animated authoring world. XObjects and XCMDs were also included in the new version and these allowed for development by third-party companies. Once Shockwave appeared in 1995, a compression format was available for real-time "video" playback within a stand-alone application. Shockwave files were generated via an internal program named Afterburner. In 1996 Director 5 was released; the new version contained a list of added features that allowed for debugging of Lingo scripts, multiple casts, a watcher – to see how you were doing via the scripting and a host of add-ons called Xtras. Now the updates were coming a little too fast and way too furious. In 1997 Director 6 was released. This new version introduced a redesigned development window, integrated Shockwave (for the web) support along with a library of drag and drop Lingo scripting – behaviors. Coming close on the heels of Director 6.0 and 6.5 was Director 7.0. Said to have numerous upgrades it was in general a disappointment to most designers and developers. Although it did increase capabilities – such as a Multi-User Server function overall it was slow and a bit buggy with its new dot syntax for programming. But just wait until next year.

In 2000 Macromedia released Director 8.0, a stronger, more sophisticated program although still built to look primarily the same as Director 4.0. More stable and more reliable than 7.0 Director 8.0 – but wait there's more, later that same year Director 8.5 was released. Made to be more web-friendly than ever before it was totally compatible with the new kid at Macromedia – Flash 5.0. We will talk more about this true web-based animation program in the next paragraphs. By the fourth quarter of 2002 Director released DirectorMX, the current version being used by designers and developers worldwide. Made to be compatible with a number of options – three-dimensional navigation (from Director 8.5), total OS X compatibility and some major authoring enhancements – all in all MX is pretty similar to Director 8.5.

Whether you agree with the marketing of this product or with how quickly it was rushed to market at times – ala Dilbert – Macromedia Director changed the way two-dimensional interactive computer-generated animation was used throughout the world (Parker, 2010).

2.6.10 Not Really a Flash in the Pan

The story goes that the inventor of Flash started with an interest in the children's toy Lego – bits of colored plastic that connect. Jonathan Gay is the inventor of one of the interactive animation world's premier authoring programs-now known as Macromedia Flash. In an article about his life, Gay stated, "Lego was an early way to express my passion for building things." Obviously, this desire to build carried through to his invention that has revolutionized the way the world communicated.

Gay started his journey toward computer programming while still in high school (approx. 1984). He entered his graphics-editing program into a high school science fair. This was just about when the Macintosh computer hit the scene. His father brought him to a Macintosh users group meeting and told a few people how well his son had done in the high school science fair. There he met Charlie Jackson who was starting a Macintosh software company called Silicon Beach Software. Jackson had just purchased a $10,000.00 Macintosh Lisa computer. This is exactly what Gay needed to start his programming career. Gay's first creation was a Macintosh game called Airborne! It was a hit. It was one of the first animated games that used sound where the animation did not flicker. After this success Gay hired a professional artist to design screens and Gay wrote the program for his new game Dark Castle. It was a big hit and subsequently paid his way through college. After this success Gay worked on adding a Postscript-style drawing option to SuperPaint II (a software drawing program for the Macintosh).

After college Gay worked for Silicon Beach Software full time. Here is where he started working in C++, a programming language, which would allow for cross-platform authoring of games and programs. This work later became a drawing program called Intellidraw in hopes of competing with Adobe Illustrator and Aldus Freehand. What made Intellidraw different from Illustrator and Freehand was that back then you could add behavior that would allow the user to change the images based on information that would be entered. (Believe it or not the first computer drawing program – Sketchpad, developed by Ivan Sutherland, already had this option.) Another company called Visio was able to use this concept in its organizational aspects of its lead program – Visio. From watching other companies create marketable products, Gay figured that he knew enough to venture out on his own.

At the same time a company called Go was busy creating an operating system that would allow the user to work with a pen directly on a computer screen (foreshadowing of all PDAs). Here was an opportunity to create new software for a new computer company. As it turned out Go didn't! It ran through its capital faster than it could turn out work. Subsequently, Go and its products were purchased by AT&T. AT&T had other ideas for the software, but right before ship date they also pulled out. So Gay and his team were stuck with their new product SmartSketch and no means to distribute. But as good fortune has it Gay sold a copy of SmartSketch to just the right person – the architect working on Bill Gates' new home.

Even though you would have thought that would open doors for SmartSketch, it didn't. Gay was still stuck in a tough established drawing market with a product he knew could be infinitely better. In 1995 he shared a booth at SIGGRAPH and was able to display his product. Many who stopped by thought it would be better suited as an animation development program. Being a vector-based tool the product was focused on the two-dimensional animation marketplace. From here he started to give SmartSketch animation capabilities that would work on the web. Writing through Java Gay started to develop the SmartSketch animator. After a while he realized that the true marketplace would be in a product that could animate across the web. He changed the name to CelAnimator but soon realized that the name limited the marketplace. He changed the name to Future Splash Animator.

In October 1995, Gay attempted to sell his new technology to Adobe Systems. John Warnock of Adobe was interested in how the product worked but not with its slow animation playback. Adobe said no. However, Gay's luck continued in the summer of 1996; he began shipping Future Splash Animator and interest soon peaked. One of the first clients was Microsoft. They wanted a web animation program to aid in the launch of their new product for the web – Microsoft Network – MSN. Gay's other high-profile client was none other than Disney Online. Disney started to use Future Splash Animator to build online animations and user interfaces for their subscription-based online service Disney Daily Blast. Around November 1996 Macromedia heard about this new web-based animation product and approached Gay and his team about working together. The next month Gay sold Future Splash Animator to Macromedia and Flash 1.0 was born.

With the birth of this program an entirely new level of two-dimensional animation was born. Now web-based computer-generated animation could be achieved. Once Flash 1.0 hit the marketplace the entire web community was a buzz with its ease of use and infinitesimal files size. Soon after Flash progressed to versions 2.0 and 3.0. Flash 3.0 brought a simpler interface design along with greater capabilities but it wasn't until Flash 4.0 did we get the addition of the much-needed Actionscript. Now users can truly interact with the product, which set Flash on its course to be the web development tool that designers and developers had been waiting to get their hands on the section on authoring later in this book, but for now you should know that Action Scripting, like Director Lingo, allowed users to animate their objects by virtue of programming again cutting down on the work necessary to create animations. Although it is no walk in the park, animating with Actionscript was the way to go. Finally, with versions 5.0/FlashMX everything was in place to make this animation tool the predominant product in the web design and development arena.

2.6.11 The WOW!!! Component – 3-Dimensional Animation

2.6.11.1 Whitney to Butte to Jim Clark to Pixar

The move from two-dimensional animation to three-dimensional animation came about first very slowly then once technology caught up with the needs of three-dimensional animation it leapt forward at super natural speed. The early animators – Whitney, Butte, etc. needed the intelligence, financing, and ability of people like Jim Clark, John Lassiter, Ed Catmull, Andrew Stanton, Pete Docter, Steven Spielberg, Jeffrey Katzenberger, David Geffen, Steve Jobs, and George Lucas. Most of these names are well known but who is Jim Clark? If you ever heard of Netscape, WebMD, Healtheon, or Silicon Graphics then you know the name Jim Clark.

2.6.11.2 Jim Clark – From Dropout to CEO

James Henry Clark's research work in computer graphics led to the development of Silicon Graphics Incorporated, known for its fast rendering of three-dimensional computer images. Born in 1944 in Plainview, Texas, Clark did not do well in school and got family permission to enlist in the Navy at

16. After some issues with testing, Clark went on to do very well in aptitude tests for engineering. He later returned to school earning his PhD in computer science from the University of Utah. While teaching at Stanford in 1979 his group developed the Geometry Engine, an early hardware accelerator for rendering computer images based on geometric models. This led to the future development of the Silicon Graphics rendering system, which became the backbone of animation for the motion picture industry. Once systems became more sophisticated SGI's command of the market dropped. However, Clark departed SGI before this happened in 1994 and used his newfound wealth to fund new ventures like Netscape, WebMD, Healtheon, etc. (Lewis, 2001).

2.6.12 Leading 3D Software

There are many diverse versions of 3D software available to the end user. These packages range from the simplest like Strata to the extremely advanced like Maya and Unity. The rise of 3D animation and rendering will only grow with the advent of Artificial Intelligence. Here are a few quick descriptions from the websites of some of the leaders in this area.

Strata: Strata 3D was one of the first desktop 3D graphics applications, releasing its first application, StrataVision 3D, in 1989. (InfoWorld, 1991)

Lightwave 3D: LightWave is a software package used for rendering 3D images, both animated and static. It includes a fast rendering engine that supports such advanced features as realistic reflection, radiosity, caustics, and 999 render nodes. (Murdock, 2005)

Maya: Maya was originally an animation product based on code from The Advanced Visualizer by Wavefront Technologies, Thomson Digital Image (TDI) Explore, PowerAnimator by Alias, and Alias Sketch! (Kumar, 2022)

3D Studio Max: Autodesk 3ds Max, formerly 3D Studio and 3D Studio Max, is a professional 3D computer graphics program for making 3D animations, models, games and images. It is developed and produced by Autodesk Media and Entertainment. (Murdock, 2005)

Unity: Unity is a cross-platform game engine developed by Unity Technologies, first announced and released in June 2005 at Apple Worldwide Developers Conference as a Mac OS X game engine. (Blackman, 2011)

To finish this section on 3D animation and not to speak clearly about the effects of Pixar and DreamWorks is illogical at best. Since the early days of "Luxo" and "Tin Toy" to the "Toy Story" franchise, the mesmerizing power of 3D animation has captivated audiences. Its value to the gaming and entertainment industry is indisputable. However, to do a full history of all that in detail would take an entire book. But one thing should be noted: if it wasn't for Steve Jobs putting $150 million toward the purchase of Lucas Films' Graphics Group and renaming it Pixar (Pixel Arts), many of our childhood memories may not exist (Sito, 2015).

2.6.13 Section Summary

To quote a well-known 20th-century pop icon "What a long strange trip it has been." From the days of Egyptian carvings to Roman storytelling to the aspirations of Winsor McCay and right through an amazing bag of tricks, the journey to replicate life in sometimes-surreal conditions has finally made it to almost every media available for distribution. As I have stated before many names were lost and many omitted obviously for brevity's sake. The chronology presented in this section is not meant to reflect the end all of lineage toward animation; it was put together to merely create some kind of linkage from date to date, from accomplishment to accomplishment. But the truly amazing part of this journey is no matter what new technology comes along there is one thing that stays exactly the same – the narrative.

Storytelling dates back to prehistoric times, with the goal of every generation and every civilization being to tell that story in the best way possible. In most cases the story has been one-sided, hence the term his-(s)tory but now with the ability to tell that story around the world in an instance the stakes for attention get higher. The new media we are choosing to use has truly changed the nature of storytelling. Now we want the listeners to interact with the story. In some cases (Margot Lovejoy – Whitney Biennial, 2002), the work they do as artists requires input from the viewer/user. This will change things for sure. But it is not anything new, it is just being used more today than ever just remodeled to work anew. Storytelling is always at the base of every piece of work whether it is the slapstick of the MGM studios or the heritage of Disney, the story is premier above all else.

2.7 BIRTH OF DIGITAL AUDIO

2.7.1 From Edison to Taylor Swift

If you have ever watched a movie with the sound off you know what intensity and fullness audio can bring to any form of communication. For centuries mankind has raised its voice in praise or disillusion. From Gregorian chants to today's pop music the nature of sound has motivated the masses whether it was to war or to peace. Whenever we heard the reverberations of the human body or an instrument of any nature, we felt our spirit move. Sound, very simply, is defined as "mechanical compression or longitudinal waves that propagate through a medium (solid, liquid or gas)" – http://www.wikipedia.org. This section is focused on how we came to capture that function of the human condition and how it came to capture us.

2.7.2 Edison's Little Lamb

One of the first individuals to delve into the study of audio capture was none other than the "Wizard of Menlo Park" – Thomas Alva Edison. Edison first started to play with capturing or playing back sound with his work as a telegraph operator in 1862 at the age of 15. By 1877 Edison was working on a machine that could transcribe telegraphic messages through indentations on a paper tape. This then could be sent over telegraph lines repeatedly without the need for an operator. Basically, this was recorded sound – with this idea came another. Edison figured out that the human voice could be able to be recorded in a similar way. He experimented with a number of configurations from paraffin paper to wax cylinders to needles and embossing points to diaphragms and such. Edison finally gave a sketch of his idea to one of the workers in his Menlo Park Laboratories–John Kreusi and supposedly the first working phonograph was built from scratch in 30 hours. The legend goes that Edison immediately tested the machine by saying the old nursery rhyme "Mary had a little lamb." The rhyme was recorded and played back.

The patent on the machine was issued on February 19, 1878. The invention was just one in a long line of inventions by Thomas Edison that has obviously changed the world. The only other evidence of any other machine for recording and playing back sound was by the French scientist Charles Cros. Since his machine remained only a theory with no working prototype, Edison's work stands. On January 24, 1878, the Edison Speaking Phonograph Company was established in Menlo Park, New Jersey. The machine was a success instantly, first as a novelty then later as an office apparatus for recording dictation. Edison conjured up a long list of uses for this new device such as dictation, elocution, speaking to the blind, music recording, family records, clocks, educational recordings, connection with telephone for recording of messages, and conversations. I guess that's why they called him the Wizard of Menlo Park.

Cylinders quickly gave way to the concept of revolving discs. Edison's staff worked in secret, even from Edison himself. Dr. Jonas Aylsworth, chief chemist for Edison, was responsible for overseeing the project to create a suitable substance to use as a disc for recording and playback. The overall goal of the experiment was to produce a disc that revolved around a center shaft that would allow a needle to ride in grooves and literally bob up and down. The records could now run for five minutes per side, quite an improvement over the cylinder-style recordings.

The Edison Disc Phonograph debuted to the public in 1911 in Milwaukee, Wisconsin, at an industry fair. There were three basic models by the end of 1912 ranging from $150 to $250 – (remember this is 1912!). Edison was not the only company to be making discs and players. Columbia Records was turning out discs as early as 1912 as well as Edison's. Although the marketing and acceptance of these new products was strong the fluidity of content seemed to drive the audience. In order to consolidate efforts, more songs were featured on just one disc – creating an album of the artist's work. In 1924 the record business started to weaken primarily due to the advent of radios in the home. (This could be seen as another example of content over distribution. Could be similar to the web versus archival discs – DVD or CD-ROMs. History always teaches us something no matter how distant the time frame or subject matter.)

Once Edison could see the future of the home radio and phonograph market was finite he started to focus on a new device. In the second half of the 1920s Edison started to work with his moving pictures and accompanied narrations. The first effort in this direction was called the Ediscope. (Remember Edison had already created a motion picture system and full studio by the early 1890s. This will be covered in the next chapter – digital video.) The Ediscope gave way to the Cine-Music Phonograph whose basic purpose was to supplement movies with musical accompaniment.

But once Edison realized that the combination of a radio and phonograph seemed to be the best solution to the marketing part of the problem, he capitulated and took over the Splitdorf-Bethlehem Electrical Company of Newark. With this acquisition and through the development and acceptance of diverse forms of records – long playing discs and portable players – Edison's new venture would seem to have found profitability.

Edison, along with many other inventors of the time, had been attempting to link motion pictures with sound since the late 1890s. In 1895 he released his first solution to this new problem – the Kinetophone. Similar to his Ediscope and Cine-Music Phonograph, this device didn't synchronize with the motion picture as much as it offered a musical accompaniment to the action on the screen. The one system that seemed to really start everything rolling however was invented by Lee Deforest and previewed in 1923 in New York City. DeForest called it the Phonofilm system. Amazingly the system photographed the voice onto a filmstrip. Although extremely progressive for his time, the system didn't expand and flourish due to poor financial backing and the lack of adequate patents. As always there were some positives that came from this invention's life span. AT&T (American Telephone and Telegraph) would secure the rights to DeForest's Audion tube and through their manufacturing division – Western Electric. By combining this tube with other components, they were able to create a complete sound on disc system for the movies by 1924.

By this time everybody in the motion picture industry was starting to see the worth of adding sound to movies. One of the largest studios of the day Warner Bros. stepped forward to eventually become the main innovators in the group. With financial backing from Wall Street, Warner Bros. made a deal with Vitagraph Corporation and a limited time deal with Western Electric to develop this new technology. In 1926 the Vitaphone Corporation was created as a subsidiary of a merger between Western Electric's experimentation and Warner Bros expertise. The Vitaphone Corporation released its first short on August 6, 1926. Almost 100 Vitaphone systems were installed in theaters by the end of 1926. At the end of 1926, Western Electric was being asked by numerous other companies to develop motion picture sound systems and dissolved any ties they had with Warner Bros. Western Electric went on to form ERPI – Electrical Research Products Incorporated. ERPI's main goal was to establish the Vitaphone system in as many motion picture houses as possible but ERPI was also established to develop anything that fell outside of telephone communications.

However, it would be Warner Bros. that had the last laugh or maybe the first? In 1926 Warner Bros. released the Jazz Singer starring Al Jolson. Although this was not the first talking picture it was the first popular talking picture released to the general public. A tremendous hit many more "talkies" were released soon after.

Although a hit, the method behind the Jazz Singer was still in need of improvement. As Warner Bros. continued with the sound on disc system Fox Studios worked toward incorporating the sound directly onto the film system. Sound on film used an optical method to actually record the sound waves onto a thin portion of the motion picture film reel. This system was created by trial and error from a myriad of developers such as Harold Arnold at Bell Labs to E.I. Sponable to Theodore W. Case. Each system in development had its pros and cons but it was William Fox of 20th Century Fox who licensed Case's system. By 1927 the Fox-Case name became synonymous with the concept of sound on film. As Fox increased its theater empire in the United States, they soon found themselves in direct competition with the Warner studio system. Since Warner was well experienced at this marketplace, they had already contracted most of the talent leaving Fox without bodies to actually put in their new motion pictures. Remember the current day prophecies of Esther Dyson "Content is King." Due to this lack of actual acting talent Fox focused on making informational-based movies – newsreels.

The next mogul to get into the sound on film mix was Radio Corporation of America or RCA Victor. Their system was titled – Photophone. Exhibited for the first time in 1927 at the Rivoli theater in New York, RCA found that the marketplace had been cut up pretty well by Vitaphone and Movietone (Fox). Distribution, always being a major problem in the entertainment marketplace, RCA found that it needed an empire of its own through which it could display their new sound on film pictures. In 1928 RCA purchased a small theater chain – Keith Albee-Orpheum and RKO pictures was born. This was the studio that would create such incredible masterpieces as Citizen Kane with the 28-year-old Orson Welles at the helm (Sobel, 1986).

As it is with any new technology there was much debate whether sound on disc or sound on film would be the format of choice for this new entertainment delivery system. But many of the caveats attached to sound on disc were just too obvious to ignore. Sound on disc was fragile (as the discs could easily break), often out of sync, and the disc did not have the capability for a great length of play. Sound on film was more difficult in the manufacturing aspect of width of film and necessary playback technology but as we see now the quality of synchronization won out over cost and flexibility. By the late 1930s almost all theaters used what was now called optical sound and despite the troubles the adoption of sound into the motion picture industry moved rapidly. Although it was now commonplace the industry and the audience didn't truly understand that it "ain't seen nothing yet!" – (Lyons and Plunkett, 2007).

2.7.3 The More Channels the Better

In 1932 Bell Labs and RCA began developing a multi-channel recording process for the motion picture industry. By 1937 this process had worked out to the development of a nine-channel system that would be mixed down to four final tracks for use as an optical soundtrack on the edge of the release prints of movies. One of the first movies to be released with this new multi-channel soundtrack was Disney's Fantasia. Released in 1940 this multi-channel extravaganza was mixed down to four master optical channels, which were used in conjunction with a speaker system theater installation named "Fantasound." At a cost of $85,000.00 a theater to install, the movie was a huge success, but the sound system didn't take off very well (Baxter, 2023). Many other picture and sound systems formats have been tried over the years from Cinerama to Cinemascope to ToddAO to Perspecta to THX (Lucasfilms) to Digital Disk Playback to Dolby, today in its Dolby Digital version 5.0. Each system improved upon the previous one finally bringing us to a time where home theaters are promising the equivalent of a movie experience.

Sound and image have always been a critical part of the entertainment field. Whether it was full orchestrations or simple dialogue the only way to get the audience to truly believe that they are there is to let them hear for themselves. This usually required a team of technicians and a slew of equipment. But all that was about to change with the advent of magnetic tape. Again the story of magnetic tape starts with the Wizard of Menlo Park.

2.7.4 Sound on a Wire

After the development of multi-channels the next real big invention is the use of magnetic tape on which you can record. As most of the major inventions of the time this also started out from Edison's labs. In 1878 Oberlin Smith was inspired by a visit to Edison's labs in New Jersey and conceived the idea of how to record electrical signals onto a steel wire. Now at the time many different recording devices were coming in to play for various uses. But this innovation would actually create another entire industry. Smith decides not to pursue his idea (for whatever reason) and writes a paper on it and has it published in the journal *Electrical World*. Later in 1889 a Danish inventor Valdemar Poulsen reinvented the principle of magnetic recording. Poulsen's patented idea on how to record on steel wire and tape is manifested in the development of a recording machine called the "Telegraphone." Amazingly it is described as an answering machine for the telephone. The product's first international display came in 1900 at the Paris International Exhibition. Poulsen's invention is heralded as superior to the phonograph and soon he has investors.

In the early 1920s German Curt Stille improved upon Poulsen's device and modified it to use electronic amplification and started to market the rights to other countries. One such developer is Ludwig Blattner Picture Corporation of England. These machines use steel tape instead of wire. By the late 1920s machines like a Blattnerphone or the Marconi-Stille were being used to record radio programs. The sound waves are recorded onto huge reels of steel tape that spin rapidly. The machines were huge and the tape was very expensive.

By 1930 Bell Labs had an interest in this new form of recording and they initiated a research effort headed by C.N. Hickman. By the early 1930s designs were finalized and many devices were prototyped for mass production. Alas they never made it into the hands of the general population due to issues over recording of telephone conversations. But the Marconi-Stille Company goes forward with production of their versions of recording on magnetic tape. Marconi-Stille managed to sell several steel tape recorders to the British Broadcasting Company. But in 1928 AEG, a German electrical manufacturer, obtained the patent rights of Fritz Pfleumer for a system of paper-coated tape with a magnetizable steel layer. AEG then sets out to make a recording device to work with the new tape system. In 1935 a new version of the recorder called the "Magnetophone" was used in London. Soon sales of "Magnetophones" rose to approximately 1,000 units by 1943–1944. Once the quality of recording improves, the "Magnetophone" is heralded as no longer a novelty but is regarded as an excellent recording device. Magnetic tape has made a strong and lasting impression upon the world of sound recording. But how does it work? The magnetic tape recording system worked by use of an electromagnet that applied a magnetic flux to the oxide of the tape. The oxide permanently sees the flux and in essence records a signal. The signal is then amplified to drive the speakers and voila sound is put forth.

The New British Broadcasting System in the mid-1930s truly brought magnetic tape to its early pinnacle by discovering it through the work of two German recording engineers Drs. Otto von Braurmuhl and Walter Weber. These two innovators discovered that when you used a high frequency signal with the audio during the recording, the final output was very close to a live recording.

Soon after World War II, 3M (Minnesota Mining and Manufacturing) and Brush Development Company of Cleveland, Ohio, worked together to develop a tape, which had an emulsion containing a uniform dispersion of ferromagnetic powder. The two companies along with the Navy Department were about to introduce magnetic sound recording tape to the industry. Now mass produced and ready for distribution the next step is to reduce the cost of the recorders. And who else to do this but a historic figure in American song and movies. During the time from 1944 to 1947, 3M almost gave up on producing the magnetic tape, but one of its employees Dr. Wilfred Wetzel saw the future of magnetic tape as being a vehicle for not only sound but possibly for recording television images.

During this time an Army Signal Corp technician Jack Mullin had been scavenging for German recording devices. Mullin would later demo a tape recording system to the Institute of Radio Engineers in San Francisco. The demonstration was a tremendous hit. As luck would have it an employee of Bing Crosby Enterprises was in attendance. Crosby was a high-level recording artist of the day. Since most recordings were done live in the 1940s Healy figured that this device might be the solution to the problem of recording these performances. To help expedite Crosby's time at the microphone, Healy decided that since the quality of the tape recordings had been brought to the level of "nearly" live that Crosby could pre-record his songs and broadcast them at later dates. Previously to this method Crosby's recording had been done on discs and when they were recorded over and over to mass produce, the quality diminished and not to mention the time and expense.

At the same time 3M was working on a commercially viable recording tape. The problem was that the tape quality was too good for Mullin's old scavenged machines. So Wetzel and 3M had to dumb down its quality to work on the older machines. Ampex Corporation was also developing machines after Mullin's scavenged devices; again the new tape was too good so Wetzel and company had to use the older lower quality tape to record some of Crosby's shows for the Kraft Music Hall.

In the late 1940s companies such as Ampex Corporation and Sony Corporation were beginning efforts to manufacture higher-quality tape recording devices. With these machines obviously came a need for mass production of this new "magnetic tape." For a while many in the industry did not think that 3M could meet the demand. In 1947 3M introduced the Scotch 100 magnetic tape. And the magnetic tape industry was born in the United States. These same tape developments would later lead to the use of tape as a means to record images. But more on that when we get to our next chapter on video.

By 1958 the stereo tape was introduced to the general public. Requiring a special player the product does not do well and soon leaves the marketplace. But later in 1962–1964 the Phillips Company of the Netherlands introduced the Compact Cassette. This was a portable tape recorder that used a special cartridge cassette tape. The tape cassette was born. Soon after in 1965 Ford and Mercury car companies started working with Motorola and RCA Victor to introduce the "eight track" tape system for luxury automobiles. This medium first developed by William Lear in 1962 became a success as the first consumer tape distribution model. Of course it went out of production by 1980. But the cassette tape has lived on much longer going from answering machine to stereo system to home tape recorders to small pocket recorders. The next step in the life of "Mary and her little lamb" was to be digital (Aaron Foisi Nmungwun, 2012).

2.7.5 Harmony in 0s and 1s

The launch into the next stage of audio recording was going to be explosive and, as is the case with all innovation, directly launched from tradition. The term "analog" actually comes from the word "analogy," meaning that something is related to something else. Digital relates to the concept of individual pieces, in this case 0s and 1s. But how do you do that? The first thing you'll need is a PCM or pulse code modulation. Well lucky for you we just happen to have one – the only problem is it's 1937. An employee of International Telephone and Telegraph invented pulse-code modulation in France; his name was Alec H. Reeves. He received patents for the device in France 1938, England 1939, and in the United States in 1942. Bell Laboratories used the new modulation theory in telephone encryption during World War II for conversations between the U.S. President Franklin D. Roosevelt and British Prime Minister Winston Churchill. By the 1950s the first synthetic sounds from a computer were being heard by Max Matthews at Bell Laboratories. But it wasn't until 1962 that telephone companies started to use the PCM commercially. Since the world had not gone digital just yet, the space for storage and development of these PCM-based systems was tremendous and costly.

Later in 1962 T. Stockham from the Massachusetts Institute of Technology started to create digital audio tape recordings. Although these were digital audio recordings they were still done to tape. In the late 1960s he left MIT for the University of Utah and in 1975 worked with Malcolm Low to develop a 16 bit digital recorder. During this time magnetic drum storage was being developed for use as a storage medium for computer information.

A lot was happening around the world as well, in 1971, Heitaro Nakajima left NHK's technical research division to join Sony Corporation. Nakajima had been working on sound digitization for almost four years and within that time he and his team had developed the first digital audio recorder. Nakajima became one of the first people to produce digital sound. Even though he had multiple years of setbacks he could not give up his quest to produce digital sound. In 1973 Nakajima became the chief of Sony's Audio Technology Center. Here with other technologists he began working on what would become the development of digital audio for the masses. The only problem was that the recordings were digital but the tape storage necessary (like the others before) was approximately 100 times that of analog. Since he was using the typical PCM method the problem wasn't how to record, the problem was how to store for playback? Regardless of this obvious problem Nakajima and his team built Sony's first PCM digital audio recording machine the X-12DTC in 1974. Even though it did represent a breakthrough in digital recording the device was approximately the size of a home appliance, therefore, it was never marketed.

Nakajima and his team were undaunted and continued working on a smaller cheaper version. Sony had, however, been working on a video recording system – Betamax. Another technologist at Sony, Akira Iga, had an idea. He wondered what would happen if he made a digital audio recording using the Betamax VCR. Iga was told by Nakajima to get to work immediately on the idea. The result was a PCM circuit that would record large-scale signal processing. The system was a Betamax and a PCM processor together made up one of the first commercial digital audio tape recording systems. It was marketed in 1977 as the PCM-1 and was intended for home use. The PCM-1 met with both praise and criticism. The technologists praised it for its innovation but the music world criticized it for its (minimal) lack of clarity and quality. But what was learned here would help advance the world's digital sound and storage capabilities (Lucarelli, 2015).

2.7.6 Invest in a CD

In 1980 the compact disc was brought to market, the child of an international marriage between Sony Corporation and Phillips of the Netherlands. The development efforts began earlier in 1969 with the work of physicists Klaas Compaan and Piet Kramer. The original prototype worked with a laser beam and was intended to record video images. It was modified to work with audio as the laser beam read a track of pits used this time to code an FM video signal. To make it a digital audio storage device a digital PCM code was substituted for the FM signal. In 1979 Sony collaborated with Phillips to develop a set of standards. Once these standards were formatted Sony began production on a compact disc player. By 1983 both Sony and Phillips started rolling out over 500,000 units a year. The cost to the customer of the actual compact audio disc was approximately $15.00 with the manufacturing cost being approximately $1.00 – good profit margin.

Later in the mid-1980s Sony introduced the D-5 portable compact disc player developed from the previous larger versions. This then led the way for an entire world of portable personal music players. In 1985 Sony and Phillips again collaborated to produce the standard Compact Disc Read Only Memory for use in computer disc storage. In late 1988 Sony introduced the D-88, a pocket Discman that used 3-inch compact discs. These smaller discs were used as digital 45s having enough storage space to hold just one typical pop music song.

In the early 1990s this technology was taken to the music recording industry with such devices as an ADAT that could record eight tracks of digital audio to a standard S-VHS videocassette thus making multi-track digital recording affordable for the smaller studios. With all of these innovations now in place the world of digital audio recording, storage, and consumer playback truly came of age.

2.7.7 How It Gets in There

There is no question that the CD revolutionized the world of audio and computer storage but how does it do it? The answer is all in the 0s and 1s. Back in 1939 John V. Atanasoff and graduate student

Clifford Berry built the first electronic computer with a drum storage device. The device used capacitors to store information temporarily on the drum. The two programmed the computer using the binary algebra of George Boole (hence the term Boolean), a 19th-century mathematician. The 0s and 1s allowed for a sequence of information to be transferred into simple on or off stages. Although a complicated concept it worked as a means to minimize structure down to a simple sequence.

Compact disc recording and manufacturing use this same method but they transfer the sequence to a physical state. The storage structure of a compact disc is different from a traditional computer disc in so much as the information is stored optically in a spiral fashion consisting of approximately two billion small shallow pits on the surface of the disc. The spiral itself consists of approximately 20,000 windings. This is a lot of space for information.

The information has been translated into these 0s and 1s and they are now ready to be translated to the disc. The land area of the surface meaning no pit or indentation is read as a 0, the pit or an indentation is read as a 1. The sequencing of these Boolean numbers – 00001110101000 1001 is read by a laser beam as the signal and thus read as the information. Problems occur when the sequence is too tight. Modifications have been made to adjust for this limitation. In an audio CD the data is read as a stream of 16 bit values that come to represent the voltage of two audio signals. This is then referred to (again) as pulse-code modulation data. Right back to 1937 – amazing! There are many other factors to understand before this process can be completely understood and that is for another book and definitely for the experts. For now just understand that there is constantly a link between previous discoveries and what we find fascinating in today's world. Now this is all good for recording the sounds but who came up with the concept of creating in a digital format? And how did they do it? (Smiley, 2010).

2.7.8 MINI, MIDI, Compact Disc

In the 1960s early analog synthesizers were popping up in popular music. Robert Moog was able to design synthesizers for the mass market by the use of transistor technology analog but much smaller than previously needed. By the early 1970s the first one channel digital audio recorder was born, this started to allow for a crossover between the analog and digital worlds. Once these crossovers occurred the music world knew it had to start paying attention to the echoes of digital audio. One of the first standards to arrive was MIDI (Musical Instrument Digital Interface). Now MIDI is a system of creation but not the same as an audio compact disc storage system. With a MIDI system you can create and store music whereas a compact disc is merely used for storage. But where they are similar, are the time and the world they lived in.

Electronic commercial synthesizers were around well before the MIDI standard. They were commercial analog synthesizers and designed primarily to be stand-alones. Soon after, once the cost of microprocessors, integrated circuits and other components started to decrease the manufacturers of synthesizers started to experiment and incorporate the components into their systems. Typical analog switches or patches could now be replaced with small control panels and software made specifically for MIDIs. Along with these new digital enhancements came digital control screens. The problem with these interfaces was that they were each unique, resulting in there being no standard. The solution to this problem was MIDI 1.0, a formal published standard that became a reality in August 1983. These computer-based music systems were part and parcel of the entire digital revolution and more realistically part of the concept of computer networking.

The basis behind the MIDI system was the creation of audio within a computer environment but it was also about connectivity. The MIDI port structure allowed for output and input but also allowed for instruments to be "daisy chained" in a form of network. The MIDI In port to MIDI OUT allows the device to issue messages to other devices. The MIDI THRU port allows devices to pass these messages directly from its IN port. This allowed for the "daisy chain" effect. And because there was such a standard, engineers felt freer to develop or purchase diverse commercial products without the fear of incompatibility.

What this created was the ability for computer-based algorithms to drive musical instrumentation. This means that musical instrument sound could be reproduced by anyone as long as they had the compositional skills. So now an entire band could be contained within one simple machine. This was possible due to the keyboard model upon which synthesizers were built and for which MIDI was invented.

MIDI was not the be-all and end-all for the digital synthesizer marketplace, there were many problems; lag time from input to sound out, an erratic transmission rate and the sound generated may be a bit homogenous when it comes to truly distinguishing between instruments. This tends to allow for a loss of human music-making quality. It also does give ground to non-Western scale intonations. Even though all of these caveats are true MIDI still exists as one of the leading electronic music systems in use today (Warner, 2017).

2.7.9 Even Stevie Wonder Could See It

One of the leading names in electronic music is that of Raymond Kurzweil. Kurzweil started off by trying to design a machine that would help blind people scan written material and have a human synthesized voice read the material back to them. One of his first customers was the blind musician Stevie Wonder. Wonder challenged Kurzweil to develop a synthesizer that could capture acoustic sounds without allowing them to sound mechanical or computerized but yet have the ease of use that comes from a digital interface. Kurzweil and his engineers accepted the challenge and developed the K250. The goal of this device was to capture the range of acoustic harmonies while allowing for a ROM-based sampling keyboard structure. As with most of Kurzweil's production the K250 was a huge hit in the music industry and set a new and higher standard for all the rest to follow. He first developed it for musician Stevie Wonder who is blind (Lemelson, 2020).

2.7.10 Formats for Everyone

There are many file formats available for storing or compressing digital audio files from the basic -.aif (audio information file format developed by Apple) to the new Mp3 and Mp4 formats. But first it would be best to understand how a digital audio file is structured? DAF are structured in three parts, similar to that of an image file. There are two main parts; the header, which is used to store information about the file and the actual audio data – the 0s and 1s that make up the audio signal. Included with this is a third part called the wrapper. The wrapper is used to add features such as streaming capabilities, license information, creator file type, etc. Some of the major file format types are AIFF –. aif (Macintosh), AU .au (Sun/Next – Unix based), MP3 .mp3 (MPEG Audio-Layer III), Windows Media .wma (Microsoft), QuickTime .qt (Apple), RealAudio .ra or .ram (Real Networks), WAV .wav (PCMworks with other codecs).

Each format has its pros and cons but in each case compression versus lack of quality is the key issue. The real question is why do we need formats and what do they do for the end user? The main purpose behind formats is to make them a standard or to allow for compatibility issues. Well, if that's the case then why all the different ones? That is actually a better question. The reason for all the different formats is mainly because of the three most popular operating systems: Windows, Macintosh, and Unix. As these systems start to combine functions and allow for crossover the format issue may disappear to really only be left at an Mp3 or ? level.

Where did they come from originally? Formats evolved because of a need for (as was said operating system compatibility) and a need for compression or what is called a codec. Codec stands for compression decompression (co-dec). The same was that modem stands for modulation-demodulation (mo-dem). There are two basic kinds of compression that we deal with in computer technology. They are lossy and lossless; lossy compression is typically found in simpler compression algorithms and delivers a lower-quality end product. In other words, the quality suffers as the compression level is increased. In a lossless compression quality becomes a greater concern than final file size. This is not to say that the files do not reduce greatly in size but they do end up being a bit larger than a lossy compression but yet the quality stands out. Examples of lossless compression are LA (Lossless Audio), FLAC, LPAC, OptimFROG, and Monkey's Audio. These are primarily used in a recording studio to mix with as opposed to sending out over the web or done through a home system. Examples of lossy compression are Mp2, Mp3, MPC, Vorbis, AAC (Advanced Audio Coding {MPEG-2 and MPEG-4), and a WMA (Windows Media Player). Another form of lossy compression that was primarily used for the Internet was RealAudio. RealAudio was developed by RealNetworks and made its debut in the late 1990s. The difference between RealAudio and other forms is the fact that it is a streaming audio setup. There are other streaming audio formats but RealAudio was the most popular of that configuration.

But what actually happens when we compress audio and who figured this out? The main codec used today throughout the Internet world is Mp3. Most compression formats are based on relatively simple mathematical transformations such as the modified discrete cosine transform. This transformation then converts sampled waveforms into their component frequencies. In other words, compressing them down to their relative bases with a key to unlock them for playback. Some compression formats reduce down the file by only using the areas of the file in which a change in frequency occurs. Not ignoring the other values, but compressing them into a sequence and only noting the changes – kind of like bins or suitcases. One suitcase contains all the sweaters and the next one contains all the shoes. How many suitcases with sweaters and how many with shoes you will have all depends on the individual who is packing for the trip. A little crazy but I hope you get the idea.

A group known as the Motion Picture Expert Group started the compression format MPEG in the late 1980s. This group was started to construct standards for the coded representation of moving pictures, audio, and the combination of both. Starting as a relatively small group it has grown into a consortium of over 350 experts representing some 200 companies and organizations worldwide. Through meetings at least three times a year the group has generated such standards as MPEG-1, MPEG-2, MPEG-4, (consisting of versions 3, 4, and 5, MPEG-7 and MPEG-21). But how did the popular Mp3 format really get started?

In 1987 the German company Fraunhofer Institut Integrierte Schaltungen started to work with MPEG-1 Layer 3. Under its given name the EUREKA project EU147, Digital Audio Broadcasting (DAB) Professor Dieter Seitzer of the University of Erlangen developed an algorithm that eventually became the standard known as ISO-MPEG Audio Layer 3. Starting into the 1990s, Fraunhofer was granted patents on the algorithm and then contacted independent developers of Mp3 to create encoders and decoders. Any developers who wished to use Fraunhofer's algorithm needed to apply to him for a license to use it. The Mp3 format has proved to be the most popular to date with industry designers and manufacturers having released a mass-market approach to Mp3 delivery vehicles and portable players.

But how does this Mp3 thing really work? As with all sound the basis for compression (or just regular singing in the shower) has to do with something called psychoacoustic model. What this means is that the human ear cannot hear the full range of frequency of a recording or for that matter anything. The range of a typical ear is between 20 Hz and 20 kHz and is responsible for most sounds between 2 and 4 kHz. What Mp3 does really well is that it uses an algorithm that can compress CD quality audio – usually 16 bit and 44 kHz – down to a drastically smaller file space between one-tenth and one-twelfth the size with a minimal to the audio quality. This is achieved by eliminating those certain frequencies that the human ear cannot hear.

Remember the more that is compressed the more range is eliminated. So with compression final quality is the big issue! The fact that Mp3 can compress files to such a drastic reduction and still allow acceptable quality playback is what has made it so popular. Also this much smaller size allows for download times to be greatly reduced as opposed to other formats such as .aif and .wav. Today these formats are a necessity given the emphasis on the Internet as a delivery system (Vic Costello, 2017).

2.7.11 Determining Playback Rates

Once you have implemented your information architecture to the project you need to be sure that the final deliverable of audio is constructed correctly for its final delivery medium. Compression formats and quality of audio files will be different when you are going from web to archival media. For example, a CD-ROM or a DVD can accommodate higher levels of audio than the web. On a CD-ROM, audio files can finalize as high as 16 bit–48 kHz or the quality that is associated with Digital Audio Tape. Red book quality compact disk levels are 16 bit–44 kHz. This makes for a large final file size and is not easily compatible with typical web delivery methods.

Ways of getting around these limitations are by buffering the download of the file. This is accomplished sometimes by using systems like RealPlayer. RealPlayer uses a buffering technique in order to let the end user get a greater quality playback than is normally associated with the web delivery format called streaming. Other forms of streaming are ASF or Advanced Streaming Format for Windows and other players like RealAudio or Beatnick can usually cost the end user a fee. This has greatly hampered the adoption into the overall multimedia market as always the simpler the better. Typically, when compressing files or even when just creating them for digital output you used to have to choose between .wav or .aif formats. The .wav is associated with a Windows platform (Windows Media Player) and the .aif is associated with a Macintosh platform. Now since the adoption of the Mp3 format everything seems to mesh quite nicely.

2.7.12 Keeping It Quick, Real or Mp3

At the start of this century there were many different audio formats for use over the web or for use in multimedia in general. I personally will always trust QuickTime over all others. It has proved for me to be a consistent standard for over a decade. Mp3, however, appears to be the one that will stay for at least a few years. It does require license fees for developers but it has the best sound in relation to the smallest file format. See size does matter.

2.7.13 Section Summary

From "Mary had a little lamb" to Taylor Swift streaming everywhere, quality has always been at the center of audio recording and formats. Whether it was a wax cylinder or disc or the ability to optically stamp audio onto celluloid or onto oxide-covered paper or to a droplet of plastic–audio rules. And the ability to make this phenomenon accessible to the masses is what has brought the world of audio reproduction to become one of the world's largest revenue-generating industries. What is bigger today than the music industry and its ties to marketing, clothing, television, movies, automobiles, consumables, etc.? It influences our culture like nothing before. In the immortal words of Paul Simon, "Every generation throws a hero up the pop charts," but what new technological advantage will they bring down with them?

Probably before we were actually born, we first recognized sound. Even though we couldn't experience the image of the world we recognized the sounds of heartbeats and movements. This is why audio is such an integral part of any presentation or project. It is the most common level of enhancement (actually necessity) of any work. If you have ever watched a movie and turned the sound off you know exactly what I am talking about. Could you imagine the movie "Jaws" without the telltale theme that played every time the shark got close to another victim? Video and animation are great enhancements to a story but nothing changes the mood like audio.

The reason that audio is critical to a story or presentation is that we as a species store audio tracks in our psyche. For example, there are sound associations that we make and have stored that we carry with us at all times. Color has a similar reference structure: apples are red, the sky is blue. These things change but we all have a common starting point. And we use this when developing a storyline in multimedia. When a window changes, we might utilize a click sound or if a screen is dissolving up we may use a futuristic sound, it all depends on the image. All in all what we are looking to do is to generate an emotional response from a wide array of pre-established connections.

2.8 BIRTH OF DIGITAL VIDEO

2.8.1 From Columns to Cellular Towers

2.8.1.1 First Italian Miniseries

In an interview for his film *Gangs of New York*, the film director Martin Scorsese made direct reference to the Trajan column. The column is over 100 ft from the ground and it has displayed on it the complete story of the Roman Emperor Trajan, since approximately 100 AD. It tells his story from the first battle up to the end of his reign. Each image is hand-carved and full size, they wind around the column in a spiral. The viewer needs only to walk around the column to get a sort of filmstrip of the life and times of Emperor Trajan. Scorsese referred to the column as being the first item that actually was made to tell a story and not for any other reason. As a filmmaker he made reference to the concept of the motion necessary for the viewer to participate in the message. Scorsese thought it was the first time in the history of man that motion was used to tell a story, but obviously not the last time.

The concept of use of motion has its start back in 65 BC with the philosopher Lucretius (mentioned in Section 2.4). Lucretius proved that for a set amount of time the eye and brain would hold an image. This allows time for humans to create the illusion of motion from an inanimate basis behind animation. But with film or television or video or DVD, the objects are not inanimate, therefore, the problem was how to generate a series of moving images.

For this solution we go directly back to the Wizard of Menlo Park for (what maybe) one last time. In 1888, a man named Eadweard Muybridge visited Edison. As was stated earlier in this book, Muybridge had flung open wide the world of motion pictures. Muybridge approached Edison with his own invention, the Zoopraxiscope. The device did similar things as the magic lanterns and zoetropes of the 19th century. Edison was not greatly impressed with Muybridge's device but he was impressed with his idea. So much so that soon after the visit, Edison filed a caveat in the patent office for a machine that would do for "the eye what the phonograph had done for the ear." He named it a Kinetoscope after the Greek words for "kineto" motion and "scopos" to watch.

Although it has been reported that Edison had limited input into the invention of the motion picture projector, he is credited with the motivation behind its development. One of his assistants W.K.L. Dickenson was given full charge of the assignment and worked with photographer Charles A. Brown to fully develop this new product. The mechanism development problem was simple to fix, but the ability to reproduce images on a horizontal strip was the difficult part. Edison, having seen the experiments of Etienne-Jules Marey and his Chronophotographe, while in Europe, was influenced toward the solution of using strips of photographic emulsion. This produced a better flow to the series of images. Prior to this Edison was trying to use a rotating cylinder of images to project motion – something he may have learned from his time with Muybridge and his Zoopraxiscope. The process used by Marey worked but due to the fact that Marey lacked sufficient amounts of celluloid, the Chronophotographe never made it to the popular market. Instead, Dickenson found a supplier of this horizontal celluloid in the Eastman Company (Eastman Kodak). By around this time Dickenson was given a new assistant William Heise. A version of the final Kinetoscope made its debut in 1891. It acted as a viewing machine and a camera. The camera worked by using the basic premise of the persistence of vision. The shutter mechanism would expose images at a rate of 24 frames per second and then once the filmstrip was developed the sequence of images could be played back through the same mechanism and shutter series to play back the images at 24 frames per second. The patent was filed in 1891 and one of the specifications is true today – the film was specified as 35 mm. The first public demonstration was in May 1893.

Once the mechanism was in place Edison set out to create one of the first motion picture studios, dubbed the Black Maria. W.K.L. Dickenson was Edison's first director and his first copyrighted film was *Edison Kinetoscopic Record of a Sneeze*, in 1894. By the mid-1890s Dickenson and Heise had produced nearly 80 films. In 1894 the Edison Manufacturing Company was given the right to manufacture Kinetoscopes for popular use. The Kinetoscope was typically available in parlors at a price of 25 cents of viewing. These parlors opened countrywide (Ball, 2013).

2.8.2 Everybody Gets in on the Act

The Kinetoscope slowly left the American landscape as the motion picture industry evolved. Edison was also involved in the beginning of this industry as well, creating a motion picture system and getting directly involved with the beginnings of sound (through the use of his phonograph cylinders) in motion pictures as was discussed in Chapter 5 of this book. Soon after this the major studios started to flourish, Warner Bros., Metro Goldwyn Mayer, Radio Corporation of America, United Artists, etc. The movies started to become and stay part of the American landscape. But what was possibly a bigger explosion is what occurred on a much smaller screen.

The story goes that in 1921 a young Idaho farm boy named Philo T. Farnsworth was out in the field and noticed how the crops grew in rows but what amazed Farnsworth was how each row was seen as a whole image when viewed from a distance. As the leaves moved the image changed. This is truly a form of brilliance but it becomes true genius when you realize that he was only 14 years old when it happened. The story continues that Farnsworth went to his high school physics teacher with a sketch of his idea. The teacher concluded that it was feasible.

By 1927 at the grand old age of 21, Farnsworth demonstrated this new medium as an electronic wireless system television in his Green Street Lab in San Francisco on September 7, 1927. Prior to this on April 7, 1927, Farnsworth had originally demonstrated the system with a television image of then U.S. President Herbert Hoover via a closed circuit using a 185-line system. This is the same year that Farnsworth applied for a U.S. patent on the product. Many people argue that Farnsworth was not the inventor of television. It is credited to an evolution of ideas starting from the inventions of numerous individuals such as Ives, Alexanderson, Jenkins, and Baird all working on spinning disks and mirrors culminating the principles first proposed by Paul Nipkow, a German in the 1880s. Although these devices were landmarks, they in no way are equivalent to the breakthrough created by this young lad.

As it is with history there is always an argument as to who did what first. The next name that comes up when you talk about the invention of television is Vladimir Zworykin. Zworykin is credited with inventing a device that was called an Iconoscope for the Radio Corporation of America in 1923. Zworykin's patent for the idea was issued in 1938, a full 15 years after the date given to his invention. The main problem with the dates is that although Zworykin is credited with inventing the product in 1923 there are claims that it simply did not work back then and it was only after he visited Farnsworth did he get a working version by the later date of 1938. Arguments and legal battles over the date and the original inventor went on for years with RCA finally giving in and licensing the product from Farnsworth. David Sarnoff then a major player at RCA and felt that he had enough might to combat the young Farnsworth and eventually get the patent rights for RCA. But for once it didn't turn out that way. RCA had to pay royalties to Farnsworth for use of his idea in their televisions. RCA continues to make television sets to this very day. Farnsworth continued with his inventions debuting the first color television broadcast in 1939 at the New York World's Fair (Edwin Brit Wyckoff, 2013).

From screen to tape and back to screen again. Although television is a major invention of the 20th century, what proved to be more influential was the invention of the video tape recorder. The VTR is the predecessor to the modern VHS or Video Home System. After the development of magnetic recording tape in the late 1940s, the concept arrived to record images on the magnetic surface. Once the late 1940s came around television had taken its place in the American household replacing in some cases the radio and possibly even the fireplace. But back in 1948 audiotape had its first big year being implemented in recording studios around the world. But later that year Bob Herr an engineer at the 3M corporation came up with the concept of recording images onto the magnetic tape that 3M was manufacturing for the audio industry. Herr's invention came alive when another 3M employee Dr. Wilber Wetzel demonstrated a simple black-and-white recording in 1950. The recording took nearly 7,000 feet of magnetic tape and only lasted for approximately 15 minutes. It was at the 31st annual convention of the National Association of Radio and Television Broadcasters in Chicago that videotape made its worldwide debut. But as science, invention, and luck would have it, the Ampex Corporation's Mark IV system (that the tape was supposed to make its debut on) couldn't handle the lower frequency tape produced by 3M and Wetzel. One of the staffers called Dr. Wetzel who they thought knew how to fix the problem. (Wetzel who was actually being purposely left in the dark by Ampex, in fear that 3M might rush to produce a competitive recording product understood enough to come up with a solution.) After a full night of work, 3M engineers came up with a solution and rushed it out to the airport for Wetzel to fly to Chicago in time for the demonstration. Needless to say, the demonstration was a tremendous hit and the videotape era started (Abramson, 2003).

2.8.3 Hitting the Airwaves

By the late 1950s videotape and a video tape recorder had become part of the standard equipment in many television studios. Each of the major networks at the time made sure they had an ample supply of the tape. But due to the rush to market all the networks combined only had a total of 50 reels between them, with each one costing approximately $250.00 apiece. The problem wasn't the price; however, the problem came about with the network technicians not knowing how sensitive the tapes were to dust and humidity. They found out quickly once reels became damaged or unusable from the day-to-day wear and tear. Even with these problems the use of videotape grew rapidly in delayed broadcasts or news usage.

The most memorable use of videotape came about in the famous "Kitchen Debate" between then Vice President Richard Nixon and Soviet Premier Khruschev. In Moscow, on July 24, 1959, Nixon and Khrushchev had a debate in a model American home exhibit set up in Moscow as part of a U.S. Information agency show. The two argued over the merits of capitalism versus communism. The two knew that they were being televised and recorded. Nevertheless, the two continued with their debate to a point of heated argument.

Once the debate was over Khrushchev knew about the recorder and wanted the tape to be translated before it was released. Ampex executive Philip Gundy smuggled the tape out of the Soviet Union and broadcast it the following day. The newspapers had declared the debate to be close to starting World War III, but once viewed by the public it was found to be more of an opinionated sparring match between two international leaders. Proving once and for all that written history succumbs to visual recording (Phillips and Hamilton, 2014).

2.8.4 Taking It to the Home

After the first network broadcast by video tape recorder in November 1956, the concept of home use became apparent to the electronics industry. By 1964 Sony Corporation (already producing audio tape recorders and televisions) developed the world's first video tape recorder for home use – the CV-2000. Nobutoshi Kihara, who was trying to realize the dream of Masaru Ibuka, a Sony icon, led the development team for the CV-2000. Although the CV-2000 displayed extremely high-quality black-and-white images, was as small as an audio tape recorder, and was only a tenth the cost of its cheapest predecessor, the problem was with the reel-to-reel action of the device. The Sony Sales Department wished for color television playback and possibly a cassette tape similar to that of the audio recorder. They felt that consumers couldn't successfully use the reel-to-reel action and that the cassette would eliminate any problem by the user in operating the device. The color aspect was a different problem. This took the Sony development team to devise a new system for recording color signals and luminance. Conceived by Toshihiko Numakara it was called the Color-Under System.

Subsequently called the Numakura Patent, this system became the standard for all other video tape recording systems (Lucarelli, 2015).

2.8.5 The Package Is the Real Issue

The color problem had been solved but what about the cassette? What about the quality and the subsequent size of the cassette tape? What about loading the cassette into the VCR? Engineers Akinao Horiuchi and Yoshimi Watanabe were assigned to develop the new cassette tape delivery system. At the same time other companies were also investigating the development of the videocassette. Companies like JVC and Phillips had developed solutions to the problem of reel-to-reel consumer use. These two diverse solutions led to one of the biggest battles in the home electronics marketplace. By 1975, Sony introduced the Betamax system to the home user. Betamax's strengths were its very strong image quality, color display, and quality audio playback. The JVC tape system was called the VHS – Video Home System. During the mid-1970s until about the early 1980s the Betamax system enjoyed a strong market showing but in the early 1980s that market share started to slip and the VHS market share increased. Some economists cite Sony's inability to cover the worldwide market demand for its Beta version that allowed the VHS system to enter the market and subsequently meet the demand left by Sony.

With the introduction of two different systems and realizing that the consumer would be left to choose, both Sony and JVC proceeded to develop a VCR with a ½ inch wide tape. The first VHS VCR was introduced in the fall of 1977 featuring a two-hour recording time-perfect for feature length movies. This stabilized the home consumer marketplace and allowed for the development of a prerecorded marketplace to grow. And the market did grow at a bit too fast a pace for the assurance of quality and stability. Many other manufacturers came to market with similar products, the major player in this being Matushita (the parent company of JVC) whose other brand names are well known: Panasonic, Technics, National, and Quasar. Seeing this as a major competition Sony (it is believed by critics) made one crucial mistake. They did not partner with other companies to develop the superior Betamax brand. They tried to monopolize the market and subsequently lost the home consumer market (although my father-in-law holds on to his). JVC had allowed other companies to refine the VHS product thus gaining a market advantage of 49% by the end of the 1970s.

The lesson to be learned by this is simple; just because it is a superior product does not mean people will buy it. If the choice is limited then that is what the market will use. If you control what you supply as a global cartel then that is what the consumer will use (Greenberg, 2010).

2.8.6 It's What's Inside That Counts

Now that videotapes were commonplace, what about the camcorders used to make these tapes? In 1969 two engineers from Bell Labs developed what was to become a major breakthrough in all forms of digital imaging. The device was an improvement on the Bucket-Brigade Device or BBD invented earlier that year by F. Snagster and K. Teer of Phillips Research Labs. The device that Willard Boyle and George Smith of Bell Labs came up with was the Charge Coupled Device or CCD. The device worked with a semiconductor that would enable designs that were geared for imaging, digital memory, and forms of signal processing. This device is still used today in machines such as scanners, facsimile machines, cameras, camcorders, and video digitizing boards. The principle behind the device is that it reads light emissions more so than just color. This all gets back to the concept of frequency. The CCD has sensitivities equal to 100 times that of other devices. Within a few years the CCD replaced many other forms of sensors. Being that it is a digital format the device carries information from one component to another without loss of information.

Since the images captured by the CCD get electronically distributed over the chip and not deposited into the chip the electrons can be distributed into neighboring pixels. This allows for greater tonal values and resolution control. A stronger final image and much less storage capacity would be needed. Truly a breakthrough in digital imaging! What does this have to do with digital video, well just about everything! CCDs containing pixel grids are used in digital camcorders to measure and record the images based off of the light emitted from the object thus converting analog information to digital information (Janesick, 2001).

2.8.7 Digitizing Video

Digitizing video was originally dependent on specific hardware and software configurations. By the mid-1980s companies were developing systems that would allow for commercially acceptable digitized video to be finalized for broadcast. One of those companies was Avid Technology. Avid remains as the leader in digital video for the broadcast arena. As a pioneer of non-linear editing, Avid introduced a personal computer-based editing system that used software compression. Truly revolutionary for the time!

From this pioneering effort simpler video digitizing boards were born. These boards (still used today, although iPhones are slowly replacing the need for this hardware) enable an image to go from an analog state (video recording tape) to a digital state (hard drive, CD-ROM, DVD). The converter needed to do such a thing is called an ADC – analog to digital converter. The ADC's primary function is to convert an analog signal to a series of discrete digital information packets. Most ADCs convert on a one-to-one basis – voltage is proportional to input. Some work in a logarithmic state, this means that the relationship can be modified to increase the input. Since these values are described as discrete means that they are sort of indigenous to the device or the logarithm, the final input may never be equal to the quality of the original image. This is typical of any compression format. Where this doesn't change as drastically is in a DV Cam format.

Typically, to digitally convert a video what is needed is an analog device to record the image, an ADC and a digital editing system to edit and compress the final output for playback. In most cases, an RCA cable with an input and output receptor jack is used to connect the analog device to a digitizing board. Upon capturing the video what actually happens is that the digitizing board relays the images frame by frame to the digitizing software and hardware system. Each frame is rendered to be digital information and stored on the computer. Enormous amounts of RAM and storage space are needed, in order to do this function properly.

Compression within video editing is a necessity. Similar to the problems faced by engineers in the analog world, the amount of space necessary for video is tremendous. Also the playback rate of the microprocessors was originally too slow to handle the video playback necessities. What was needed was a compression scheme that would allow for acceptable quality while accommodating the then current hardware restraints of the late 1980s and the early 1990s. Compression for video works by dropping unnecessary repetitive frames and only showing in the final output those frames that contain pixel redraw or movement of the image. Once video is captured a compression format is used to finalize the editing process (Poynton, 1996).

2.8.8 Watching the Compressors

A consortium known as the Motion Picture Expert Group or MPEG was formed in 1988 to create a set of standards for video compression. One of the first forms of video compression to come out of those meetings was MPEG-1. Subsequently, MPEG has been evolved to include a professional level of acceptance for video (and audio) compression formats. The list of accepted levels of compression accepted by MPEG are MPEG1, which includes layer 3 Mp3 audio; MPEG-2, which includes audio and video standards acceptable for broadcast quality, MPEG-3 developed for HDTV but since abandoned; MPEG-4 developed to support three-dimensional objects and movements; MPEG-7 developed to work with multimedia subjects; and currently MPEG-21, which is being established to deal with multiple forms of "multimedia" structures of the future.

About 1989 a format for working with these digital video files came about. The format is known as QuickTime. This product was developed by Apple Computer to allow for playback of digitized video without the need for special extra equipment. The original version only supported small 120 × 160 pixel width playback windows. The images played back were typically jerky in motion and digitized in resolution. QuickTime quickly became the playback format of choice due to its ability to play cross-platform. Currently, Apple has released QuickTime 6, which works with an MPEG-4 format.

One of the original compression decompression (codec) formats used with QuickTime is Cinepak. Cinepak is a codec developed jointly by Apple and then a company called SuperMac. Used primarily to work with slower rate CD-ROM drives, Cinepak has the ability to playback at full screen (640 × 480) resolution but it was developed partially for and was primarily used with slower rate microprocessors (Solari, 1997).

Shortly after the release of QuickTime (originally a Motorola chip-based product), the Intel chip side of the equation released its own version of video compression techniques called Indeo 4.0. This codec was developed for optimal playback with Pentium chips. This allowed for smoother playback and scalability of the playback window. Another Windows-based codec for video is called Audio Video Interleaved (AVI); this format was released in 1992. This is a Microsoft format developed to allow for playback on almost all machines. Similar to QuickTime, AVI allows for playback without special hardware.

There are many other codec formats in use today, some require special hardware and some do not. In some cases, the incompatibility of computer compression formats can wreak havoc on playback systems. Sometimes the projector-digital or not will not allow for the audience to actually see on the projection screen what is on the computer screen.

A newer form of compression in the Apple community was Sorenson 3. The Sorenson codec was originally created by a joint venture between a company called Sorenson Media and Apple Computer. Sorenson 3 compression was originally developed for use with video destined for playback on the World Wide Web. Sorenson 3 is used in conjunction with programs like Media Cleaner Pro as a compression format. The compression ratio with Sorenson 3 is roughly – 12 to 1. The newest version of the Sorenson codec to date is called Sorenson Spark and is needed for use when inserting video into Macromedia's FlashMX. The Windows version of this compression format is called DivX. DivX is a compression format originally developed primarily for the Windows format by the company DivXNetworks. Earlier version of DivX (not related directly to the company) has relations directly linked to an illegally obtained copy of Microsoft's MPEG-4 layer codec. The most widely used today is h.264 – this was generated because of YouTube (Apple Computer, Inc., 1993).

2.8.9 Straight Digital

The newest thing to have come along for video recording was digital video or DV. DV encoded the video directly onto the tape with internal compression allowing it to be used or transferred directly onto a computer. Originally developed in the mid-1990s by Sony and other electronic manufacturers, this technology is placed directly in the home as well as commercial broadcast companies. Companies like Panasonic had moved up the DV codec to work with high-definition television. In a straight digital environment like in a DVCam, the signal is mainly relayed through a port to the editing systems. Here is where the CCD comes back to haunt the developer. Since a DVCam is completely a digital environment, an ADC is not needed. What is needed, however, is, what has been called commercially by its developer Apple Computer, FireWire. FireWire is a version of a standard sanctioned by MPEG and known commercially as IEEE standard 1394, for example, it is also referred to as Sony iLink. FireWire allows for digital video to be transferred in a compressed state. This allows for a faster transfer rate and is now incorporated into most desktop and laptop computers. To capture video in a full digital environment the only things needed are a camcorder, a computer (with FireWire installed), and an editing software system.

Once the images are captured, video-editing software is the last stop for this footage. Currently, the desktop video-editing marketplace is flooded with software products. Everything from Apple's simple-to-use iMovie to Adobe Premiere and After Effects to Apple's upscale product FinalCutPro. FinalCutPro was a hot commodity. Having its roots in Macromedia's development labs it was purchased by Apple in the early 2000s. Since being retooled by Apple, FinalCutPro with its inclusive audio, video, and special effects components has come to take its place as the program of choice for amateur and professionals (Watkinson, 2001).

2.8.10 Determining Formats

There are many formats for compression and presentation of video files. The web is different based on the user's access. Selecting a final format is crucial to producing an effective product. The most widely accepted format today is H.264, originally developed for YouTube.

Until the advent of streaming the two most common formats for compressing or finalizing digital video files originally were Sorenson 3 and MPEG. Simply put compression allows for the video files to be played back at a visually acceptable frame rate. Which format and compression rate we select to use comes about by knowing our marketplace and our install base. How we start out is usually indicative of how we will finalize. For example, if you start out with a digital video camera you will almost need to finalize with a DV compression format; this will lead to a smaller final files size and a multi-purpose end product. However, if you started out with the more common VHS (Video Home System) format you will finalize typically with a lower-quality end product and a compression ratio that generates a larger file which is typically made up of lower quality.

This also is brought about by which non-linear editing system or software you have chosen to use. For years the leader in digital editing has been systems like the Avid and Media 100. These systems use proprietary software and are quite expensive. Since 1991 software products like Adobe's Premiere have brought non-linear digital editing to the individual. Premiere until recently had been the leader in the lower end non-linear editing market. But over the last three years Apple's Final Cut Pro has taken a decisive lead in the market. At the time of this printing, it appears that Adobe has suspended support for the Macintosh platform due to Apple's development of a competing product (Watkinson, 2001).

2.8.11 Section Summary

As with any of these sections many, many names have been lost due to page constraints. Many individuals were involved in the creation of the technology that allows for us to produce these new digital videos. Now that almost all computer systems are produced with FireWire connections, DVD burners and ample storage space, the entire home computer community is becoming the next D.W. Griffiths, W.K.L. Dickensons, or Spike Lees. The one thing that remains the same is the telling of the story. And with the advent of new technology the telling of that story changes. Now the endings of movies are becoming interactive. Users can select which way they wish a story to finish. Now users can determine their own path. This allows for greater freedom of expression and greater freedom of choice for the end users. This was always a problem not of the production of the movies but of the distribution of those movies. Now that the distribution of video files is making its way onto cellular phones and portable digital assistants and the web, the world of digital video has become the Kinetoscope of this new millennium and the cost is still roughly 25 cent per minute.

Whichever editing format you choose to create with the final format is almost more important. True you can only compress or work with what you have captured but you can only compress what actually exists. In other words, if you are looking to finalize a small throughput world your compression needs to be extensive. This is where you need to be sure that a professional videographer is involved in the process. Compression formats are brutal so a great start is what is needed. Quality lighting can solve or cause many problems. You need to be sure of your shoot before you get to the edit or compression stage. Your compression formats need to be proper for your final delivery method.

2.9 AUTHORING, DESIGN, AND DEVELOPMENT

2.9.1 From Codex to Codecs

2.9.1.1 Codex's and Kells

Individuals scribing information onto first clay tablets and later papyrus started as far back as 5,000 BC. Rolls or scrolls of this recorded information were used until approximately 800 AD. The papyrus was folded into quarters and a cover was fashioned from wood and the "book" then would be stored as documentation to an event or a decree. This is called a codex. Once the codex became the accepted form of document recording or storing "books," archivists found it easier to store recorded materials. The codex or folded paper that created a series of pages still stands today as the basis for all recorded print information. These "books" were stored in "libraries" as far back as the ancient library of Alexandria considered to be a landmark in civilization.

One of the most heralded codices of all time is the *Book of Kells*. As the Middle Ages took hold papyrus was replaced by parchment and vellum as the main substrate for "books." The *Book of Kells* was written in approximately 800 AD. What makes this work unique is that it is a premier example of a vehicle that contains a combination of exquisite illustrations and calligraphic text. This was one of the first complete presentations of image and text viewed by modern man (Hall, 2022).

Obviously, books have become our main source of information distribution. So we are not going to list the history of publishing at this time (that was covered in Chapter 1). What will be discussed, however, is the ramping up from books to the new media form of publishing. As part of that transition, one individual needs to be mentioned most of all. Bob Stein was the founder of one of the world's first multimedia authoring companies The Voyager Company. Stein's foreword thinking and vision helped to start the world of interactive multimedia authoring. In an introduction to his 1984 lecture at the University of California at Los Angeles, Stein was called "one of the pioneers of expanded publishing" and as the founder of the Voyager Company "has explored new ways of presenting books in digital form through their (Voyager's) Expanded Books Project. Stein previously acted as a consultant to Alan Kay (PARC) and to the *Encyclopedia Britannica*.

Stein was also a computer devotee. Michael Cohen, one of the original Voyager staff, in an article "The Teachings of Bob Stein," states,

> the first Apple Powerbook had just come out, and Bob Stein… realized it was just that, a powered book. He put a bunch of book-loving geeks (including me) in front of some Macs running HyperCard, and we created what were probably the first commercial e-books.

Cohen goes on to state how a simple interactive authoring program – HyperCard – could handle their needs (at the time) with "grace and flexibility" (Virshup, 1996).

But Bob Stein is just one of thousands of stories of the early digital publishers and we will get back to him at the end of this section. But the real question is what was the starting point for all this authoring and how did we go from paper to pixels and from codex to codecs.

2.9.1.2 HyperText and HyperMedia

To understand the need for UX, we need to understand what makes a project necessary to be authored in multiple media? What makes something truly necessitate an interactive or multimedia environment? Typically, it is if there are many diverse forms of media contained within the one presentation then that structure could be called multimedia. But how many levels of multimedia are there? Is it just enough to contain multiple visual-level formats? What determines a level of interactivity?

If we go back to the start, we can see how those levels have been determined. In most cases, the first level is a simple slide show like presentations with minimal interactivity just typically a forward or back. Although these level programs typically allow for audio, video, slight animation, and navigation they might not be built for heavy authoring or any form of true programming. The next level up, our second level typically allows for audio, video, animation, playback of various formats (.swf, .avi, or .qt). But what it really does is allow for true interactive programming and not just simple navigation. Programs at this level should allow for some form of customization of the final application. In other words, they should allow for a stand-alone projector. This should also contain the ability for cross platform authoring of files. The third level is determined by whether or not a program allows for all of the previous concerns plus greater levels of interactivity. At this level what is usually needed is whether or not the developer can rely on his or her ability to work within the scripting language of the authoring program to create time or event reactions. The concept of time and event response is truly everything to multimedia.

John McCarthy begins developing LISP (LISt Processing, or in a meaner way, Lots of Irritating Superfluous Parentheses:-)). This language is, thanks to its features, the base for Artificial Intelligence applications. It's used, amongst others, in the game Abuse, which levels and functions are determined by LISPcode, which is compiled when starting the game.

The basic definition of ALGOL is published. It isn't, though, any user's manual; instead, it deals only with the theoretical part. ALGOL58 is also known by the name IAL (International Algebraic Language).

The concept of user interaction coupled with time and event is what may be considered an example of hypermedia. Hypertext and hypermedia is a concept that belongs primarily to two individuals. They are Ted Nelson and Douglas Engelbart.

Hypertext started as a prophecy by Ted Nelson in 1962. Nelson's project was called Xanadu and is still online today at http://www.xanadu.com. Xanadu's main goal was to create a venue for selling copyrighted materials online. Nelson believed that the distribution of information was at the core of this new hypertext world. Imagine what he thinks of Napster and Kazza! Nelson's book *Literary Machines* defined hypertext as "non-sequential writing." By this it appears that he meant any form of writing where the document being created could be succeeded in versions easily or so that the sources or any form of research could be quickly brought to the front when desired. His goal, it would appear, was true seamless integration of information, but meant to be done in a non-linear format. Although this sounds like the World Wide Web that Nelson has rebuked, the quality of the current web is not being done in the correct manner of his vision of hyper-connectivity. Nelson believes simply, "it's being done all wrong" (Wright, 2014).

Douglas Engelbart among many other things helped to solidify the terms "hypertext" and "hypermedia." But it was his essay "Augmenting Human Intellect: A Conceptual Framework" that truly put his theories to light. Engelbart felt that we could actually use a comprehensive approach to solving problems, not an isolationist approach in other words compiling as opposed to separating. Engelbart set up his own research facility in 1963 to elaborate on hypermedia. The research went on through the 1960s and 1970s developing into a "groupware" system called NLS (oNLine System). The goal of this facility was to help generate digital libraries and diverse forms of storage and retrieval of electronic documents. (Basically, how the book came to conquer the concept of scrolls storage and retrieval.) He also invented the mouse primarily because of the need to click on a GUI (graphical user interface). The GUI environment helped users to "surf" his NLS for their e-mail and to work between windows with varieties of word processing options.

Engelbart's research and his NLS brought about what has been referred to as the "mother of all demos." Through his network he was able to, in 1968, showcase a 90-minute multimedia presentation that contained a live video conference from 30 miles away. The demo resulted in a few attendees considering it to be a hoax. Engelbart's vision has truly influenced the world of multimedia in a visual sense like no other pioneer of his time. Engelbart later worked with ARPA and their ARPANET (later the Internet) through his NIC (Network Information Center). The goal of this center was to use a hypertext system that allowed for document retrieval and integration. ARPANET used this network – NLS – as their central location for all documentation and online publications, truly the center of all knowledge. So what have these two giants wrought? (Landau, 2018).

2.9.2 Starting with the BASICs

To truly understand the path of interactive authoring you need to understand what you are authoring and why. Are you creating an actual application or simply creating a stunning multimedia presentation? Once you have decided what you are doing then you can pick your poison. There are two routes that developers usually take, one is the simplistic user-friendly route initiative via a program like Visual BASIC or in some cases they went the path of rich content systems – Macromedia DirectorMX or FlashMX.

2.9.3 Getting Down to Basic

Visual Basic is for many the only way to author applications and or programs. Its history is deep in the world of actual computer program creation. Starting in 1963 Beginner's All-purpose Symbolic Instruction Code (BASIC) language was developed by mathematics professors John G. Kemeny and Thomas E. Kurtz at Dartmouth College, New Hampshire. The goal of this new language was to make it easier for students to learn programming but be powerful enough to actually do something. It also provided (according to George Mack – *A History of Visual Basic*) "direct user-computer interaction and clear and friendly error messages, give fast response for small programs, minimal amounts of computer resources and require no hardware knowledge and allow the user to have no knowledge of the operating system." It was these capabilities that made BASIC the first high-level language available on new mini and microcomputers. It was BASIC that was used in 1975 by Microsoft's Bill Gates and Paul Allen to create a BASIC compiler for the MITS Altair. Remember this is what gave Microsoft its start in the computer program and software business. Since that time BASIC was used in some way in all early DOS-based computers of the late 1980s and very early 1990s. Even though BASIC was basic enough for typical computer programmers it still wasn't basic enough for the average programmer or user. Enter Visual BASIC. The difference between BASIC and Visual BASIC is that in VB the program will actually write the code as long as you visually create the new application. For example, if you want a dialogue box just use the toolkit and draw one. The code is created for you and the dialogue box exists. This is very similar to interactive programming today. Alan Cooper created Visual BASIC in 1987 when he worked for Coactive Computing Corporation. The program was originally called Ruby and was developed in BASIC. Its goal was to allow for visual programming for the average programmer and user. By the start of the 1990s Microsoft had acquired the rights to develop Visual Basic and in 1991 Microsoft released its version of Visual BASIC 1.0 for Windows at the Windows World convention in Atlanta, Georgia. Later in 1992 Microsoft released Microsoft Visual Basic for MS-DOS the Professional and Standard editions. Both the Windows version and the MS-DOS version allowed developers and programmers the ability to use visual elements to create applications. This is interactivity with the concept of functionality at its very core.

Visual Basic version 4, 5, and 6 increased the users' capabilities and works directly with a host of other Microsoft development tools including Microsoft Visual C++, Visual J++, and Microsoft Office Developer plus many other Microsoft development systems. Development systems like Visual BASIC are focused on interactive application-based authoring and are very professional methods of creating actual applications. There have been many other watershed moments in the history of interactive authoring as well. One of the first was Janet Walker's Symbolics Document Examiner in 1985. This system allowed for book marking and was one of the first support manuals to go totally hypertext-based. It wasn't so much an application development tool as it was simply a first of its kind hypertext document. Many others would follow but before we get to that we need to recognize the birth dates of many other systems. Some other structures in which to develop simple interactive applications or presentations are NoteCards, OWL-Guide, HyperCard, SuperCard, MetaCard, PDF, Persuasion, PowerPoint, Formula Graphics, mTropolis, Director, and Flash. In the next section, we will deal with the history behind some of these programs and their integrated lineage (Ignotofsky, 2022).

2.9.4 Interactive Building Blocks

The early 1980s were strong years for interactive authoring with the development of systems such as NoteCard, a LISP-based hypertext system created by Xerox in 1985. Some unique features contained within the NoteCard program were scrolling windows, pre-formatting, and browsing and navigator windows. Intermedia, members of Brown University, created a Macintosh A/UX system hypertext application, the same year. But the one to last for the past almost two decades has been HyperCard.

2.9.5 HyperCard

Started in 1987 by Bill Atkinson of Apple Computer HyperCard has come to reign quietly as the mother of all hypertext-based authoring systems. Still in use today by thousands of devotees, HyperCard is looking for an upgrade. Having not received a true update in almost 15 years, HyperCard is respected as one of the simplest and relatively powerful authoring software to ever come to market. Its beauty is in its simplicity. Organized in "stacks" of "cards," HyperCard allows users to create simple animations that react to the users selection. The developer can organize the "stacks" anyway they desire and can use HyperTalk – with its simple user-friendly language authoring system – to create or organize the playback of the "cards." The "cards" can be linked via HyperTalk to act just like hypertext links do on the web. Still in a black-and-white format many users would like to see the product upgraded to real color capabilities and made compatible with Macintosh OS X.

It is still reportedly being used in numerous countries around the world to do everything from manage building systems to create gaming environments. It is said that HyperCard has been the engine behind such popular games as Myst and Riven. It has also been heralded as the impetus behind such software language giants as Java and most web browsers. Within the last ten years it has been superseded by an array of interactive authoring programs like SuperCard, MetaCard, Revolution, FileMaker Pro, Director, and Flash. Many companies, both large and small, find that HyperCard's main benefit has been its' easy-to-use scripting language. Using an English language style for scripting, HyperTalk, the language allows for a quick comprehension and a short learning curve. These are two very important issues when just starting. But the next generation seemed to have some quality upgrades (Cheifet, 1987).

2.9.6 SuperCard

In 1991, the next generation of hypertext-based authoring systems was introduced to the marketplace. Its name was SuperCard, an obvious takeoff on the concept of HyperCard; SuperCard one-upped its predecessor by allowing for color graphics. Originally marketed by Aldus Corporation, SuperCard is now owned by Solutions Etcetera and (at the time of printing) is in its 4.0 version. Some of its new features are its function called shell (). This new function allows for the execution of Unix shell commands under Macintosh OS X operating system.

2.9.7 MetaCard

Created in 1992 MetaCard was used by a strong portion of the application and presentation segment of the authoring community. MetaCard's user interface works conventionally with each operating system. In Windows it looks like Windows and in other OSs it looks indigenous to the system. But MetaCard itself was written in C++ (a product of Bjarne Stroustrup). Its final projects are called "stacks" just like HyperCard and it uses a simple English-like scripting language called MetaTalk. But one of its strongest advantages is its ability to link around these "stacks" by working in a doubly linked list. This allows for a greater form of relational database structure where address books that relate to input can be formatted and developed. MetaCard was also developed to include support for vector graphics, raster images, audio and video files. Although MetaCard is not what would be called a full-fledged interactive authoring environment, it still can deliver a usable product.

2.9.8 Portable Document Format

The concept behind the PDF was to create a paperless office. John Warnock, founder of Adobe Systems and an alumni of Xerox PARC, started working on PDFs to solve the problem of too much paper and to create a means by which anyone working for Adobe could read their documents regardless of the workstation they were on. To solve the problem Warnock created a system he called "IPS" or "Interchange PostScript." It made its debut at Seybold Conference 1991 in San Jose. The final release of Version 1.0 was at Comdex Fall in 1992. Very simple in its first release the prepress world had no use for it. But soon it became a standard method of transmitting and transferring files to an image-setting device.

By 1994 Adobe Acrobat had the ability to hyperlink content pages. With its built-in search engine, article threading, security features, and note-taking abilities, a PDF could now act similar to a HyperCard stack. Adobe itself created all its manuals and internal documents in a PDF format. By 1996 PDFs were ready for the commercial print world. Now including support for a multitude of color formats and being able to link to HTML pages this simple application was set to take on the interactive world. But it seems for now that the release of the newest version of Adobe Acrobat 5.0 in 2001 is satisfied with PDFs mastering the world of print and usable in the interactive world. Version 6.0 is ready for release at the time of this printing.

2.9.9 Making the Point

When you think of a computer-generated slide presentation, one word pops into almost everyone's mind: PowerPoint. In the early 1980s slide presentations were all the rage. Every middle management hero was sending their work out to third-party companies to create slides or transparencies. The process was expensive, slow, and usually ended up with poor results. But all of that changed, when Forethought Software of Sunnyvale, California, published PowerPoint 1, in 1987. The program was originally named "Presenter" and was designed for Microsoft Windows 2.0. Ironically, it was produced using Apple's Lisa and ran on any Macintosh with a minimum 512K of RAM. By August 1987 Microsoft purchased Forethought Software in an attempt to add to their growing line of office products. At the same time the Aldus Corporation, makers of PageMaker, released Aldus Persuasion 1.0. Soon after Microsoft released PowerPoint 2 for Windows, the main problems were fonts compatibility. Past this point the versions would come fast and furious with enhancements being cross-platform compatibility, stronger navigation, importing of sounds and video, and inclusion of third-party plug-ins. With its newest version PowerPoint is totally web compatible and used widely throughout the presentation portion of all industries. But now you might just need a bit more power to make your point.

2.9.10 Director

To any game authoring developer or web designer, the name Macromedia conjures up both disgust and unbridled excitement. Started in 1984 by Mark Pierce, Jamie Fenton, and Marc Cantor Macromedia – originally Macromind – has become the industry standard for interactive game and web development. Whether the product is DreamweaverMX, Peak, Freehand, FireworksMX, DirectorMX, or FlashMX you are sure to have visited http://www.macromedia.com once in your career.

The original product generated by the team was called VideoWorks and displayed only in black and white. In 1987 VideoWorksII was released this time with full color capabilities. Later VideoWorks evolved into VideoWorks Interactive, which was based on the programming language BASIC. Although never released to the general public many software companies used the product to develop tutorials and tours of their own software products. Macromedia created for Mac-based users one of the first CD-ROMs in 1987.

Later in 1988 VideoWorks Interactive Pro morphed into MacroMind Director 1.0 via the programming genius of John (JT) Thompson and Erik Neumann. This switch in authorship replaced the simple BASIC language programming with the more powerful Lingo Scripting engine. Once Director 2.0 was released with its new scripting language, inventor John Thompson said, "...I was researching object orientedness in general – SmallTalk, Objective C.... so those ideas got synthesized into Lingo and XObjects." John Thompson is someone who needs to be looked at a bit closer than just a casual mention. Thompson – with his invention of Lingo – is truly a major force in the field of interactive programming. Johnson is a MIT graduate with degrees in engineering and computer science. He does, however, cover more ground than that. Thompson appears to have a wider range of interests,

> I don't have a single interest, and my interests cover math, science, art and culture.... I've always tried to create technology that makes it easier to interact with computers...the computer is an incredible canvas; and I'd like to keep playing with that canvas....

VideoWorks/Director has been rumored to have been used to create the computer graphic displays for StarTrek: The Next Generation. Once this adoption into the world of commercial motion pictures had been made multimedia was just about to launch to the general public. With the advent of Director 3.1 the program crossed platforms with a Windows-based player for its final projects. In the fall of 1993, MacroMind released Director 3.13 and MacroMind became Macromedia. Now this release is supporting the use of Apple's new format QuickTime. Now videos could be played directly in the window of the Director, along with the audio and Raster image motion that it always possessed. Once the three visual capabilities were rolled into one interactive program, true multimedia was coming of age for the typical developer.

In 1993 Director 4.0 was released and quickly became the state of the art for the interactive animated authoring world. XObjects and XCMDs were also included in the new version and these allowed for development by third-party companies. Once Shockwave appeared in 1995, a compression format language was still used widely at IBM. The same company also publishes the RPG language (Report Program Generator), which is specialized in making commercial reports. It was available for real-time "video" playback within a stand-alone application. Shockwave files were generated via an internal program named Afterburner. In 1996, Director 5 was released; the new version contained a list of added features that allowed for debugging of Lingo scripts, multiple casts, a watcher – to see how you were doing via the scripting and a host of add-ons called – Xtras. Now the updates were coming a little too fast and way too furious.

In 1997 Director 6 was released. This new version introduced a redesigned development window, integrated Shockwave (for the web) support along with a library of drag-and-drop Lingo scripting behaviors. Coming close on the heels of Director 6.0 and 6.5 was Director 7.0. Said to have numerous upgrades it was in general a disappointment to most designers and developers. Although it did increase capabilities – such as a Multi-User Server function overall it was slow and a bit buggy with its new dot syntax for programming. But just wait until next year.

In 2000 Macromedia released Director 8.0, a stronger, more sophisticated program although still built to look primarily the same as Director 4.0. The goal was to make the release of Director 8.0

more stable and reliable a version than Director 7.0 – but wait there's more – later that same year Director 8.5 was released.Made to be more web-friendly than ever before, it was totally compatible with the new kid at Macromedia – Flash 5.0. By the fourth quarter of 2002 Director released DirectorMX, the version currently being used by designers and developers worldwide. Made to be compatible with a number of options – three-dimensional navigation (from Director 8.5), total OSX compatibility and some major authoring enhancements (Reeves, 2005).

2.9.11 Actionscript Wasn't Really a Flash in the Pan

It wasn't until Flash 4.0 did we get the addition of the much-needed Actionscript. Now users can truly interact with the product, which set Flash on its course to be the web development tool that designers and developers had been waiting to get their hands on.

We will go over Actionscripting in the section on authoring later in this book. But for now you should know that Action Scripting, like Director Lingo, allows users to animate their objects by virtue of programming, again cutting down on the work necessary to create animations. Although it is no walk in the park, animating with ActionScripting is the way to go. Its similarities to JavaScript make it a great start toward understanding higher levels of programming necessities. Finally with versions 5.0 and FlashMX, everything was in place to make this animation tool the predominant product in the web design and development arena. Flash lasted for close to 20 years as the premiere product for designing and delivering interactive products. Now it has been reincarnated as Adobe Animate, and Adobe XD (Bedingfield, 2019).

2.9.12 Java and JavaScript

First things first, Java and JavaScript are at best distant cousins related by marriage twice removed. In other words, there is no direct link between these two items. Java is a software platform and JavaScript is an object-oriented scripting language, two totally different things. Let's stop the comparisons and explain the history behind each.

2.9.13 Java Starts Brewing

Java is a software platform that was developed by Sun Microsystems beginning in December 1990. The main individual behind the development of this new platform was Patrick Naughton. Naughton, becoming frustrated by the limitations and complications involved in Sun's current C++, C, and APIs tools, started what's called the Green Project. The Green Project was started in Menlo Park, California (ironic to be so close to Edison's Menlo Park, New Jersey) near the world of venture capital monies of the early 1990s. C++ was the original choice for the platform's main language but Bill Joy (chief scientist at Sun Microsystems) felt that there were some issues with using C++ as the base. So the story goes that Joy had a concept of integrating the best features of one language MESA with another C. This integration was proposed in a paper to the Sun group. The paper's title was "Further," appropriately named. The main structure behind the new platform was to be its object-oriented environment that was based on Stroustrups' C++. Many others got involved as well. James Gosling was one of the other key players in the development of the Green Project. However, once Gosling started to work on it the project took another form. The team focused on new applications for the platform. The first attempts focused on a PDA device that dealt with an agent called Duke. The device was called Star7 (based off of a telephone feature). This feature allowed for users to answer a telephone anywhere. Later Green Project, now called Oak after a tree outside of Gosling's window, was redeployed once again to become a subsidiary of Sun Microsystems called FirstPerson, Inc. And guess where the team relocated – Palo Alto. It seems that the circle never stops! (Javier Peris, 2022).

2.9.14 Java and the Net

After a three-day brainstorming session, the Oak platform or now as it was called Java was destined for the Internet. But why use the name Java? Well, there are many recounts of the reason from the team members visiting a coffee shop to an acronym taken from the team member's names – JAVA – James Gosling, Arthur Van Hoff, Andy Bechtolsheim. The Internet seemed to be a perfect venue for an object-oriented software platform especially since the release of Mosaic. Naughton wrote a small web browser called WebRunner – this would later be renamed HotJava. By the end of 1994, Java and HotJava were ready to be presented to the executives of Sun Microsystem. The first release of the Java platform was available by 1994 with the HotJava web browser by 1995.

At the 1995 SunWorld conference an announcement was made by John Gage, director of Science for Sun Microsystems, who made the announcement for HotJava's release. At the same conference Marc Andreessen announced that Netscape would include Java support in its Netscape Communicator browser. By the beginning of 1996 Sun Microsystems had formed a JavaSoft business group to further develop the Java technology.

Later the Java platform has been adopted by many companies worldwide but is still looking to establish itself as the de facto network platform. Jonathan Schwartz, executive vice president of software for Sun Microsystems, Inc. states, "Java technology has become the open standard for network computing and Web services everywhere" (Biancuzzi and Chromatic, 2009).

2.9.15 Java Script – DeCaf?

JavaScript or what Microsoft calls Jscript is an object-oriented language developed originally by Andreessen and the staff at Netscape Communications under the name LiveScript. It was renamed to "JavaScript" after Andreessen's announcement to support Java in Netscape's browsers. A smart move public relations-wise but a bit confusing to most. Since it is a bit confusing the European Computer Manufacturers Association (ECMA) standardized the language to be called ECMA-262 Edition 3 (well that's certainly easier). You just need to know that JavaScript and Java (the platform) have almost no relationship whatsoever. The only similar aspects are they both are used extensively on the web and share some forms of syntax.

JavaScript and ActionScript do, however, share similarities. The syntax used in FlashMX's ActionScript and DirectorMX's lingo bear great similarities to the syntax of JavaScript. Since JavaScript's main function appears to have been to establish a dynamic web structure, it makes sense that it is used primarily to embed functions into HTML pages. These functions make it possible to do certain things that HTML just won't allow such as opening a window from another window, setting the window's size, closing the window, inputting values, changing cursors or images. It is one source, which has helped to make the web truly dynamic (Keith and Sambells, 2010).

2.9.16 HTML – The Big Little Invention

The invention of HTML – HyperText Markup Language, is credited to Sir Tim Berners-Lee. This version of HTML was based on SGML (Standard Generalized Mark-up Language), which was an internationally agreed-upon method for marking up text into structural units such as headings, paragraphs, list items, and body copy. SGML could be implemented on any machine especially now through his browser the World Wide Web. Traditional SGML does not include hypertext links. Berners-Lee brilliantly based his version of HTML on SGML, which was already used in traditional digital typesetting machines. Additionally, by adding the use of the anchor element the "link" or AHREF used to create the hypertext links made Berners-Lee's invention incredibly useful (Longman, 2000).

2.9.17 Pioneers Let the Games Begin

In the 1960s Bell Laboratories was the breeding ground for all things multimedia. Electronic music was being made for the first time. Images were being relayed to hard drives through scanning devices. Storage systems were being invented. Individuals were developing digital video capabilities. The question was not so much the hardware or even the software but what was to be created? As Esther Dyson said, "Content is king." But who is going to create this new form of the content?

2.9.18 Pong and Spacewar

It would be remiss if we did not at least mention the advent of the computer gaming industry. In a recent interview it was stated that the gaming industry represents an $8–10 billion revenue stream. Well how did this all start? We go to the Massachusetts Institute of Technology and the year 1962. A young programmer named Steve Russell created what was to become the first computer video game. Russell worked on a Digital Equipment Corporations' (DEC) PDP-1, which was hooked up to a cathode ray tube display. The system was donated to MIT by DEC probably not intending for the world of computer games to debut. What made the PDP-1 perfect for game developing was that the operating system allowed for multiple users. The perfect play station!

But before the entire advanced computer gaming industry was initiated came Willy Higginbotham and Tennis for Two. In 1958, Dr. Higginbotham was the head of the Brookhaven National Laboratory's Instrumentation Division. As a sample of each respective division's research, they were required to produce a project for tour reviews. Of course Higginbotham decided to do something new and unusual. Along with technician Robert Dvorak, Higginbotham set out to create a display. Using an analog computer (1958) and an oscilloscope for a display the two scientists developed the simple tennis game that we have come to call "Pong." Simple beginnings for an $8–10 billion industry but again it started at MIT.

2.9.19 The Media Lab

In 1980 Nicholas Negroponte and professor at the Massachusetts Institute of Technology (and also a founding member of Wired magazine) along with the late Jerome Weisner (former science adviser to US President John F. Kennedy and former president of MIT) founded the MIT Media Lab. The Media Lab's mission, at its inception and to this day, is to pursue the convergence of computing, publishing, and broadcast systems.

By 1985 the Media Lab was a reality and the work started. Emerging from MIT's Architecture Machine Group, the Lab's work focuses on how to take traditional concepts concerning content and rework them toward a fully interactive multimedia frameset. It is held to fact that in 1978 Andrew Lippman of MIT's Architecture Machine Group developed the first true hypermedia experience with the work Aspen Movie Map. Videos to videodisks allowed the users to jump to different views of Aspen from an interface screen. It was a landmark in the lineage of hypermedia that it displayed what could be done in the future with increased capabilities.

The Media Lab has also continued the tradition of encouraging a full collaboration between industry and academia. Bringing together the best and brightest from around the world to work together on the future of digital communication and experimentation. Negroponte is just one name to be taken from MIT's Media Lab, other groups (especially the Aesthetics and Computations Group) and individuals who need to be noted and researched for their huge contribution are too numerous to mention for this particular section but I must attempt to name a few: John Maeda, V. Michael Bove, Jr., Glorianna Davenport, Janet H. Murray, Benjamin Fry, Megan Gailbraith, Nanako Ishido, etc. These are the new pioneers of multimedia – now too simple of a word. But who was taking these new concepts and introducing them to the common man? (Brand, 1989).

2.9.20 Next Turn

By the early 1980s the world of interactive authoring took a serious turn. As we spoke before one name comes to the forefront when we discuss the creation of electronic books and interactive multimedia. That name is Bob Stein. The founder of the Voyager Company Stein is responsible for making his and subsequently all of our vision of interactive multimedia a reality. Originally involved in laserdiscs and other forms of hypermedia, Stein's Voyager was responsible for bringing together many worlds of content and the interactive marketplace.

A Renaissance man by trade Stein's past included working for a number of high-profile companies before starting the Voyager Company in 1985. His most fortunate meeting and prior industry experience (in my opinion) was working with Alan Kay at PARC. Kay, if you remember, is the visionary behind the laptop computer and many other computer technologies have come to influence the world forever. But many products have survived those times and have found almost a cult following. Such products are the works of artists like Laurie Anderson and Peter Gabriel (Virshup, 1996).

2.9.21 From Archive to Interactive

As all of these archival methods slowly gave way to the full-fledged web a new method of design and development became prominent. The term "Information Architecture" became part of the overall term "Interaction Design." Tools that had their heritage in software like HyperCard, SuperCard, Action, Director, and Flash.

2.9.22 UX Design Software

The concept of design and development have now become intertwined like never before. Gone are the days of the wall between the coders and the designers. Now the main goal is complete seamless collaboration between teams. As the stakeholders make decisions and changes to the prototyping and wireframing of the project, those changes are quickly deployed to the designer and reviewed by the development team in order to redirect development. No longer do teams get knee-deep into useless coding before they are made aware of changes and new goals. Now usability and functionality are built into the actual visual design. In some cases, pseudo interactivity can be developed in these early stages and tested before it gets into production. All forms of media and UI elements can be inserted. Wireframes become prototypes and become working apps in a matter of hours as opposed to weeks. These new tools have made major advancements in just a very short time.

2.9.23 Adobe XD

As a product of Adobe Inc., Adobe XD or Experience Design is a grandchild of Adobe Flash and a great, great, grandchild of Macromedia's Action, Director, and Flash. The lineage of this product came about once Adobe purchased Macromedia in 2005. Slowly as this evolves and utilizes most of the Flash's interactive qualities it seems that it will evolve into the premier Experience Design tool.

One of its major features is its Content-Aware Layout tool. This tool allows for the user to Design and edit components without the typically frustrating nudging or the tinkering. This tool will align and evenly space elements as you add, remove, or resize objects. Designers can input any media they need to from images to videos to animations to text to typography. Button stages can be developed and linked to internal screens or even to outside web pages.

As stated before, Adobe is in the final stages of purchasing Figma and this appears to be how Adobe will cement its place as the leader in this field. Currently 20% of all UX designers use Adobe XD over other software choices.

2.9.24 Figma

Currently Figma is the leading UX design software in the marketplace. Dylan Field and Evan Wallace developed Figma in 2012 while studying computer science at Brown University. Their goal was to develop a collaborative web application for interface design that also had offline features enabled by desktop application. Figma's main focus is on developing the user interface and user experience design, with an emphasis on real-time collaboration. This is achieved by utilizing a variety of prototyping and vector graphic tools. Viewing on both the Android and iOS platforms are built into the Figma mobile app which allows for viewing and interacting with Figma prototypes in real time on multiple mobile devices.

Adobe (which produces AdobeXD) made an offer to purchase Figma in 2022 for $20 billion but this was denied by the Department of Justice. Sixty percent of all UX designers use Figma over other software choices.

2.9.25 InVision

InVision's main product is its app. The app is much bigger than simply a UX design tool but a complete collaborative planning platform. It boasts numerous awards from work and adoption by large Fortune 500 companies.

According to their website,

InVision is the visual collaboration platform powering the world's smartest companies. More than 7 million people across global enterprises and small teams come to us when they are looking to transform the way they work together. That includes tens of thousands of organizations, including 100% of the Fortune 100. We exist to make every kind of work more collaborative, inclusive and impactful so that companies can get the most out of their most valuable asset: their people.

Its main design tool is called Freehand AI. Please note that Freehand was originally the competitor to Adobe Illustrator back in the 1990s and was truly a horrible malformation of vector tools; however, it seems that InVision has reinvented the software tool and it is working. As of 2023 almost 20% of all UX designers use InVision over other software choices.

All of these design software choices seem to have one thing in common – they are all dedicated to the grid structure. This is just the way design works best, whether online or print.

2.9.26 Section Summary

One thing that is sure to be learned by these chapters is that all roads lead to Edison, PARC, MIT, or Microsoft. It is kind of like playing six degrees of separation. It seems that everyone who has contributed to the making of interactive multimedia authoring systems is interrelated or hyperlinked to each other. Many of the names have been left out again simply due to time and page length but I need to type a few more here so that they are at the very least mentioned: Frank Halasz, Andy Van Dam, Jeff Raskin, Todd Rundgren, Peggy Irish, George Landow, etc.

Each section of each chapter could be an entire book by itself (which probably has been written and published). It appears that the story of authoring doesn't die off or reach a pinnacle, it just continues to mount. At the beginning of this chapter the question was raised as to what does a book possess that makes it a better vehicle for interactive content? As we can see from our journey through this chapter almost nothing besides flexibility, convenience, size, and expense. But we do need to work toward the advancements of interactivity over some of these existing virtues and if that is true then how can we utilize new technology and yet never enter the quagmire of technology for technology's sake. We must be careful because the advancement of technology never seems to end, it just seems to always allow us to rise up and carry the message forward. So is it always about the story or content whether it is a book or a laptop? Esther Dyson was correct and our future appears to lie somewhere close to Socrates and Shakespeare – more so than Sagan and Segal always story over technology but how can we change that story to better integrate with the new technologies? Should we bother? Why do we bother improving technology if the story really never changes? What is our point? Is it just some morbid curiosity as to how much confusion, pain, and anxiety we can endure as a race or is it truly for a function? We cannot expect an answer to that question since we are still in the dark ages, technologically speaking, for it would appear that we will need time, vision, and chance to give us that answer.

2.10 REPLICATION AND DELIVERY

2.10.1 From Quills to Quasars

2.10.1.1 What Is Available and Possible?

Nothing was more fearful than when you signed off on a print run or on a replication of CD-ROMs or DVDs. You've checked and double-checked and triple-checked but you were never quite sure. Some call it Obsessive Compulsive Disorder, some call it perfectionism or possibly professionalism, but sometimes it is never enough because any way you slice it once you replicate you're stuck or are you?

Whether or not that is the case the focus of this section is on detailing the history of replication, media archiving, and diverse forms of information delivery. Who started recording events and why? How have we received the information? How have we chosen to store it? What has changed and what can we expect in the future?

2.10.2 Books for Everyone

Replication is best associated with the printing press or the monastic scribe. In most of these cases, it was a one-off scenario with each replication taking the same amount of time as an original. A German artisan Johan Gutenberg changed all that for the modern world with the use of movable letters in conjunction with the printing press. However, the concept of a printing press had been around for centuries with the Chinese Tang Dynasty (618–906), which used block cut from wood to print textiles and short religious texts. Moveable type was used during the Song dynasty (960–1279) and furthered the world of printing. The invention of movable type in printing is credited to Pi Sheng (died 1051). The use of two-color printing was seen as early as 1340 with the first dated example of block printing being the Diamond Sutra 868.

It is thought that through the travels of Marco Polo and other explorers of the Far East, the concept of the printing press and moveable type made its way to the Western world and ultimately to Gutenberg. Gutenberg introduced the concept of the printing press and movable type to the Western world with the advent of his printing of the Bible on February 23, 1455. Gutenberg is also credited with the use of oil-based ink and diverse forms of substrate such as "rag" paper. Gutenberg's methods helped to spurn the advent of the delivery of the written word to the masses. Any work that was produced between the start of Gutenberg's printing career and the year 1500 are collectively referred to as incunabula. One of his first works after the Martin Luther translation of the Bible was a continuation of Luther's writings, which helped to expand the growth of the Protestant Reformation (Lemelson, 2020).

One of the reasons for the acceptance and growth of the printing industry and the use of movable type was its ability to reproduce books in an inexpensive manner. Although started by the Chinese and the Egyptians, what made Gutenberg's method revolutionary was the ability to rework the content, which is directly related to Gutenberg's concept of moveable type.

The next step for printing was the Industrial Revolution. Now that the structure of replicating text and image were in a sense mechanized the power of steam and electrical power would push that to the max. Now the mass production of printed works was getting into full swing. But this is where we stop our discussion of print. The only reason for analyzing the advent of print was to discuss how a media originates and how it influences society. What the mass distribution of the printed word and the birth of the printing industry did for society is more important than a further discussion of lithography, chromolithography, flexography, etc. What is important is how society responds to a new media and its potential for the distribution of knowledge. A question that I believe to be fascinating is "Where will the next lightning bolt hit?" Who will have the next big idea for the distribution of knowledge to the masses?

2.10.3 Maxwell's Hammer

As we discussed briefly before, in Section 2.1 of this chapter, the discovery of electromagnetism was a monumental occurrence in the history of science and industry and art. James Clerk Maxwell is credited with presenting a unified theory on the topic (Sutter, 2021). Now in case you are unaware, electromagnetism is a relatively new theory that was slowly perfected and unified by Maxwell throughout the course of his work during the 19th century. What does all this have to do with the replication of CD-ROMs, DVDs, computer storage materials, and the Internet? Well just about everything, for without Maxwell's theories on electromagnetism these devices and their properties may never have been utilized so quickly. Also without Gutenberg's press maybe Maxwell would never have had the opportunity to be educated? Also Maxwell's theories are directly related to Einstein's theories of relativity. So how important of a figure was James Clerk Maxwell? Einstein described Maxwell's work as the "most profound and the most fruitful that physics has experienced since the time of Newton." One thing always relies on another.

2.10.4 Electromagnetism

At the base of Maxwell's theories is the concept of an electromagnetic field. The field travels from its origin in the form of a wave. These waves exist in wavelengths and these wavelengths are contained with a spectrum. These fields come about via two distinct formats, stationary and dynamic. Fields that stay stationary are considered magnetic fields, which can be found around a current carrying wire or the field of electricity between the two plates of a capacitor. But the real big deal is with the dynamic fields. Electromagnetic fields that change or move are called dynamic fields. Examples of dynamic fields are things such as the spectrum of color, radio waves, microwaves, X-rays, and gamma rays. See where we are going with this?

What this ultimately led up to was the ability to create lasers, masers, basic electronic components, and obviously the computer and disk storage systems that we use today. A computer's disk storage system, as do most other parts of a computer or any other form of digital communication, rely upon the use of lasers and obviously electromagnetic fields. But let's just talk quickly about some precursors to that time.

2.10.5 De Forest for the Trees

Lee de Forest is the perfect example of not being able to see the actual worth of an invention. De Forest, working in the early 1900s, could have been considered the true Father of Radio if he would have concentrated on the long-term effects of his inventions. However, in his biographies it is written that he was more so concerned with the concept of getting a shot at quick fame or fortune. De Forest's theories on radio waves and vacuum tubes helped to spawn many ideas including the birth of sound in motion pictures.

De Forest, while working on improving wireless telegraph equipment, created the Audion. (The Audion was later used in developing sound in the motion picture industry.) The Audion was a triode, which incorporated a filament and a plate similar to other vacuum tubes. The difference with this version was that it also contained a grid located between the filament and the plate. This allowed for the signal to be strengthened and also allowed for amplification of radio signals. Later other engineers and inventors, ultimately a group of Bell Lab engineers, used de Forest's theories to develop the transistor.

Transistors transcend the world of electronics. Transistors are almost solely responsible for the miniaturization of the computer. Because of the work of Lee de Forest future inventors could work off of his principles of vacuum tubes. De Forest stated that within a vacuum tube alternating current could be converted to direct current and an electronic signal could be amplified. This premise was used throughout the development of television and by 1945 the ENIAC (Electronic Numerical Integrator Analyzer and Computer) ("Lee DeForest | Lemelson," n.d.).

One of the leaders of the future teams that used de Forest's theories and inventions was William Shockley. Shockley attended two famous institutes of advanced learning and both directly related to the world of computers, CalTech and Massachusetts Institute of Technology. Shockley teamed with two other Bell Laboratory engineers, Walter Brattain and John Bardeen, to develop what was to become a major contribution to the electronics industry. This was brought about first by applying quantum theory to the development of semiconductors and later utilizing a silicon substrate for developing a component called the transistor (transfer and resistor).

The problem that started all this research for Bell Labs was that the telephone industry was using vacuum tubes as line switches. The tubes were troublesome, expensive, and very volatile. The transistor on the other hand was simple to manufacture, solid, sturdy, inexpensive, and worked instantly. What this did for the telephone company was outstanding but what it's done for the computer industry is unparalleled. The ENIAC (first commercial computer) utilized over several thousand vacuum tubes. With this revolution in computer technology (remember – originally developed for the telephone industry) systems could be built that were less expensive, smaller, lighter, faster, and more powerful. But where did this need for speed come from and what about storing all this information?

2.10.6 Hollerith, Hopper, ENIAC and Univac

How did we start computing and storing information? The earliest recorded item used to tally or store information discovered so far is the Ishango bone circa 6000 BC. It had tally markings incised in its surface. Obviously other forms of markings, writings made their way through history. But when did we really start archiving data through modern-day machines?

2.10.7 Hollerith and Watson

One of the first to do this mechanically was Herman Hollerith. Hollerith was the originator of the punch card. An alumnus of Columbia University, Hollerith became a statistician for the US Census Bureau where he first conceived of the idea of using punch cards to tally information. From this concept was born Hollerith's company – the Tabulating Machine Company (1896) and later a bit larger company was also spawned.

After 28 years of incorporation and development TMC became part of something called International Business Machines Corporation or IBM. IBM was actually a conglomerate of three corporations, the Tabulating Machine Company, International Time Recording Company, the Computing Scale Corporation. Originally called the Computing Tabulating Recording Corporation – CTR was headed by founder Thomas J. Watson, Jr. (Bashe et al. 1985).

2.10.8 ENIAC and Beyond

The ENIAC (Electronic Numerical Integrator Analyzer and Computer) was originally developed in Philadelphia in the 1940s as a solution to develop firing tables for use in the military during World War II. It contained 17,468 vacuum tubes, 10,000 capacitors, 1,500 relays, and 6,000 manual switches. One of the original developers of this machine was John Mauchly of the Moore School of

Electrical Engineering at the University of Pennsylvania. Mauchly was working on using vacuum tubes to speed the calculations of the current computing machines.

A section of the military called the Ballistic Research Laboratory was responsible for determining the specifications for developing firing tables for each specific weapon that the military had developed. The computations were laborious and time-consuming. In order to expedite this process, the military instructed BRL to develop a method to quicken the process. Lieutenant Herman Goldstine of the BRL had heard of Mauchly's work and commissioned him along with J. Presper Eckert to develop a new high-speed computer.

The project started in 1943 and took until the end of World War II to develop. But it was worth the wait. The ENIAC could compute the cube root of 2589 to the 16th power in a fraction of a second. If given a bit more time – a whole second – it could execute 5,000 additions, 357 multiplications, and 38 divisions. Although this was astounding, the problem came in with changing the basic program of the machine. This could easily take days of rewiring the system with punch cards and switches. The next step was to use some form of stored memory – enter the Whirlwind. But before that we need to see what happened after the ENIAC.

2.10.9 Hopper and the Bug

One of the foremost names in computing technology is Grace Murray Hopper. Working with Presper and Mauchly, Grace Murray Hopper, who was a Lieutenant in the Navy's WAVES (Women Accepted for Volunteer Emergency Service), had been working with computations and computers since her days at Vassar College in 1928 through her PhD in math from Yale.

In 1943 Hopper worked with Howard Aiken at the Bureau of Ships Computation team at Harvard to develop a machine that would calculate tasks for determining such strategies as creating a minefield for the Navy's ships. In essence what Hopper and her team was doing was starting to create the first digital computer called the Mark I.

Hopper is also infamously credited with the term "bug." The story goes that a computer had failed to do an operation and when Hopper investigated, she found that a moth had gotten inside the machine. She supposedly taped the moth to a page in the notebook and had written, "first actual bug found." Hopper is credited with the computer terms "bug" and "debug." But Hopper is known for much more than this.

In 1949 Hopper started work on a new machine called the Univac. Working with the team of Presper and Mauchly (original ENIAC developers), the new start-up would develop a computer that recorded information onto high-speed magnetic tape. This was the first step toward recording the base program of a computer. In other words, this started to establish the operating system concept and the concept of programs. The success of this process caused the Sperry Corporation to purchase Univac's parent company.

Hopper stayed with Sperry and through her efforts important advances were made in the development of automatic programming and the transition from computing scientific data and processing toward the use of this technology in the world of commerce. Hopper was also instrumental in developing the language that would allow for this transition Flow-matic which would later be incorporated into the standardized programming language called COBOL (Computer Basic Operating Language) (Yale University, 2000).

2.10.10 Memorable Memory

Now that the use of a computer was becoming accepted in the military and in industry the need to store this information and the necessity for speed was becoming apparent. But before all this could be accomplished the next step in computing history had to take place.

Memory storage was the next key step in the history of computers acceptance into the mainstream of business and academia. Storing a computer's instructions had always been the one component of the process that was the most time-consuming. Remember the ENIAC could accomplish tremendous computations in a fraction of a second but to change its main function could take days. To do this many transitions had to take place.

2.10.11 Banging the Drum

Drum memory was at the start of computer storage. Drum memory was a large metal cylinder that was coated on the outside surface with a ferromagnetic recording material. This is obviously very similar to recording tape. A row of read-write heads runs along the long axis of the drum. The heads are aligned with each track of the memory drum. The drum moves in order to seek a piece of stored information. Although fast for its time the method still reeked of hand manipulation.

2.10.12 Wang to the Core

The next step for memory storage and transfer was magnetic core memory. Simply put the magnetic core memory process relied upon the small magnetic rings (the cores), which stored the information within the polarity of the magnetic field they contain. And that was the simple version! Remember how important James Clerk Maxwell was!

An Wang, an American physicist, created the pulse transfer-controlling device in 1949. This device was some of the earliest work done on core memory systems. Basically what this device did was enable the computer to read information stored on the core and transfer it to 0s and 1s for the computer to translate into instruction. Wang was part of Harvard's Computation Laboratory at this time but luckily for Wang, Harvard did not have an interest in controlling any research invented in their labs. Wang owned the patent and started Wang Laboratories. Wang enjoyed a strong share of the office computer market until approximately the early 1980s (Lemelson, n.d.-b).

2.10.13 Enter the Whirlwind

In 1944 Jay Forrester was an MIT graduate student who was recently transferred to the Servomechanism Laboratory from MIT's High Voltage Laboratory. Forrester who was looking for greater responsibility was transferred to work for the US Air Force on project Whirlwind. The goal of Project Whirlwind was to create a flight simulator that would work in real time. The project started in 1945 and would finally go online in 1951. Whirlwind, which used a full video display, required faster computation and output than the drum memory could generate. After Jerry Crawford, a team member of the Whirlwind project, saw the ENIAC in action he also concluded that it would be necessary to use a digital computer instead of attempting the design with an analog model. This required the use of additional code in the computer's program. More code and more computer programs meant a need for faster access to the information.

After trial and error and much discussion and planning the team decided in favor of core memory. Core memory as described above was the first step toward internal disk drives and subsequently external, CD-ROM, and DVD storage systems. After the end of the design process Kenneth Olsen converted the Whirlwind structure to a transistor format. This was successful and led to other even smaller versions of the mechanisms. Whirlwind was used as an air defense system until the 1980s (Church, 2016).

2.10.14 Hard Disk Storage

In 1948, approximately in the middle of the Whirlwind project timeline, the University of Manchester ran the first stored program on vacuum tubes with a drum memory system added later. Although Forrester had moved on to core memory, the development in Manchester meant that other projects were headed in the same direction. In the late 1950s memory systems were taking jumps from core memory to magnetic disk, called RAMAC ("random access method of accounting and control," this method was developed by IBM. The system could store 5 million bits on 50 aluminum discs coated with magnetic iron oxide. This device weighed over one ton.

Everything, however, started to change once Arthur L. Schawlow published his famous paper "Infrared and Optical Masers." Masers were the precursors to lasers. Masers basically referred to semiconductor light amplification and modulation. Schawlow worked with the inventor of the laser Charles H. Townes also of Bell Labs and Columbia University. Their respective inventions changed the way information would be "seen" by a computer. But it wouldn't be until years later that the world would see a "floppy disk" to be used for storing information.

2.10.15 Floppy Disks

Forever being concerned with staying ahead of the field, IBM introduced the first 8-inch flexible plastic "memory disk" in 1971. Al Shugart originally developed the "memory disk" with an initial capacity being only 100 kilobytes. The disk had originally been developed by Shugart to hold the program code for the IBM 3330 Merlin disk pack. Later in 1974 Gary Kildall for IBM originally designed the CP/M operating system, using a floppy disk storage system. This format replaced the punched cards still being used by large computer systems.

The progression of disk storage continued with the first sealed hard disk drive with the development by IBM of the two 30-megabyte disk platters. The system was called the "Winchester" drive due to project manager Ken Haughten's 30–30 rifle. This was a stationary system obviously limiting portability. Al Shugart made the next big step, for disk storage, again in 1976 with the development of the 5.25-inch floppy disk. This disk was developed for the desktop systems being sold by Wang Laboratories. The floppy disk drive was fast becoming the method of information storage. Steve Wozniak of Apple Computer fame developed his Apple I computer at the Homebrew Computer club of Palo Alto in 1976. The Apple I was upgraded in 1978 to include, among many other things, a floppy disk drive.

Sony introduced the 3.5-inch "floppy diskette" in 1980. This quickly became the industry standard for portable personal computer storage and has lasted until just recently. Most computers today do not even include this storage device or any form of internal storage preferably using the cloud for storage and access.

2.10.16 The Past as Prologue

Many forms of archival media have taken shape over the years, always relying on physical standards or mechanization to work properly. Today the physical nature is still dominant but the structure is based on infinite growth not an end-to-end system.

Previously standards such as CD-ROM, Laserdiscs, DVDs, Flash Drives, CD-I (CD-I–Compact Disc Interactive), CD-R (Recordable), and CD-RW (Re-Writable-circa) were heralded as state of the art. At the writing of this book the standard jumped ahead by light years.

2.10.17 Web Storage Starts

Today archiving on the web is commonplace but what we may need to understand is how this concept evolved. The use of the web as an archive is basically what it was intended for from the start. The web can become an archive for any of our work or it can be used for what Ted Nelson and Douglas Engelbart felt it should be used as – an archive center. But how will we use this new format? What will be the problem with copyright issues and fair use legislation? Where did Napster start and Kazaa end? What about a peer-to-peer network? In this next section we will discuss the archiving files on the web.

2.10.18 P2P

Ted Nelson envisioned the concept of sharing hard drive space through the network of interconnected computers. This concept of a peer-to-peer network started basically at the same time as the web itself. Simply put, a peer-to-peer network is made possible by file-sharing software administered through each "node" on the "network." Users download the necessary software and put it onto their stations to create a repository for file sharing throughout the "network." Users can then upload or download the desired files. It also allows for the users to create listings of their favorite files and to seek out answers and FAQs around the "network." A P2P is probably the best use of a computer's hard drive space. The only problem is that this software is being used sometimes, without the user actually knowing about it. Sometimes hackers are setting up files-sharing nodes within your hard drive and you're not aware. This is a detriment to the overall concept of a P2P.

Once you have downloaded these files you need to store them somewhere for easy transport. Enter the portable document assistant (Dechow and Struppa, 2015).

2.10.19 PDAs

Portable document assistants are not really that new. The first PDA to be made available to the consumer market was the Newton released by Apple Computer at the 1993 MacWorld. The Apple Newton was a bit like a brick. It weighed nearly as much and did just a little bit better in the usability area. But actually it was quite the breakthrough for the portable market. In this next section and as a wrap up to replication and delivery we will discuss the use of handheld devices for the delivery of information.

2.10.20 The Newton Was at the Core

Apple's first foray into the world of PDAs was the Newton. And at the center of that core was one gentleman Steve Capps. Capps had been originally at PARC and was drafted by Jobs to work at Apple during the original days of the late 1970s starting with the apple Lisa. Having worked on the original Macintosh OS, Capps was enlisted to work on the original Newton OS. As of recently, however, Capps after being recognized as an Apple Fellow (Apple's highest honor) has started to work for Microsoft's Internet technologies division. Another person involved in the Newton development was Walter Smith. Smith was responsible for the unified data model that ties the Newton software together, the compiler, interpreter, and runtime library for NewtonScript (this with Packet Interface-5).

Michael Tchao is a name that should be mentioned when it comes to the creation of the Newton. Tchao was involved in Apple's advanced R&D and had been working on slate computers that read handwriting and beamed data across rooms. It was Tchao and Capps who originally came up with the idea of incorporating these high-end technologies into portable handheld devices.

Once the device had been designed and an operating system developed, the next step for the Newton was developing applications. Typically referred to as a "killer app," the next step for the Newton was handwriting recognition. Stepan Pachikov and Larry Yaeger are the two technologists responsible for the development of handwriting recognition. True these two gentlemen should be incorporated into the chapter on content selection and input but the concept fits better into this chapter on replication. But the one component of this device that truly made it rise above all others was the hand recognition. Dr. Pachikov was from a company called ParaGraph International of Moscow. Pachikov took a group of start-up engineers from Russia to create technology that was to be licensed by Apple Computer. Quite the slick move!

Pachikov originally developed that handwriting recognition software to teach children to write. After understanding its overall benefits, he started to shop it around Silicon Valley to numerous software and hardware companies. Finally landing at Apple, Pachikov was granted a meeting with

Larry Tesler, Steve Capps, and others. Pachikov gave a demo to the group. Of course like all good demos nothing worked properly but, nevertheless, the group was impressed by the potential.

Larry Yeager enters the picture after he left work at Alan Kay's laboratory in Los Angeles. Yaeger helped to develop the Newton's 2.0 Print Recognizer. There were many revisions to this program before release and it has never truly lived up to its hype but what was accomplished by this simple device has truly started a revolution in the transportability of computers and their overall ingestion into the mainframe of society.

A lesson to be learned by the history of this device is that many times "defeat is snatched from the jaws of victory" and Apple has always been the company to do so. After the Newton Systems Group was bounced around for years it finally was supposed to be spun off into Newton, Inc. This company was to be made up of Newton researchers and marketing people. After a few incarnations the spin-off and the company were finally killed by Jobs in 1998. As Jobs is supposedly quoted as saying, "the stylus restricted the use of the four other ones – your fingers" (Mat Honan, 2013).

2.10.21 iPods – $10,000 in Steve's Pocket

Maybe the iPod was the reason as well? The reasoning behind a product is not so much its effect on the world as much as it is its effect on cash revenue. The iPod was originally released in 2001 with a price tag of approximately $750.00. Quite an expensive toy! But it was obviously a step up from the Newton to a more functional handheld device. The iPod was Apple's first serious venture into capitalizing on a new revenue source generated from the Internet. But even though it was originally created for downloading music, the iPod is in reality an external FireWire capable hard drive. Whenever it is plugged into the FireWire port to download music, the iPod's lithium batteries are recharged. Now with the strength and acceptance of this new device Apple hoped to cash in on the music revenue possible from the Internet but mainly it hopes to start to be an integral part of the whole handheld movement. It was heralded as putting 10,000 songs in your pocket – may be but it has definitely put billions into the coffers of Apple.

Many individuals predicted that the handheld world will come to be the driving force behind computer usage. Today with thumb drives capable of over 1 gigabyte of storage and museum artists using these handheld devices to actually deliver artwork to viewers/users, the age of the walking computer is upon us. One forerunner in this marketplace is Palm Pilot.

2.10.22 Palm Pilots

In March 1996 Palm released the Pilot 1000 and Pilot 5000 organizers. By March of the following year, the Palm Pilot platform had inspired over 2,000 developers worldwide to create hardware and software for the devices. By the end of 1997 Palm had licensed its OS to IBM, 3Com, Symbol Technologies, and U.S. Robotics. Obviously a strong revolution was underway!

Through a series of strong and ever-expanding alliances, Palm plants a very strong foot into the future of portable computing and telecommunications at the start of the 2000s by integrating technologies with companies such as Sprint, VeriFone, and Pacific Bell. Palm followed up this consumer and industry market invasion with entrees into the academic market as well. By the middle of 2002, Palm sealed its fate within the industry market by entering into a reseller agreement with IBM.

2.10.23 iPhone

The iPhone was launched in 2007. Prior to this many incarnations of the mobile cellular phone had been developed, finalizing in 1973 with the first commercially viable model developed by Marty Cooper and his team at AT&T Bell Labs. To say it changed the world is obviously an understatement. Smartphones in general have shaped a generation. One that might be in need of heavy chiropractic services later on. The ability to communicate globally – immediately is truly insane as Steve would say. The ability to listen, watch, play, and maybe even learn with no immediate boundaries. The only boundary seems to be our self-control. As of 2023 there are more than 1.46 billion active iPhone users worldwide. iPhone users account for 21.67% of the world's total smartphone users. The AppStore boasts over 1.6 million iPhone apps available. Android numbers double that to over 3 million. This current generation lives and dies just a mere 5 inches from their face.

2.10.24 Cloud Storage

The concept of storing on the web is relatively new. Systems like P2P started this initiative. Please remember visionaries like Ted Nelson were the pioneers of all that connected hard drive space. But platforms like "the Cloud" have come to remain a mystery to most lay people. Its omnipotence works hand in hand with all forms of mobile devices, especially those accessing the Internet via Wi-Fi. Simply put it is off site storage with access via an Internet client. This form of storage is fluid, unlimited, but can be somewhat inexpensive for the individual user. It is a great concept for back up and for remote working operations. Access can be limited or granted depending upon the administrator of the account. Many companies have Cloud storage available; the three leading ones are Amazon, Apple, and Google. A simple definition of cloud storage is listed below (*What Is Cloud Storage? – Definition, Types, and Benefits*, 2021).

> "Cloud storage is a broad term for cloud-based data storage services.
>
> Cloud storage works by letting users store data on remote servers. Users can access their data from anywhere via the internet.
>
> Cloud storage providers cater to individual consumers and businesses alike."
>
> *(Posy, 2021)*

According to Amazon Web Services, there are three kinds of Cloud storage – Object storage is data storage architecture for large stores of unstructured data. File storage is widely used among applications and stores data in a hierarchical folder and file format. Block storage is used for Enterprise applications like databases or enterprise resource planning (ERP) systems, often requiring dedicated, low-latency storage for each host (Amazon, 2023).

2.10.25 Next Step Forward

Soon after the emergence of archival media came into the mainstream the speed at which these technologies changed was a true expression of Gordon Moore's Law. Moore's Law predicted that technology would double in capability every 18 months. Today that law is now restructured to be every six months or less. The archival format structured from the mid-1980s to today included at first formats like CD-I (CD-I–Compact Disc Interactive), CD-R (Recordable), and CD-RW (Re-Writable-circa).

2.10.26 Quantum Leap Forward

Today Moore's law is running at full speed. Recently in 2023, quantum computing has been unleashed into industry. It works with qubits not 0s and 1s. This employs a multi-dimensional approach to computing. The new hardware part of this is now capable of working with these new computational structures (Voorhoede, 2022).

Quantum computing works within a completely different form as opposed to current processing. Quantum computing works directly with the laws of quantum mechanics where the solutions are achieved via a multi-dimensional format. Where classical computers encode information in bits – 0s and 1s. The qubit (or quantum bit), which encodes the information, is a two-level system – a linear combination of 0s and 1s. This can then be extrapolated out. The name of this phenomenon is superposition (Voorhoede, 2022).

2.10.27 Section Summary

In this section we ended with quantum computing as the next leap forward along with Artificial Intelligence. All of this travels across the Internet and its reach is growing every minute into all of our lives and occupations.

It will exist everywhere from homes to offices to trains and planes and automobiles. The concept of replication is actually a question of archive or delivery. Archival methods have now advanced to the cloud terabyte level of content storage. The more storage space we have the more we fill. Nature abhors a vacuum. The real methods of archiving have become omnipotent and the outgrowth of Ted Nelson and Douglas Engelbart's vision – online systems of storage. A bit scary at times but nevertheless factual. It is really all about delivery. For what good is storage if you can't get to it to use it?

The trip has certainly been a long one and for some of our future Gen-i individuals a trip that appears seamless and endless. What has been the constant from these quills to quasars? What has kept us on this journey – the need for information or the profit that is to be made from getting us there? It would appear that the information we exchange is the same from generation to generation. The only thing that changes is when and where we receive it and how it is customized for our needs. That is what makes it valuable to us and to marketers globally.

Maybe it is not in the delivery method but in the timing of the delivery. After all the ink and the wires, the presses and the mold makers, the inventions and the inventors, it still appears that the efficiency and the nature of the message wins out over the replication or delivery method. A strong course to be changed, but assurances to the speed of the delivery or the permanence of the media will allow for us to concentrate more on what is truly important – the content.

2.11 THE FUTURE OF INTERACTIVE MEDIA

2.11.1 From Joy to Jevbratt

2.11.1.1 Good News/Bad News: We Have Absolutely No Idea Where We Are Going

For years everyone has been hypothesizing on the future of interactive multimedia. How will it work with the new operating systems? How will it integrate into our lives? What will be the new stages of development? How would new computer architecture affect interactive multimedia? How small will actual computers get? How integrated into our lives will they become? Will we be wearing computers and dealing with their capabilities? Will learning and all forms of information distribution only be done this way? Will advertising and entertainment be analyzed and homogenized before we even get a chance to go near it? Will the individual be considered when developing the interface or will they actually be the interface structure?

Interface development is one area of concern and rightfully so since it is the first area in which you come in contact with the intended product or service. However, there are many schools of thought on the interface as metaphor and maybe the arch of interactive multimedia will bring us to a point of generating an instant personal metaphor for all situations. According to Brenda Laurel, "The working definition of the interface has settled down to (is) a relatively simple one how humans and computers interact but it avoids the central issue of what this all means in terms of reality and representation." The term "metaphor" may become a metaphor for actually just using something. Sometimes I believe there is too much hypothesizing and not enough trying and actually doing. Nothing – not even a metaphor – actually replaces work. There is an old joke about the future of interactive multimedia – whatever we think it is will be virtual reality. We really don't know until we try! A couple of things that need to be brought up again here are Bill Joy's article "Why the Future Doesn't Really Need Us" and Nicholas Negroponte's theory on "Being Digital." Joy discussed how we would work in the future and Negroponte discussed how the future will work. It needs to be mentioned before we progress to discuss the future of interactive media. Negroponte's title reflects his entire point – we need to understand how to be digital. How to realize that we are dealing with bits not atoms? And how will we manipulate that change?

So with that in mind we will not focus on the design aspects of interactive multimedia, but more so on the future use of interactive multimedia as it pertains to four basic areas – education, entertainment, information, and promotion.

2.11.2 Education: Learning and Teaching Interactively

The discussion of how we can use this new-fangled form of computing and intelligence to teach and learn interactively and to evolve to the user's needs has always been an odd question for consideration. How else have we been teaching? I hope we have been teaching interactively. I hope we have been considering the student demographic in our classrooms every time we present a lesson. As a teacher for over 15 years, I have come to learn that, you need to know the room. A classroom is a living, breathing organism, one that needs to be cultivated – not separately but as a cohesive whole. Students need to be given their roles in the classroom. It is almost a matter of psychological safety. How can interactive media address these concerns?

We also need to know our rights and obligations if we decide to use this information on a website or even on any form of archival media. Because according to the DMCA (Digital Millennium Copyright Act) of 2000, every time we archive we are publishing and publishing brings with it a list of copyright issues. When preparing the information for a lesson today we now need to be sure of copyright infringement even when we speak. But yet our students may be committing huge acts of intellectual vandalism even as they click. It is sort of like leading by example down a dark hallway in a new house. How we teach in this new world to this new Gen-i will be a unique problem to correct. We will probably rely upon the methods of antiquity to coerce out a solution. For no matter what method we use, it should always be about developing methods of critical thinking.

There are many schools of thought on teaching – but really just two main ones however. There is the oratorical style – lecture format, and the other the Socratic method or philosophical approach – generate questions to be answered. Both of these can be employed when looking to create an interactive multimedia learning experience.

2.11.3 Oratorical Education

A presentation environment – sort of a PowerPoint on steroids approach, may best serve the lecture format. The use of any form of digital video presentation software is also a solution but the only problem with this is how the students arrive at the information. If the lesson is purely instructional – how to change a tire – then yes this is an appropriate solution. But what happens when you are changing a tire and the jack starts to slip. What happens if the lug nuts are stripped or almost soldered on by the tire store where you just purchased the tires? What happens if the spare is flat? If you have ever had this happen you know that it usually happens at the worst time and sometimes – jack slipping – at night on a hill with no one else around. These problems need to be addressed in the presentation as well. The presentation of information needs to evolve and flex to the event. So even in the simplest learning environment, there are still options and opinions. Here is where the learner might prosper if curves can be thrown at them? How would they solve a problem? The lesson goes from purely step-by-step to a "what-if" scenario. What happens when you take this to extremely serious levels? This is where interactive media in education needs to be going. But who is designing these systems and will an oratorical fashion suit them for long? This is where the philosophical approach is necessary. Let the user ask the questions and motivate a response. This works more so with the model of Think, Listen, Speak, and Add. Once we adapt this methodology to developing interactive media is when we will start to truly use interactive media in an educational format.

2.11.4 Philosophical Education

Once we start to adopt the method of Think, Listen, Speak, and Add to our Information Architecture and UX design then we will actually be using a philosophical approach to developing educational interactive media. We need to use a "what-if" approach to all educational products. The problem is that the "what-ifs" become exponentially enormous. "What-ifs" can go on forever? How to decipher or create a dynamic structure for deployment of these potentials is the problem. But a bigger problem is how to actually create a system that is flexible enough to handle the dynamic structure of a philosophical debate? For once you put a limit onto the sequence or to the list of probabilities you have stifled the debate and, therefore, debilitated the philosophical format. It is not a simple solution. It has all the potential problematic structure of a mythological two-headed serpent. If you ask the question you have negated the ability to think about it as a possibility. The one solution to presenting this kind of an educational structure in the future is to utilize a form of chat room forum. This allows for an ongoing structure to develop. But the problem with this is who moderates the theory and the direction of the discussion. Who calls time out and sends in a curve ball to sway the discussion toward an end result? Obviously, the perfect moderator would be a teacher, but isn't this distance learning and not a use of interactive media? Or will it become both – distance learning with a form of interactive multimedia? This seems to be the best solution and seems to have a better chance then the accepted oratorical fashion.

2.11.5 Nodes on the Horizon

If we accept the concept of education being dependent on constant growth then we need to also stand up for Ted Nelson's 50-year-old belief that the use of HyperText and HyperMedia is meant to continually expand the growth of content. This is true education and pedagogy development with validation above ownership.

2.11.6 Education Summary

Education has been the pioneer of interactive media. And as Engelbart said, "Pioneers are the guys with the arrows in their back." So with that said, attacking the development of educational interactive media is not noble. The method in which it is used today is, however, in need of an infusion of ideas and vision. The world of the Gen-i is here. Our recent pandemic and all of the school closing should have taught us how ill-prepared we are for this new age, and it is not slowing down to consider limitations. We may want to consider this new lack of attention span as a decline of Western civilization – either way – it is here and we may not be ready.

2.11.7 Entertainment: Net and Gross

Nothing but net is where the entertainment community seems to be focused in regard to interactive media. The future of media and entertainment seems to be completely focused on how to use broadband and or wireless technology to deliver content to the user and how to generate workable revenue from this model.

From games to movies to iTunes to infomercials, the Internet is the avenue of dreams and total ambition. And at the center of this dream and nightmare was Bill Gates. An article in *Wired* called Gates "the entertainment god" and the reasoning is who else controls 90% of the desktops in the world and through things like Xbox, WebTV, and MSNBC and the distribution and sometimes the actual content of everything that is on those televisions.

Recently visionaries, such as Bezos, Dorsey, Musk, Altman, Zuckerberg, Gates, Jobs, Ellison, Joy, Clark, and Andreessen, have been likened to the Carnegies and Rockefellers of the previous century. These magnates saw the future and purchased it before anyone else even woke up. Where railroad lines once directed traffic and goods, now Internet lines will do the same. What stopped Hollywood from jumping on the Net-wagon and taking it for a big ride? The answer was simple: it appeared to be the 12-year-olds from Norway to Nebraska and all lands in between. Piracy was the key issue.

Twenty-five years ago a then 15-year-old allegedly decided to help out the world's hacker base by releasing via the Internet the encryption-cracking program DeCSS. This program allowed hackers to break down an encryption code, which kept individuals from copying certain content. Cracking is as common today as turning on a light switch. This one act of piracy or benevolence almost halted the reign of a new delivery method for entertainment. The 15-year-old was brought to trial but eventually acquitted. But since Norway does not entertain the "double jeopardy" concept in its law, the prosecutors were able to appeal the verdict and attempt to retry him. What will happen to this young man and others like him is still up in the air. At the time of this printing, the RIAA (Recording Industry Association of America) was considering dropping its suits against ten-year old children for downloading music. Meanwhile at the same time that this Napster-like whirlwind is calming down a new piracy issue is starting up. It will always be something new from the days of the Ivory Coast to the days of Sun servers. Methods by which content can be stolen and redistributed are going to constantly be in development. However, this will probably never stop the progress of content distribution.

2.11.8 The Future of Content via Distribution

At this point of the discussion, I think we need to agree upon one fact. The method for content distribution is not the issue, broadband and convergence is here, the issue is how will content change with the distribution method? What will happen when the Gen-i population starts to be the requisite demographic (and in case you haven't seen that is happening as we speak)? What will happen to content structure when reality television teams up with blogs and roaming cameras or when guerilla television is a channel and cinema verité is sponsored by mouthwash and toothpaste? When influencers roam the plains. That is a question that can only be answered by time and vision and chance. Now is the beginning of the growth spurt.

Who is controlling this growth spurt? Right now it appears that this is partially in the hands of companies like Microsoft, Amazon, Netflix, and Apple. And now that smart TVs are under $100 the invasion of your home is on big time. The goal of these systems is to create a center point in the home for all media to enter and then be distributed. This appears to be the hook and you my friends are the bait. And who is looking at the issue of security? Not us.

2.11.9 Music: Apple's iTunes, Spotify, Pandora, etc.

Music is the key to the beginnings of content delivery. Think about it – first you start small with audio Mp3 then you go large with video files. Apple (and many others) are going after the home and portable entertainment marketplace. Their products currently are the iPhone, iPad, and the AppStore, and iTunes-delivery method and device. This is going to be a constant issue for the media for now and the future. What impact will this have on the field? Companies such as listen.com were purchased by RealNetworks in order to utilize their model and structure. The purchase price was US$36 million for what industry experts called a US$100 million operation. Sounds crazy but not when you realize that the purchase was a bargain basement price 36 cents on the proverbial dollar. It was actually a great deal, but what about the big fish? Another concept that is the swaying point in these discussions is whether to use an a la carté method where users can download what they want and store accordingly or pay a flat monthly fee. The market will work that one out. But what about the bigger deals – the Hulus and Netflixs and Prime Video who are delivering movie content and building this content. How will the content and the marketplace continue to change with the delivery media?

2.11.10 Video: Netflix and Prime

They started with DVDs delivered fresh and piping hot right to your door. Sounded great and it was by most accounts. Netflix is that company and that was their original product contribution to the world of entertainment. Remember delivery determines the media. For if you can't get the "boat out of the basement" then what good is the boat? Just like the Phoenicians they control the language.

Reed Hastings started Netflix, in 1997, with the goal of delivering content to the people by whatever means necessary. The company refocused and really developed streaming as the standard for entertainment delivery to the masses. Now they are winning awards for their content development. Just like Amazon's Prime and all of its original content. Razors and the blades (Randolph, 2019).

Surprise, surprise, it has actually worked the annihilation of the competition. The real question is how will these kinds of companies and their original content providers react to perpetually feeding this beast? If content structures are changed what is going to happen and what about delivery and the enticements and competition? Remember media is not just about showing images and sound it is about creating an environment to promote and use that content as well. And some of that entertainment space is primarily dedicated to the ever growing, hugely popular, and profitable gaming industry.

2.11.11 Gaming: Example: Broken Saints and Sim(ulated) Dreams

Many games are now entering the world of full-fledged movies and series. For those of you who do not know, www.brokensaints.com was a cult phenomenon in the early 2000s. Utilizing animation this site told an interesting and intriguing story of a hardcore hacker, a Muslim soldier, a Shinto priest, and a mystery Shandala who are all drawn together by an apocalyptic dream. Easy reading stuff! But what will happen when these kinds of games and RPGs or role-playing games are infused with the ability to have live action actors intertwined with the story line? Imagine a daily dose of the motion picture "Final Fantasy – The Spirits Within."

Professor Steve DiPaola, a leading entertainment futurist and developer of the facial system used in Sims – the interactive game, stated that gaming development products would have greater effects on the entire interactive market than just the immediate gaming world. DiPaola foresaw how the facial systems used in gaming will be employed by markets such as the communications industry as well as the motion picture industry or the educational Interactive (multi) media industry in the years to come. The term "synthespian" has been thrown around for years. But through these systems of what is called "face space," the reality of a synthetic actor may come to pass. This may be another decade of court battles as we are seeing today with the Writers and Actors strikes in Hollywood.

The concept of synthetic actors obviously opens up the entire world of interactive entertainment to entirely new levels. If facial expressions and voice intonations are now available to change on a real-time basis what will actually happen to the entertainment field? Will actors be only needed for voice-over work? Will they sell their virtual souls for future profits? Will we ever need actors again if we can model off of Barrymore and Bogart, Pickford and Bacall, Pitt and Redford? Duchamp and Bowie had it right? Sell yourself off one piece of stock at a time. As models like Sora and all of the other AI video content producing platforms quickly evolve and quantum computing becomes easily workable at the base Internet level, how will the entertainment marketplace morph in these new times?

2.11.12 Streaming: Wait and Switch

The premise of Bait and Switch has been replaced with Free to Freemium or Wait and Switch. All of the major streaming services are using this model. The first episodes are free, we just need your credit card or iTunes account. Then based on the distraction of modern life we will forget about it and then we are charged monthly. Too busy eating the gerbil food to notice so we continue on our way. They just wait for us to switch back and we never do. Soon we will pay for multiple overlapping streaming services. STARZ on Prime on Showtime on Max, the same content just jumbled up enough for us not to notice. Now we will watch the same show throughout different devices in our homes. Imagine food product commercials right on the refrigerator! Just how morbidly obese do we wish to get? The fourth chapter in this book is entitled "Veal" for a reason.

This seems to be the mentality for the future. So if your entertainment files are streamed and fluent why stop? That is exactly what the media companies are hoping for, because once you purchase the razor they will keep selling you the blades. The episodes will follow one after another until you wake up a week later and realized you binged an entire season of that new show in one day. Oldest trap in retail and just the one they have set for us.

2.11.13 Entertainment Section Summary

For anyone to tell you what will happen to the entertainment field as it applies to these new technologies is unrealistic. Everything is changing at a very fast pace. By the time the general public finds out about it, it is five years old and has already been in and out of court. The one thing you can count on starting from the days of Al Jolson to now is "You ain't seen nothing yet!" or P.T. Barnum's "this way to the egress?"

2.11.14 Interactive Advertising: This Way to the Egress

The world of interactive advertising is poised to hit its highest revenue numbers in the following years. One of the main reasons for this is the change in the user marketplace with an abundant amount of time spent online. Today, as of this writing, a new wave of users is coming of age. Today the Gen-i nation is ready and willing to purchase online and to spend the majority of their leisure time surfing, downloading and watching WebTV or whatever connection they will ultimately plug into.

According to studies done at major universities and throughout the interactive advertising community, today's branding efforts are developed and executed for a multilevel, multi-venue approach. Some of those are major companies such as Procter & Gamble, Kraft, and Unilever are three huge powerhouses in the consumer marketplace. The approach that many of these retailers are using is the umbrella look and feel. Here they are gathering up their product lines under one umbrella site in an attempt to aid in the brand name awareness for the dispersed product lines. A good idea for brand awareness, market management, and cross-promotion efforts.

One case study brought up by Bob Tedeschi (1999) speaks about creating cyber-celebrities from product mascots. An image described by Tedeschi is a sample of the cross-marketing of products and brands. The image is that of the Planters symbol Mr. Peanut in a barbeque apron with a Weber grille. Dawn Brozek, an analyst with Nielsen NetRatings, stated, "They've found the recipe for success online." Today influencers who brag about revenue but really have minimal have replaced that.

Service industries as well are becoming accustomed to the concept of online branding through not just a secondary media but slowly progressing toward becoming only secondary to television. Credit card companies, banking, travel, etc. are starting to realize that they need to fully develop their new media presence in order to be ready for the wireless, cable television convergence. To discuss the dreaded long-term convergence would take six books, not just a segment of one chapter to fully explain. But, nevertheless, it is here – and it is ready.

A part of the market, which proves the development of this online commerce industry is not just the low-ticket items – airfare and hotel reservations, is the continued use of interactive multimedia by the big ticket companies. One such company is Mercedes-Benz. Ever since the advent of interactive multimedia, CD-ROM or Web, Mercedes-Benz has been at the forefront of brand development. Other automakers have also been heavily involved in the use of interactive multimedia to promote since the early 1990s. Believe it or not the first company known in this market was Buick.

Unbelievable when you consider their market demographic is probably 45 years old and over. Now it makes sense but it was startling 30 years ago.

Now with even 50- and 60-year-olds doing their banking online it all makes sense. But the premier market is the 14–19-year-old age group. Now with teenagers spending more and more time engaged in Internet chatting and messaging the trend for the next generation – Gen-i is going to be online. Major ad companies understand this and are finally full on dedicated to it.

Where the world of interactive advertising is headed is also anybody's guess. It would appear that as always it is about bringing products to market and to the customers' venue is paramount. As the customer comes to understand and use this new venue so will the advertisers and their marketing companies. It all comes down to finding the people where they are most likely to purchase – in a sense the points of distribution for information and advertising messages. But owning these distribution points seem to be the ultimate goal. Typically, they have been under the control of media companies but as Barry Diller mentioned in his mission statement that distinction is getting very broad. Who will control these distribution points and how will they get the set up?

2.11.15 Interactive Advertising: Interactive Appliances and Us

The smart house is here. And retailers and realtors alike are salivating. As if the housing market actually needs a shot in the arm, smart houses are the dream of the digerati and ecommerce groups worldwide. They are also the dream of the world of interactive advertising and promotion. Imagine how easy it is to reorder when you swipe the milk carton after it is empty to establish a shopping list. This is convenience but as we all know convenience comes with a price. How will ad agencies or media companies look to control this space? Imagine advertisements that now come over that refrigerator screen. Advertisers who are willing to pay for that marketing space – an evil plan but nonetheless it's coming at you. Can you blame them? It is always about selling you a beverage when you are thirsty. Impulse buying in the home is a marketer's dream.

Amazon introduced the Dash button in 2018. This simple stick-on button worked with Wi-Fi and allowed you to order instantly. With the introduction of Virtual Dash buttons in the Amazon Prime app came the end of the physical Dash Buttons.

2.11.16 Interactive Advertising Section Summary

The term "this way to the Egress" comes from the story of a P.T. Barnum display in New York City of strange and exotic animals. Throughout the exhibition there was a sign that stated "this way to the Egress." Few knew what the word meant and thought it was another exotic animal. Of course it was just an exit (egress), way out of the exhibit so that more people could come and pay. Brilliant?

Barnum arguably was this country's finest advertising executive. He made his fortune and lost it as well on the backs of the general public by selling them goods that they neither needed nor really enjoyed. The world of online advertising may also be headed that way. Do we really need 360° views of a car before we head to the showroom? Do we need to play games with our next meal before we even purchase it? Probably not but we need to realize that the individuals who own the wireless connections, who broadcast the web access, who provide the cable connections are not about to let us get by free! Somebody has to pay and it appears that, as always, that somebody is us. This is why this current generation was bred for this. So "this way to the Egress" but you can take your time as long as you're buying or just browsing – keep paying that wireless and data fee.

2.11.17 Search: Information the Real Culprit

There is an old expression – when you are up to your rear end in alligators it is tough to remember that you only came here to drain the swamp. It seems that we have almost forgotten the purpose behind the invention of the Internet. It was originally intended to transfer useful information from point A to point B.

And many businesses are betting on your need for that information and exactly how much you're willing to spend to ensure it, access it, and/or store and retrieve it upon demand. Remember knowledge is power but having it at your fingertips makes it a weapon and most businesses can't afford not to have. In this next segment we will explore the current state of information distribution, the models that are being structured and where it all may be headed.

First, we will look at the most common forms of information retrieval – search engines and portals. Google, Amazon, Facebook, Instagram, TikTok, etc. Who is the winner? Well Google-ing someone has now become part of the American lexicon so who's the winner. It matters what you are looking for. So what is the big deal about all this information?

2.11.18 Search: I'll Google for You

There is almost one least common denominator for web usage – you need information. That is pretty much the reason for using it. No bells, no whistles, no video usage – just the facts ma'am, just the facts. The question is how does Google, a free product for users, manage to lead the entire web-based world in the arena of development and worth? Advocates herald Google's pure structure and its permeations – Google Docs, Slides, Drive, Shops, etc. Alphabet now its parent company oversees these branches. Its estimated revenue for 2023 is approximately US$290 billion range ("Alphabet Revenue 2006–2019 | GOOG," 2019). It seems to have the right business plan expansion and quality. As said before it has become part of our lexicon and many times a verb. To speak of Google would literally take multiple terabytes so for now this is it.

2.11.19 Search: Yahoo!

Yahoo!, the portal, was supposed to be dead by now. Everyone knows that portals aren't worth the code they're written with so what makes this one so unique? With its revenue rising at 50% per year levels, Yahoo! was almost an enigma. The only reasons for its death-defying leap upward were a steady hand by then CEO Terry Semel, diversification and a focus on market needs. Sounds like a success plan for most businesses. But by nature portals and any conglomeration of information is the reason for the origin of the Internet and the web. Yahoo! combines its portal structure with Google's search engine and together they seemed to have had it pretty well locked up back then. How the distribution of that information is being used is really up to the audience and the companies that respond to those requests. Selling items at ridiculously low prices appears to also be part of that request. Today 50–60+ year olds mainly use it.

2.11.20 Social Media: The New News Network

Today 62% of individuals get their news and information from social media platforms (Pew Research Center, 2023a). These echo chambers are directed through algorithms that focus on your dislikes more so than your likes. As stated in Chapter 1, Frances Haugen (a Facebook whistleblower) – most of the Facebook algorithms are focused on conflict rather than agreement. Conflict raises the views and time on page. This is true for most social media networks. X (formerly Twitter) tried to bar this but once it was sold to Elon Musk and his investment group, it quickly allowed previously barred individuals back onto the site to generate more conflict via false narratives. Couple this with the new AI deepfake apps and we are soon to be faced with a very confused and trigger populace. Later we will discuss in more depth the warnings brought to us about social media and now AI from some of the original technologists who were there at the start and left as quickly as they could. Even they seemed to fear for the future they created.

2.11.21 Microsoft and Oracle – Two Giants Do Battle

If IBM is providing the network solutions, then who is providing the software solutions? It would appear that the leader and undisputed champion of software for the working world is still Microsoft but following very closely behind is Oracle. Both companies are in the business of creating information systems that manipulate information for others to use. Sounds crazy but it is really simple – reinterpreting information for use by others. This is where the world of computing is going and where it should be headed. Sure the hardware is moving rapidly but the software will always be the key. Gates' vision and dream of fully web-based network servers will be fulfilled and you will enjoy it. Ellison's focus on the world of creating cell phone mobile "Internetworks" will also be fulfilled. The question is what will be happening to these networks and their end users' machines? How will they be connected and how will they be implemented?

2.11.22 Automating Our Lives

Simpler is better. That appears to be the direction of the entire world of media. We enjoy our machines and our interfaces but what we really enjoy is the world they bring to us. In the previous sections we have been discussing the nature of the information, the storing and retrieving of that information, the delivery and the ultimate manipulation of that information to and by end users but what needs to be discussed is how all of this will work together and not through separate proprietary structures.

2.11.23 Nanotechnology – the Bots Are Here – Big Time

There is a saying in the world of nanotechnology – "it is really different down there." Dr. Jan Hoh of Johns Hopkins University spoke that in 2002 when he gave a tremendous presentation about the world of nanotechnology. Within that presentation Dr. Hoh spoke about the future of computing power via the breakthroughs in nanotechnology. One of the advances stated by Dr. Hoh and confirmed again in Scientific American October 2002 issue stated that in 30 years one laptop computer will have the capability of all of the computers currently working today. Now that was certainly an astonishing prediction but one that may be sustained by fact and research.

So with that stated and defined and coming to fruition quickly, what effect will these new computing powers have on the rest of the world? Will processing speeds have the ability to replicate the speed of the human thought process? Are we really in the primordial ooze of the computer timeline? Will we be the Neanderthals of the computer history timeline? Will the future speak kindly of us? Will we be there to witness it? Will our Supreme Overlords of AI treat us fairly?

The concept of information distribution, processing, storage, and retrieval are at the heart of the computer revolution. We may want to fool ourselves into believing that it is what we create that is really important but if we cannot retrieve it or if we can't store it then a real problem exists.

2.11.24 Artificial Intelligence

In the last chapters of this book, we will discuss the effects of AI and quantum computing on – well basically our life. How will politics, business, communications, jobs, etc. and also the arts be affected. But for now let's just have a preview.

In the early days of education the "why" was predominant over the "how." As education seeped down to the masses the "how" started to creep in to produce useful citizens. This even went into Higher Ed with the Ag and Tech colleges that sprung up in and around the farm belt. This existed and flourished for close to a century. Now with the advent of AI this concept of Applied Theory may give way to just theory again. Now the "why" may be the factor.

But bigger than the "why" today is the determining of what is important to the question or to the concept of what we are trying to accomplish. AI is quickly taking on the burden of the "how" production aspect of creating anything. Today generative AI is creating stories, images, videos, audio files, and even computer code for the entertainment and information marketplaces. In business and marketing, it is refining the efforts of companies to reach the right audience.

AI is also reaching further into overall customer service with CRX "bots" and going into the "professions." Today AI is being used in the medical arts to diagnose X-ray imaging, diverse diagnostic tests, etc. as well as review contracts and administer customer intake in legal services.

In education it is being used to construct IEPs (Individual Educational Plans) and used to evolve a traditional group coursework to match the development of the classes progress. Companies such as Khan Academy are integrating AI into its educational development process.

Creatives in diverse fields may not need the skill sets of computer software. Now they may just need to be able to conceptualize the need and to be able to properly convey that request to their AI counterparts. The concept – the "why" may dominate again? Now more than ever the formulating of strong neural pathways of strong critical thinking skills will separate the hierarchy.

CHAPTER SUMMARY

Why does the future desperately need us? After all, we are just being digital. Two names that have rocked this world of bits and bytes are Bill Joy and Nicholas Negroponte. All of today's digital philosophers really owe the birth of their discussions to these two people. Negroponte's introduction of bits versus atoms truly may have been the crossing of the Rubicon for this new interactive revolution. Ecommerce is truly the embodiment of his simple words. After all, we are not really using atoms in the world of interactive multimedia. As Negroponte stated, "GATT is about atoms." The world we are now living in is quickly becoming about bits and possibly only bits. This loss of the tangential is affecting brains and neural pathways of many young people. They are lost in this world of non-real reality, jittering in a dance of "The Anxious Generation" as perfectly said by Jonathan Haidt.

The problem with that hypothesis is that we are the end users and we are atoms. We need to determine how the future will need us. How we will use all of this information that is streaming at us. How will we determine values and will it be possible to determine markets? It used to be very simple – livestock simply meant stock that was alive. Now how do we measure our Schwab account or our AmeriTrades? Where will we stand in the formation of credible and desirable products and not just "Prompt Engineers"?

The formation of new societies via new desires may be needed. How we will do this is another question. In his article "Why the Future Does Really Need Us." Dr. Joy recounts the supposedly ultimate integration of man and machine (read cyborg) presented by Ray Kurzweil in his book *The Age of Spiritual Machines* and why this may need to happen. It is a great fear of all of us that these machines may ultimately replace human interaction – AI overlords anyone?

On the other side of the equation is a friend of Joy's, Danny Hillis. Mr. Hillis is a well-respected information and computer scientist. In a discussion with Joy, Hillis stated, "I'm as fond of my body as anyone, but if I can be 200 with a body of silicon, I'll take it." How are we measuring these advancements in technology and human interaction? How are we determining when too much is too much? Will nanotechnology, quantum computing, and Artificial Intelligence serve us or intern us? Where is all this interactivity, face time, information processing, ecommerce, and synthetic thespians, headed? Who knows? They say you have two choices in life – one is to cry and one is to laugh. Take your choice, after all life is an interactive experience.

3 The Digital Incunabula
The Economy

The changes that created this new economic and digital engine.

3.1 THE FIRST TIME

The first incunabulum is defined as:

> Derived from the Latin word for cradle or swaddling, incunabula are books printed between 1450 and 1501, the first 50 years of printing in Europe. (Chicago, 2024)

Gutenberg's invention of the modern printing press opened up the world to the age of the Renaissance and the Enlightenment through the spread of knowledge via books.

So denying this sequence or relative connection seems a bit odd, and it seems, so does the denial of the changes created from what some call the Digital Incunabula. So many disruptions have come about from the advent of this new means of communicating, archiving, and distribution that it is likened to a large stone being thrown into a pond. The closest ripples have the most power but the ones that reach out to the edge seem to have the greatest impact.

Given technologies, hardware advancements and today's advent of Artificial Intelligence, these ripples seem to have a life of their own. This time with an innate power of change whose course and effect seem to still be undetermined by the authors or the users.

While discussing this event students have responded to the question of which has had a greater impact – the printing press or the Internet. Below are some of their responses.

> As anyone could acknowledge after reading this paper, the Internet has affected our society immensely. Although I did not go into detail with the impact of the printing press, I believe that discussion only needs to be at a minimum. Not to mention the printing press created more employment than it took away.
>
> *Alecia Weiterschan, October 13, 2014*

> The Internet, like the development of the printing press, can easily be pointed out as both changing history immensely. Like the printing press, the Internet disseminates information, but this time on a grander scale. I feel that the Internet is the winner with negative impact.
>
> *Dr. Sequoyah Wharton, October 9, 2014*

Here are some facts concerning the first incunabula and what is being called the "Digital Incunabula."

The first incunabulum was before 1501:

- 1500 – 1,000 printing presses were in operation throughout Western Europe and had produced 8 million books (E.L. Eisenstein).
- In the Americas, the first extra-European print shop was founded in Mexico City in 1544 (1539?), and soon after Jesuits started operating the first printing press in Asia (Goa, 1556).
- In the 16th century [presses have increased] tenfold, to between 15 and 20 million (Febvre).

DOI: 10.1201/9781003404347-3

Now the Digital Incunabula was from after 1962:

- 1962 – J.C.R. Licklider introduces the concept of networked computers. Subsequently became ARPA's first head of the computer research program ("ARPA Is Born," 2022).
- 1964–1967 – Paul Baran of the Rand Corporation develops *message blocks* in the U.S. (Freeman, 2011).
- 1967 – Directing ARPA's computer research program, Robert Taylor initiates the ARPAnet project, which is the foundation for today's Internet.
- 1968 – UCLA team sends first data packets ("ARPA Is Born," 2022).
- 1971 – Ray Tomlinson sent the first e-mail message (Freeman, 2011).
- 1974 – *A Protocol for Packet Network Interconnection* is published by Vint Cerf and Robert Kahn which outlines the design of a Transmission Control Program (TCP) and for the first time coins the term "Internet" – Interconnected Network ("ARPA Is Born," 2022).
- 1984 – on August 3, 1984, the first e-mail arrived in Germany from the U.S. (Freeman, 2011).
- From 1999 to 2013 Internet users increased tenfold (Ryan, 2010).
- In 2005, the **first billion** Internet users (Ryan, 2010).
- In 2010, the **second billion** Internet users (Ryan, 2010).
- In 2014, the third billion Internet users (Pelchen, 2024).
- As of April 2023, there were 5.18 billion internet users worldwide, which amounted to 64.6% of the global population. Of this total, 4.8 billion, or 59.9% of the world's population, were social media users (Petrosyan, 2024).

Some 60 years after the invention of the Internet, it has now grown to have over five billion people online with countless websites and millions of online journals. Comparing this event to the dawn of print, it's incredible to realize it took over 500 years for a similar effect to happen.

But what are the ripple effects of this Digital Incunabula? What kind of effect have those ripples had on the world? Are the effects malevolent or benevolent? The biggest ripple will be the effects of Artificial Intelligence on every field, from knowledge-based careers to support service jobs. This will be discussed further in Chapters 5 and 6.

As with anything, there are parts of both involved with these disruptions. Again, it's for these reasons that we focus on the use and study of design because this is the means by which the new information and interaction is deployed.

3.2 THE FIRST TIME AGAIN

3.2.1 Schumpeter × Moore × Cerf

As a definition of the Digital Incunabula, we offer this simple response.

The Digital Incunabula is the time from after 1962 when the Internet and the personal computer were first made a possibility.

According to the Computer History Museum in Boston,

> This Internet Timeline began in 1962, before the word 'Internet' was invented. The world's 10,000 computers are primitive, although they cost hundreds of thousands of dollars. They have only a few thousand words of magnetic core memory, and programming them is far from easy.
>
> *(Spicer et al. 2000)*

3.2.2 Schumpeter × Moore × Cerf = Digital Disruption

Schumpeter's Creative Destruction theory is an economic premise where new advances create destruction of old technologies and hopefully bring new possibilities. The difference is those advancements had years to slowly migrate into the current economy.

Now Moore's Law (Gordon Moore – founder of Intel) predicted that technology would double every 18 months. Dr. Moore debates the strict timeline as being his quote but nevertheless we all get the idea. And it has actually now sped up to technology doubling every two to three months. Times all of this by another small innovation – the Internet (developed by many but usually the name Vint Cerf comes to the forefront here) and you can start to get a handle on why these times are so much more volatile.

New business models are created that are technology-rooted and delivered by wireless global communication systems. They appear like a meteor flashing through the sky at night. We all saw it and we all were amazed but only some had the wherewithal to follow the object and see where it landed. New fertile ground with huge potential – some will know what grows here and some won't.

A great quote from an article in TechCrunch.com. Here are some amazing facts about this new economy.

> ...Uber, the world's largest taxi company owns no vehicles, Facebook the world's most popular media owner creates no content, Alibaba, the most valuable retailer has no inventory and Airbnb the world's largest accommodation provider owns no real estate.
>
> *(Goodwin, 2015)*

If this doesn't prove the strangeness of this new economy what does? It is pure disruption – crowd-sourcing, social commerce, and divested industries – and at every level design and media have to understand how to speak to and speak through this new maze. It's all about the interface and the user experience design that makes for success. Please note this is not an easy task. And this is why we speak through design.

3.2.3 Y Design?

The question is why choose to focus on design? What does design invoke that makes it so important? For years design has been seen as an ornamental choice. The right typeface, the right color, the right proportions – have these choices really made any difference? Probably not but today design has transcended that previous criteria. Today design has moved forward to actually interphase with and determine our choices in all things.

Today when you see someone walking around head tilted down staring at that small screen they are absorbing information, some of it necessary, some of it discretionary. But how they absorb is the basis of all design choices today. It is called interaction design and as discussed before it is created not by one, but by many – each stakeholder – user, designer, and producer has credence. Each one has a significant part in how something is designed and why each piece received importance.

Imagine creating a compass from a simple piece of metal and a thread of silk. You charge the metal with the silk by stroking it in one direction. Then once the metal is charged you place it on a floating leaf – a movable base – that helps point the newly charged metal – now a needle to true north. This is interaction design. The metal is your message, the silk the empirical knowledge base and the leaf the ability to grow, change, and shift with the power of true north.

As we create design objects for this new millennium we need to make them malleable to the movement of the user. As the user shifts the object needs to follow that shift and point to true north. As society moves forward true north stays but their path changes based on the obstacles in their way. Our job as designers is to keep them on true north.

The problem is that any platform not malleable, not shift capable, not able to change direction toward true north effortlessly – like the floating leaf – tends to force the needle to point toward its knowledge base and conform to those limitations. Today that model has become extinct because it has lost its ability to find its way. No compass – no future. Designers point.

3.2.4 Going Digital and Interactive

In August 2015 (years before COVID-19 disrupted education), McGraw-Hill was still making strides toward its original goals of taking the industry digital. McGraw Hill Education CEO David Levin published an article on HuffingtonPost.com entitled "Dear Students and Faculty: Please Go Digital." In this article he espouses the virtues of a digital publication – not simply an eBook format but a truly interactive learning product.

> Print has been an effective way of sharing information since the time of Gutenberg, but it simply hasn't kept pace with the opportunities and demands of teaching and learning today. ... by combining digital content with software...we can work with faculty to create experiences that make learning more effective and efficient.
>
> *(Levin, 2015)*

3.2.5 Analog versus Digital

A main problem with some who have not embraced these changes seems to be just a general misunderstanding of what the criteria is for this incunabula. Numerous industry experts use the terms "analog" and "digital": Steven Heller, Paola Antonelli, and Maria Popov from the *Atlantic*.

Here is a quote from Steven Heller in his *New York Times* Sunday Book review article – "Down to the Letters 'Graphic Design Before Graphic Designers,' and More."

> Those who generate art, animation and even new typefaces from algorithms would not necessarily have been considered graphic designers in the analog environment, but they are becoming a new breed of designer/engineer in the digital era.
>
> *(Heller, 2012)*

Some experts define analog as something that attempts to mimic the real world – a paper dummy of a product is a simple example. Digital is defined as something that can only exist in a digital realm. This needs to be established upfront in this book to give the reader a better platform from which to view it. In all cases, we need to establish if a product or service could exist realistically to the best of its desired needs in either state.

Example:

Q: Can a website or app exist without a digital platform?
A: No, it can only exist via a digital platform.
Q: Can your package and deliver consumable products virtually?
A: No, they are physical items and therefore can only exist in a physical state. Any prototyping of this form of packaging and delivery is merely an attempt at an analogous display.

Seems a strange point to bring up but this is mentioned to establish what is or isn't analog or digital.

3.3 DISRUPTORS

3.3.1 Winners and Losers

"Down here it's just winners and losers. And don't get caught on the wrong side of that line" – Atlantic City – B. Springsteen – Nebraska Album – 1980

The disruptions are everywhere. They happen all the time. From the scribes getting the boot once movable type became popular to the workers at Disney and Kodak getting their pink slips once Jim Clarke and Steven Sasson finished their work. It is always going forward. The trick is, like a world-class surfer, how do you know which wave to ride? Like surfing, it comes with experience. So know your history – current and past.

What is the difference between disruption and innovation? Here is a strong excerpt analyzing that difference.

> Innovation and disruption are similar in that they are both makers and builders. Disruption takes a left turn by literally uprooting and changing how we think, behave, do business, learn and go about our day-to-day. …
>
> *(Howard, 2016)*

What will affect what? And what is connected to what? Like a good programmer, you must play the – if this – then that – else if – end game. The ripples affect everything in the pond. What did the Digital Incunabula disrupt? Here are just some examples that every family in America knows.

> "They [Kodak] were a company stuck in time," said Robert Burley, an associate professor at Toronto's Ryerson University …
>
> The company's credit deteriorated as revenue tumbled from traditional film, and the inventor of the Instamatic cameras was slow during the past decade…
>
> *(McCarty and Jinks, 2012)*

And even the banking industry has felt the effects and benefits of these disruptions.

> A fundamental shift is happening as more Americans move towards digital financial services. Eight in ten consumers use a fintech product to manage their finances, and businesses benefit from access to a wide range of services to improve their financial operations.
>
> *(Lee, 2023)*

Another simple and direct description of digital disruption is given in the *Harvard Business Review*.

> The much-used term simply means adapting an organization's strategy and structure to capture opportunities enabled by digital technology.
>
> *(Furr and Shipilov, 2019)*

At this point the discussion will break up into six sectors:

- Information
- Printing
- Promotional
- Entertainment
- Education
- The arts

Each one of these sectors has overlap, but is also unique to itself.

3.3.2 Information Disruption

At the McGraw-Hill Book Publishing in the 1980s their main focus was to produce flagship products in the Professional and Reference Division like *The Encyclopedia of Science and Technology* – 6th Edition. Thirteen thousand glorious pages of articles, figures, and diagrams. By the mid-1980s McGraw-Hill started working on putting the edition on a CD-ROM. This was totally the domain of the programming group. Typical editorial teams were consulted on how the screen could be designed. Many, not knowing a thing about designing for the screen, quickly capitulated and stated that it was fine as it was and stepped away. But what was really fascinating was that the 21 volumes on the shelf now came down to two shiny discs, inclusive of text, images, and some form of audio and video. The cost to produce the printed volumes was US$200, with the cost of the compact discs being US$4.95.

Compare that to the price to purchase printed volumes – US$2,500. Obviously, the future was staring them in the face and unfortunately many did not see the writing on the wall.

The world of information distribution and aggregating is determined today by the speed and strength of the delivery system. If you don't believe it, just look at Wall Street where information makes and breaks fortunes in a nanosecond.

Here is a quick excerpt from an interview explaining what happens when information can be accessed faster for one investor than another resulting in high-frequency traders being able to know what the market will do before their competitors. (Jain, 2023)

> As a result, program traders gave way to more sophisticated automations with much more advanced technology: High-frequency trading.
>
> High-frequency traders can conduct trades in approximately one 64-millionth of a second, compared with the several seconds it took traders in the 1980s.
>
> *(Jain, 2023)*

What does this change in the acquisition of information have to do with the world of print? A tremendous part of the print world was dedicated to distributing financial information that brokers relied upon to make sound investments. The *Financial Times* was just one publication that went for US$5 a daily copy over 30 years ago. Now that information is accessed and absorbed in that nanosecond. Any method that takes more time is literally yesterday's news and just as valuable.

The ripples of disruption will be outlined at the end of this chapter since in most cases they are similar and affect each sector.

3.3.3 Printing Disruption

Once the information sector had its disruptions, the printing sector would start to feel the ripples. If information in its most stoic form could be assembled and distributed in digital format, now it was time for popular culture to feel the effects. Entire magazine collections and special issues were being brought into the digital age. Images were being scanned and stored. Text was being aggregated onto these discs. The cost of printing the CD-ROMs was dropping dramatically and now monthly editions of magazines were being compiled onto these discs. Encyclopedias like *EST*, *Grolliers*, and *Britannica* were becoming extinct. Even one of the most popular magazines of the time, *People*, was attempting to be brought to the digital age. There were many hiccups along the way but it was clear that the plates – metaphorical tectonic ones – were shifting along with typical printing plates. Information distribution was moving and many were being left on the fading shores of that chasm.

Some of the biggest names in book printing were feeling the brunt of this new direction. Companies like R.R. Donnelly, who represented one of the largest names in printing and even down to the local printers in your neighborhood, were feeling it in lost clients and budget.

Here is an abstract detailing the health of the print industry in 1995.

> The cost pressures of 1994–95 should ease, and a stable economy with low inflation should support growth in demand for printed products. The impact of the electronic media may become further apparent in some product sectors, but a severely adverse impact on any of these sectors is unlikely.
>
> *(Lofquist, 1996)*

Another article in 2009, 13 years later after the web had risen to power.

> Taking its place, of course, is the Internet, which is about to pass newspapers as a source of political news for American readers. … As early as May, 2004, newspapers had become the least preferred source for news among younger people.
>
> *(Alterman, 2017)*

And then seven years later in 2015.

> The Printing industry is in the decline stage of its life cycle. Over the 10 years to 2019, the industry's contribution to the overall economy, measured by industry value added, is expected to decline at an average annual rate of 2.6%, due to the increasing presence of digital media. *(Printing in the US – Market Size, Industry Analysis, Trends and Forecasts*
>
> *[2024–2029]| IBISWorld, n.d.)*

And finally in 2023.

3.3.4 Printing in the US Industry Trends (2018–2023)

> Printing revenue has fallen at a CAGR of 3.0% to $82.9 billion over the past five years, including a 3.4% decrease in 2023 alone, when profit accounts for 4.1% of revenue. Digital competition diminishes consumer demand.
>
> *(IBISWorld – Industry Market Research, Reports, and Statistics, n.d.)*

The reason to study the impact on printing is because it feels the decline in revenue from all the other sectors outlined in this chapter, due to the fact that it relies on those other sectors for business. But also because it has wide-ranging implications in readership patterns, and also the effects on the brain.

Long-form reading versus short bursts of information has actual physical effects on the brain especially for young users. We will discuss this in greater detail in Chapter 4.

3.3.5 Promotion Disruption

Advertising was hit, and possibly was the widest impacted when these shifts happened. If the main delivery system of print was going to a digital format how could advertising keep up? CD-ROMs were one form but once the World Wide Web was launched back in 1993, all bets were off. You no longer count magazine distribution as a means to charge your ad rate; you need to know who actually *looked* at those "pages." Not how many entire copies were being sold, but now how many pairs of eyes looked at *your* ad? As with all disruptions a new industry was born where one just died. The magazine audit board was replaced by the IAB – Internet Advertising Board. New methods and processes had to be implemented. New rates and fees needed to be established. Automated processes were built – algorithms were devised – answers were required – basically, the earth shook and hasn't stopped. All of this has now grown toward becoming a powerful force in current advertising practices as traditional methodologies slowly lose dominance to an auction-like process that turns in a nanosecond.

One area of "corralling" the herds back to the product from their "distraction" of being online is the constant barrage from social media platforms. To combat that distraction, brands have embraced social media and learned to properly use those benefits. As discussed with Frances Haugen's "drug loop," there are statistics substantiating these effects.

- Digital advertising spending in 2023 is expected to exceed $600 billion.
- Search advertising is expected to do a volume of over $200 billion.
- The United States estimates over $232 billion in 2023.
- Per user advertising spending in the social media spending is forecasted to be $45.11 in 2023 and 2026.
- 87% of all Digital Advertising will be programmatic.

Statista, WordStream, PR Newswire.

And how has the industry reacted to this new platform currently? They are investing in its future to the tune of double digits. Here are the stats for social network ad spend in the United States during 2021–2025.

> Every year billions of dollars are spent on advertising on social media in the United States. In 2022, the amount reached 66 billion U.S. dollars, and included paid advertising on social networks as well as games and applications on social media.
>
> *("U.S. Social Media Ad Spend 2025," 2022)*

3.3.6 Entertainment Disruption

Probably no sector of industry has felt the overall deep impact effect of the Digital Incunabula like the entertainment industry. From the earliest days of Todd Rundgren building his own digital studio and releasing his albums to Sean Parker peeking into your hard drive at home, the entertainment industry – especially the music industry – quickly became the first casualty of this new age. Celebrities like Taylor Swift generate billions of dollars for the economy but mainly through touring and not album sales. No longer do record stores line the strip malls or the avenues. No longer must you save up for a $20 album; now you can just buy the one song. No longer do you have to fork over $100 to bring the family to the movies; now you swipe up the movie using services like ChromeCast, Prime Video, Netflix, Hulu, etc. on your big screen television – all for $7.99 a month. Movies have started to price themselves out of the family market and it has become a serpent eating its own tail effect. The days of the huge summer blockbuster are fast becoming memories, with the rarity like Barbie or Oppenheimer – (2023). Now music industry standards for a gold album have dropped to numbers in the 100,000 as opposed to the 1,000,000 mark.

> Since the introduction of the iTunes Music Store on April 28, 2003, music sales have plummeted in the United States – from $11.8 billion in 2003 to $7.1 billion last year, according to the Recording Industry Association of America.
>
> *(Covert, 2013)*

During the early 2010s, the movie industry felt the brunt of these disruptions from increases in digital piracy to hacks on computer systems, and a drop in summer blockbuster revenue. Summer of 2023 is the rare positive with films like *Barbie* and *Oppenheimer*. But it took a decade and a new genre like MCU to find their way back to profitability.

> Large declines aren't uncommon in the blockbuster business, where so much of the marketing push is for opening weekend. But such steep fall-offs contribute to anxiety over the ability of movies to capture and hold the attention of moviegoers in an age of so many other entertainment options.
>
> *(Coyle, 2014)*

> Between the first weekend in May through the end of August, ticket sales in the United States and Canada are expected to total roughly $3.9 billion, a 15 percent decline from the same stretch last year, according to Rentrak, a box office data company.
>
> *(Barnes, 2014)*

How will this ripple back through the overall economic landscape? Include in that the effect of AI video creation software like Sora. Well, think for a second about how many people are employed by the entertainment industry and how those same people will be looking for new jobs. The same goes for other areas in the entertainment field.

Read an excerpt from *Home Media Magazine* detailing a bit of the disruptions felt in the entertainment field.

> As has long been noted, the revenue model that has underpinned the entire Hollywood ecosystem is under assault. The theatrical business looks shaky. The summer box office disappointed, as consumers found other things to do (including watching digital content)...
>
> *(Prange, 2014)*

How has television been affected by this new Digital Incunabula? The first answer is it has been immensely.

The television industry has been fractured in the last ten years, from the effects of YouTube to Hulu to now – every social media platform and website out there is embedding video into almost every story and byline. How have the networks dealt with it? Many excerpts that follow will explain.

When Edison first started the motion picture industry, his original studio the Black Mariah was basically a hallway of dozens of small rooms all in a line. A camera, operator, and a director on one side with the scene and actors in front of them, with each of the scenes and actors dressed in storyline and acting. Each small vignette was captured on film and then distributed via the Nickelodeon or to simple movie houses.

Now that model is being replicated over a 100 years later by companies like Huffington Post, MSNBC, NBC, ABC, CBS, and every one of the major networks and news organizations. These small setups capture the content and distribute it via today's Nickelodeon – the Internet – to homes and individuals around the world.

You would think anything interactive-based would be untouchable, but of all companies – Disney felt the squeeze in the marketplace. Disney Interactive Studios was shuttered in 2016 as Disney+ its new streaming service came to life in the early 2020s. There are rumors that the interactive service will be resurrected to use its process to distribute Disney's products and services as part of its online learning division increases (Coulson, 2022).

So nothing is exempt or sacred, not even the platform, which created this beast. Take into account all of the different avenues to get your entertainment – whether music, movies, games, books, etc. All the major networks are now putting the content online and some for free. Notice how the music industry has felt the blow year after year.

Here is just a short list of the dedicated platforms and devices from which consumers can stream their desired content anytime, anywhere.

- Amazon Prime
- Hulu
- iPad
- iPhone
- iTunes
- Netflix
- Pandora
- Spotify
- YouTube
- all social media – Facebook, Reels, Instagram, Tik Tok, etc.
- thousands of apps available on SmartTVs and Apple TV boxes

3.3.7 Education Disruption

From the first incunabula to today the world of education hasn't actually changed much. Students are still taught in a rote manner – memorization is still the key to learning. Although there have been many attempts to change the model, it is basically the same since the late 1800s at best. From law school to medical school to the study of education itself, memorization is still a large part of the process. These schools are vocational. Basically, the professions have always held a place as not "vocational" but some space in between a PhD and a MD or JD? A never, never land where even in academia only recently have such degrees like a JD (Juris Doctorate) has been seen as a terminal degree. But nonetheless, such degrees do represent finality to that area of study, and they are again vocational.

It appears that the pendulum has swung again from pure theory to application, and is now hopefully settling in the middle to provide a balance of both: applied theory. Many have attempted to change this model as individual universities, school districts, and teachers, but it would seem the biggest change might have to come from the bottom up; through the desire to use and the direct response by students to these new learning techniques, aids, and requirements.

Recent events, like the COVID-19 pandemic have impressed upon the administrative levels the need for re-designing and re-deploying proper pedagogical online educational portals.

The growth of online education has been staggering in the last five years alone. There has been a 15% surge in U.S. postsecondary learners within that time span. In some areas almost 60% of all postsecondary degree candidates are taking at least one online class with over 60% identifying as female. These populations are also becoming more racially diverse as access becomes easier in urban communities. During COVID-19 these numbers accelerated.

> The Covid-19 pandemic accelerated the availability of online learning in higher education starting in 2020. NCES reports that 75% of all postsecondary students in the U.S.… took online classes in fall 2020 as the pandemic restricted in-person learning, whereas only 36% had enrolled in distance education in fall 2019.
>
> *(Hamilton, 2023)*

How far has it progressed from 2006? Obviously the recent disruption of remote learning via the COVID-19 quarantine proved that the education space was not prepared at all.

In contrast, but only minimally given that almost no one is advocating fully online curriculum and courseware, blended learning seems to be the one area of growth which allows students to be moderated and guided by – of course a teacher; see below.

> Proponents of postsecondary online education were recently buoyed by a meta-analysis sponsored by the U.S. Department of Education suggesting that, in many cases, student learning outcomes in online courses are superior to those in traditional face-to-face courses.
>
> *(Smith Bailey, 2010)*

Here is more arguing for the progression of supporting digital textbooks and overall online learning from David Levin (ex-CEO of McGraw-Hill Publishing). Levin asks for faculty and students to please "Go Digital." And to go digital via the growth of customized digital learning overall – this has been mentioned before in this book.

> Studies have shown that students using technologies designed to personalize the learning experience get better grades and have greater success completing their courses. And for the 50% of two-year college students and 20% of four-year college students taking remedial courses, these technologies can help them catch up with their peers …
>
> *(Levin, 2015)*

Some other areas of concern and innovation are listed below and will be discussed later in the chapter "The Real Ripples."

- Common Core
- Tablets in the classroom
- Use of the Internet as a learning tool
- eTextbooks
- Gen-I
- STEM & STEAM
- MOOCs

Here are some information the effects of the pandemic had on learning outcomes which further advocate the benefits of online learning. These effects could have possibly been averted if educational technology had been adopted earlier.

> For context, the math drops are significantly larger than estimated impacts from other large-scale school disruptions, such as after Hurricane Katrina—math scores dropped 0.17 SDs in one year for New Orleans evacuees.
>
> *(Kuhfeld et al. 2022)*

3.3.8 Arts Disruption

For over 50 years the arts have felt the effect of the computer. From the days of Billy Kluver and E.A.T., the arts have dabbled in the use of technology to create art. In 2002, that erupted at the Whitney Biennale. Artists like Mark Napier, Lisa Jevbratt, John Maeda, Margot Lovejoy, Future Farmers, Benjamin Fry, broke open the use of digital space and the entrance of code as an art form. Sure, artists had used computers to create video or printed computer graphics but now the user became part of the artwork and it only existed in a conceptual ephemeral space. The problem that seemed to cut at the base was – "could you actually sell the work?" Here is where a large discussion on value came into play. Conceptual thinking such as, "did the art actually exist without the user?" The prospect of value challenged the existence of the work. They tried virtual galleries but it wasn't the same, in my opinion, even in the remotest possibility. It was like trying to shove one aesthetic into the other. If art was supposed to be for the masses then how could the art world refute this attempt and manor of global distribution? And if the goal was to communicate the experience, how could making the user a necessary component challenge and invalidate the method? It was a huge shift, and of course 13 years later nothing has really come of it. Just like the economic models of many industries it will have to be simply beaten into submission by lack of dollars in the traditional marketplace or a massive devaluing of the commodity.

As examples of those economic influences just think back to Kodak's bankruptcy. Now apply that to the study of photography. Due to these breakthroughs in technology everybody thinks they can be a decent photographer – and to some point, they can. So what has that done to the area of study like pure Art Photography?

To see how hard it was for the world of interactive art to break through here is a review of the Whitney Biennale 2002 by Carter B. Horsley in *The City Review* and Holland Cotter from the *New York Times*.

Please note that this show heralded the world of interactive art and showcased the interactive art of some who now have completely changed the landscape. Even as recent as 13 years ago it seems reviewers had no idea what was going on. The show had 113 artists and was over 85% digital. In the 20 plus years since no Biennale has focused on Internet art.

> …the works of 113 artists in a variety of media and there is more architecture, performance, Internet and "sound art" in this biennial than ever before. If the overall verdict for the biennial is disappointment and ennui at what mostly is a ghastly mess, a few works, nonetheless, are outstanding and memorable.
>
> *(Horsley, 2024)*

Now here is a short list of some of those interactive artists.

- Benjamin Fry
- Lisa Jevbratt
- Margot Lovejoy
- John Maeda
- Mark Napier

And now here are two of the favorites in the world of data visualization and truly interactive art. Why? Because they do the actual work, not a fake version of it. Tim Otto Roth and Camille Utterback *(Ms. Utterback is also a MacArthur Award winner)* are both artists and technologists who inject the basis of technology into the output. The work could not exist without the specific technology.

What does all this have to do with the disruption being felt by digitalization of our lives? Remember Dr. Maeda – "art is what we do the sciences for." So if art can't embrace the change, will it go away like traditional photography has gone away? Is traditional fine art setting itself up to fail like all methods that refute Schumpeter's theory of Creative Destruction?

The expression "art is in the eye of the beholder" has been a basis of numerous discussions over the years but how could a group like this be so overlooked? My answer – the by-product – digital work – could not be packaged and sold like traditional forms of art. Therefore it couldn't or wouldn't be substantiated or promoted. Something so relevant and not that far ahead of its time was just left to kind of die off? This is important because it created a new means to do it – NFTs

Well think about Walter Benjamin and his 1936 essay "The Work of Art in the Age of Mechanical Reproduction." Benjamin's main point seems to be that there is an "aura" around a singular piece of art that any form of reproduction can never capture.

> One might subsume the eliminated element in the term "aura" and go on to say: that which withers in the age of mechanical reproduction is the aura of the work of art.
>
> *(Benjamin, 1936)*

While that's a great argument, it just labels art as a commodity – one of wealth, whether it is a cultural commodity or an actual commodity. After all, I guess it's all about the "Benjamins" – the real paper "Benjamins." But should this be true even when we deal with something as supposedly pure and clean as art? Don't we need a respite somewhere in all of this?

3.3.9 NFTs a Quick Primer

A lot of talk about NFTs – but what are they in relation to the arts? They are a means of "provenance" for the art world especially. NFT stands for non-fungible token – discussed before in this book but worthy to be noted again.

What does it all mean? It means the work can't be replicated or forged (easily). Blockchain technology allows for an encryption method by which forgery is brought to almost an impossibility. In an analog world a piece of art would be validated by a "letter" or note of provenance showing the transfer of the work from owner to owner. This validates the piece as being the original – one of a kind – work. Benjamin's theory of "aura" (Benjamin, 1936).

NFTs are then traded as commodities – the art itself contains the digital note of "provenance." However, this is only done through cryptocurrency. Some see it as a means of money laundering. Given the recent history of Bitcoin and other cryptocurrencies.

3.4 THE RIPPLE EFFECT

How big and how far do these disruptive ripples reach? Well just think about it. Craigslist destroyed the newspaper industry as we knew it. It killed the main revenue stream by allowing classified ads to be placed online for free. Classified ads were over 50% of the newspaper industry's revenue. Even a field that was once seen as very stable, like the educational publishing industry, has seen these disruptions. These changes have sent that entire business model and the rest of them; information, publishing, promotion, entertainment, education, and the arts into a tailspin from which they still haven't recovered or reinvented themselves. Many individuals have had to re-invent themselves numerous times, as have their respective fields, due to these new digital changes and shifts. In this section there will be examples of these ripple effects on each sector, and try to show how they are all interrelated.

3.4.1 Education Marketplace

Starting with education (because that is where it all starts) this excerpt shows how the education sector has suffered major shifts as well.

> Rising costs could scarcely strike at a worse time. Around the world demand for retraining and continuing education is soaring among workers of all ages.
>
> *(Vaitheeswaran, 2014)*

Now this is where the serpent starts eating its tail – in its educational demands. As education accepts more and more digital aid, is it leading itself toward the AI tutor? And in going there is it slowly reducing profits that in turn aid that quick demise?

> … the Internet has created a demand for digital textbooks, which has reduced the revenue of the established publishers. In the higher education market – primarily… – the price a publisher can command for a digital version of a textbook is lower than that of a physical copy.
>
> *(Band, 2013)*

3.4.2 Entertainment Marketplace

The entertainment industry has also felt the effect of this directly on the sales of products. This is seen through all the arguments over streaming royalties. Even such veterans and music icons like David Bowie saw how this would impact the future of the overall entertainment industry.

> The absolute transformation of everything that we ever thought about music will take place within 10 years, and nothing is going to be able to stop it – David Bowie (excerpt from the 2002 *New York Times* article "David Bowie, 21st Century Entrepreneur" by Jon Pareles)

The transference of those skills and the overall shift in the shrinking of jobs, from the disruption highlights Schumpeter's model yet again. Engineers, producers, marketing staff, session players, etc. are all being affected by these new one-person technologies. Songs and all forms of musical production is being reduced to one-person bands in people's basements. Streaming has replaced any form of archival media; therefore, the ripples touch even those ancillary support staff for warehouse and storage and trucking, and retail stores. It seems to be a never-ending conveyor belt to the end of the music business or at least for the musicians.

3.4.3 Media Marketplace

And these same disruptions are just as prevalent in the newspaper and magazine industry, with less of a lack of crossing over the existing skill sets.

> The estimated total U.S. daily newspaper circulation (print and digital combined) in 2020 was 24.3 million for weekdays and 25.8 million for Sunday, each down 6% from the previous year – though with some caveats…
>
> *(Pew Research Center, 2023a)*

Magazines, as did other media platforms, also found the loss of revenue from classifieds rolled downstream and upstream to cause disruption in the traditional ad revenue model. Agencies had to adapt to a new model as well. No longer could they bill a direct 15% for ad placement; now the ads had to actually work for them to get that commission. The old adage of "I know half of my advertising doesn't work, I just don't know which half" wasn't true anymore. Now analytics could tell you which half wasn't working and now clients weren't willing to pay for that half. Companies like DoubleClick were in control. Soon acquired by Google, the monster in the box was going to come out and drive the bus. And soon be an autonomous driver through the algorithms constructed at Facebook and the like.

Since a good portion of the advertising media buys are built on an auction structure as news trends or celebrity stories spike the ad space costs on those site spikes as well. Now an entire monthly ad buy budget could be wiped out based on one big trending story.

But what does this mean as far as an economic model, how does it affect society at large? Quick statistic from Forbes, here are the needed employees for Amazon versus traditional retailers.

> 14 [employees] – The number of workers Amazon employs for every $10 million in revenue it generates. For brick-and-mortar retailers, the average is 47 employees per $10 million of revenue.
>
> *(Bercovici, 2014)*

So now that these effects are becoming more and more a part of our everyday lives, what part of this actually affects individuals? And how can those individuals and the educational system that hopes to prepare them, properly evolve? What can be done?

At MIT there are studies to evaluate this – a short excerpt is detailed in the *Harvard Business Review.*

> "We have an educational system designed for a 20th century economy," says Brynjolfsson, with students trained to memorize and follow instructions and rules, … Instead, Brynjolfsson argues for an emphasis on creative thinking and problem-solving skills …that can only be done by humans…
>
> *(Bernstein and Raman, 2017)*

Brynjolfsson doesn't propose to have all the answers but he certainly lays out a road map and path to make this needed transformation happen. He, with other experts, has spoken directly about their respective fields and how they intertwine and affect each other. This next section presents some of the experts in these fields and their viewpoints on the future of those respective areas of study.

3.5 THE RIPPLE EXPERTS

3.5.1 Defining Design in the 21st Century

As stated by people such as David Kelley of IDEO and Paola Antonelli of the Museum of Modern Art, design is quickly becoming the hallmark of the 21st century. In a discussion with design faculty, it was stated that "the 19th century was arts and crafts, the 20th century was art, and now the 21st century is all about design." Why focus on design? Remember Antonelli's answer (more from her later).

> In truth, design has spread like gas to almost all facets of human activity, from science and education to politics and policymaking. For a simple reason: one of design's most fundamental tasks is to help people deal with change.
>
> *(Design Takes Over, Says Paola Antonelli, 2010)*

Again design is now recognized as no longer just ornamentation to be used to persuade. Now it is integrated into how we interact with products. Now that so much of our experiences are in a digital format the main experience has left the hand and moved directly to the brain. The content and the delivery method have become one. Like any other symbiotic relationship, one tremendously influences the other. Design is no longer about the curtains or the difference between khaki and beige. Now it is at the forefront of the experience. It is making the experience work or not. Later we will speak about interaction design.

It is a belief that this is due to the disruption caused by the Digital Incunabula. As discussed before, jobs, products, services have had to become more efficient due to a slowdown in economic resources. Those slowdowns are due to the changes caused by going to a digital age. From paper to pixels, as has been said for years.

So how does this all play out and where should we put our efforts? Let's turn to truly respected sources – Nicholas Negroponte, Sir Ken Robinson, Paola Antonelli, John Maeda, and David Kelley. Many of these areas overlap – for instance, the approach to design and the teaching of design overlap all areas – information, publishing, promotion, entertainment, education, and the arts.

Listed below are short bios for these individuals and some information regarding their beliefs on the needed steps to improve the design of their respective sectors.

> Nicholas Negroponte is founder and chairman of the One Laptop per Child non-profit association. …co-founder and director of the MIT Media Lab, and the Jerome B. Wiesner Professor of Media Technology. … a pioneer in the field of computer-aided design, and [he] has been a member of the MIT faculty since 1966.
>
> *(Person Overview ‹ Nicholas Negroponte, n.d.)*

Negroponte's book *Being Digital* in 1995 confirmed every belief we had in what was happening then and what would happen later. Often the question of "bits or atoms" has been raised? Negroponte's story of the value of data over physical inventory basically heralded the economic models of today. When asked the actual value of a laptop he had brought to a meeting, Negroponte posed the question of whether the value was based on bits of atoms or data? Example for you – how much would you pay for your lost iPhone?

> … asked whether I had a laptop computer with me. Of course I did. The receptionist asked for the model, serial number, and the computer's value. "Roughly US$1 to $2 million," I said. "Oh, that cannot be, sir," she replied. "What do you mean? Let me see it."
>
> *(WIRED 3.01 – Bits and Atoms, n.d.)*

This concept of bits over atoms is today called IoT or the Internet of Things, with Artificial Intelligence being at its finest and worst. This is the new way of doing business, culture, entertainment, and life. And AI is the new gasoline driving this next economic milestone.

Below is part of an article interviewing Paula Antonelli (*senior curator of architecture and design, and director of research and development, at the Museum of Modern Art in New York City*) from "AI as material for design." In this article Antonelli argues for AI seemingly understanding that it is merely a new tool. Although a major shift in technology, it is a tool we will master like all the others we have encountered.

> As designers master this tool, they will use it to expand their ability to make things better. She cautions us of the good and bad manifestations of AI, and states that these manifestations are ultimately dependent on the mind that devic[s]e or use[s] them.
>
> *(Jasani 2018)*

After hearing her speak and reading these articles one can't help but think and hopefully realize how much design has come to the forefront of this century. The reason for that in Antonelli's opinion is "one of design's most fundamental tasks is to help people deal with change."

Can anyone think of a bigger time in history when so much change occurred in such little time, and the effect that change has had on society – at the basic core level of survival? Read some Robert Reich or Thomas Friedman to see how much these "changes" have affected the everyday lives of people.

The central point is possibly – "eternal human quest to make beauty out of necessity." Now define beauty – because the question comes up – how do we define beauty? What does this have to do with the Digital Incunabula? Well, the goal of the five sectors discussed so far in this book is really connected to what we find beautiful, and is the core of beauty connected to necessity – to functionality.

So we go back to Hume versus Kant and at length Spinoza. No time for that now, but the basis of that argument is – is beauty found in objects innately or is it held to a standard of tastes? Tastes based on what? In many beliefs, it's based on whether or not beauty adds to the human experience. And now to define the core of that experience, hopefully, to function better within the society, in essence, to survive better as a large community (Spinoza), but that is another huge discussion.

To help understand that process we turn to John Maeda, Whitney Biennial artist, former president of RISD, partner at Kleiner Perkins Caufield & Byers, chair of eBay Inc. Design Advisory Board, and now VP of Design and Artificial Intelligence Microsoft.

In a recent *New York Times* article on the current disruptions, Maeda states,

> Now, Mr. Maeda said, this shift has happened to technology, be it computers, smartphones or the iPad Mini.
>
> "We have this exciting next step for design," he said. "Now that we have enough technology to do anything, design can now begin to be better than the technology itself."
>
> *(Bilton, 2013)*

This is the crux of this new time to go from the "how" to the "why." Also during this time Maeda had written numerous articles describing his experience in the educational space, the future of digital integration into learning, and states some of the obstacles he has recently come across. One major obstacle seemed to be the faculty at RISD. Just six years after coming to RISD from MIT Media Lab, Dr. Maeda was given a vote of no confidence by the faculty and asked to resign (Murg, 2011).

So here is one of the questions regarding the events surrounding Maeda's tenure at RISD, and how much this move toward embracing digital progress is being stopped. Given his prior experience, why would one of the oldest educational institutions – with literally the word "Design" in its name – not heed the advice of someone like Maeda? Obviously, we are in a time of extreme shifts and changes. So why not listen to someone who has in all probability seen and aided in these changes up close and personal and bridged the obstacles beforehand? And if individuals let that confusion guide their principles then how many others are being affected by this avoidance? How many others are having their future decided by the same individuals that families have entrusted their life savings and their children's future?

There have been record drops in enrollments at private universities in the past five years. The traditional state university systems have seen huge surges in enrollment, due primarily to price differences. So it would seem that in order to maintain their position in the marketplace these private schools would need to answer the $100,000.00 difference question. Many do, but it seems in the case of RISD and other privates that question still looms.

So with the apparent disruptions and the possibility of such potential innovation, why not adopt a different perspective? It seems some schools have invested in the difference, this opinion is coming from Dr. Maeda, whose perspective is based on industry as well as academia.

Another individual is David Kelley. Many see Kelley as the bridge between the public and private sectors.

> David Kelley
>
> In 1978, David co-founded the design firm that ultimately became IDEO. Today, he serves as chair of IDEO and is the Donald W. Whittier Professor at Stanford, where he has taught for more than 25 years.
>
> *(Tischler, 2009)*

Kelley's home is the "d.school" at Stanford and is built similarly to his corporate IDEO structure. Here Kelley melds all diverse thoughts into one cohort respecting opinions from sociologists, designers, coders, and ultimately the client's specific discipline. This process has been named "design thinking."

> Students come to the d.school with an intense curiosity, a deep affinity for other people, and the desire to gain an understanding beyond their own experience. They come from every school on campus, and beyond.
>
> *(Tischler, 2009)*

This combination of diverse users and designers, and developers seems to be the perfect blend in academia toward the profession of UX design. Kelley and IDEO's influence on the world cannot be overstated. As one of the original designers hired by Steve Jobs and many others, Kelley evokes the process of change with every assignment he and his company IDEO undertake. This new concept of "design thinkers" seems to be the basis for the next century of design and innovation. Gathering a group together to make the design process Kelley and Maeda evoke, it creates a new structure and a new order for a new century. A new structure that in many opinions is well placed to deal with the opportunities and crises that will surely come as part of the Digital Incunabula.

As stated before, this new approach to design study and the subsequent processes are in some schools called "interaction design" or "user experience design."

3.5.2 Interaction Design (UX Design)

As stated in prior chapters and sections, interaction design has become not just a "buzzword" or "industry jargon" as some design academics like to call it but obviously a real driving force behind changes in the education discipline and the economy. Therefore, to negate its position as a main criterion for overall education may seem foolish. It is like saying doctors or lawyers don't need to be educated formally in critical thinking.

Here is a description from some schools as to the focus of their curriculum.

> The MFA in Interaction Design program trains students to research, analyze, prototype, and design concepts in their business, social, and cultural contexts.
>
> *(Home – SVA MFA Interaction Design, 2023)*

These sorts of changes in academia seem to not reflect a short arc or transition but a major re-focusing. Here is a very short list of other institutions focusing on "Interaction or UX design" as both an undergraduate and graduate major.

- Carnegie Mellon University
- Institute of Design, Illinois Institute of Technology
- North Carolina State University
- NYU Graduate School
- Parsons
- Pratt
- SVA
- University of Art and Design Helsinki
- Virginia Commonwealth University

Is all education feeling the effect of these disruptions and the need for a re-design and an embracing of a new generation and a new way of presenting information? Is it the school's concern only? Well, not according to the late Sir Ken Robinson, a revered proponent of education reform (About Sir Ken Robinson, 2019).

Below are just some of his views on the future of education and the world. Heralded as one of the best minds in modern-day education and the use of technology to improve that experience.

> … The wholesale transformation of education is at the heart of the changes that are needed. It's not something that Congress, or the state governments, can get round to later on. If they put this off… , they may find that they and the whole country are left behind.
>
> *(Robinson, 2009)*

A lot of Robinson's points were what many try to reflect in class discussions. But here are more of his viewpoints in simplified form since what he has to say needs, well, an entire book.

> The fact is that given the challenges we face, education doesn't need to be reformed – it needs to be transformed. The key to this transformation is not to standardize education, but to personalize it, to build achievement on discovering the individual talents of each child…
>
> *(Robinson and Aronica, 2008)*

What is the current state of education and where should it be reformed or simply evolved into current needs? Robinson always reflected back on the early days of the industrialization lifestyle that created our modern-day educational system and how that needs to change. This quote from Robinson seems to share the belief of Brynjolfsson.

> Public schools were not only created in the interests of industrialism – they were created in the image of industrialism. In many ways, they reflect the factory culture they were designed to support.
>
> *(Robinson and Aronica, 2008)*

And what is the future of education and where should it be going in this new age of digital reformation and delivery of content? Robinson refers to the educational model as being similar to the fast-food model of quality assurance and standardization versus customization – here is where AI can be a benefit. Here is where AI can be that tutor for each child (Robinson and Aronica, 2008).

3.5.3 Summary

As a summary to these sections, it may be necessary to tie this a bit together. Why try to relate all of these diverse areas of influences and priorities to the overall use of Interaction – UX design and future technologies?

> As Antonelli says, "In truth, design has spread like gas to almost all facets of human activity, from science and education to politics and policymaking. For a simple reason: one of design's most fundamental tasks is to help people deal with change."
>
> *(Design Takes Over, Says Paola Antonelli, 2010)*

For someone as respected, intelligent, and successful as Antonelli to be saying that all of these areas of society are tied together through the practice and concepts of design truly helped validate many beliefs as well. Everything needs to be designed to function properly for the human. Some may think of innate design, as by a supernatural being (or just Steve Jobs – close enough?) – but a purpose or plan is usually in place. All changes and occurrences within a society and the market that supports those societies are influenced and then influence design choices. And please remember, when the word "design" is used we are not talking about curtains or khaki or beige… we are simply talking about critical thinking – going from problem to solution.

3.6 THE RIPPLE REFORMATION

When ripples are strong enough, they can push objects into small groupings. Undersea life, and anything else in the path of these ripples, will sort of gather together. These gatherings can be forced or formed due to where the items start or where their weight has re-positioned them. Nevertheless, they are there and seem now to go together. Maybe it's more ideal to see how they can work together, instead of trying to reposition them as they were. Maybe the next ripple will simply send them back to those points again.

As jobs fall from the bottom and the top, the middle-level jobs become basically the arena for the "Hunger Games." Think about it, the person who used to do printing is now pushed out and looking for a similar salary level. And they just keep pushing down and around. Further and further down.

Specifically in the print/publishing industry, individuals in all of these positions had to push toward other industries:

- Editorial designer
- Editor/assistant editor
- Typesetter
- Proofreader
- Photographer
- Camera lab technician
- Camera operator
- Film stripper
- Plate maker
- Press operator
- Printer/Printer's Devil – assistant
- Edition collator
- Forklift operator
- Delivery truck driver – mechanic
- Jumper – delivery assistant
- Warehouse manager and workers
- Paper companies – entire staff and facility
- Ink manufacturers – entire staff and facility
- Lumber industry – dramatic drop in need for paper and products
- And on and on…

Now apply that metric to industries that produce traditional analog products in information, entertainment, promotion, education, and the arts. The drop-off of jobs from this Digital Incunabula is astounding.

For years the question was heard at conferences, "Who can make this transition?" and the answer would be "if you worked hard and studied you could apply your skills and knowledge to this new era." The fact was – and is – not many will make the transition. Why? Because they simply weren't trained to think this new way and this goes right along with what Brynjolfsson argues (Furr and Shipilov, 2019).

> "We have an educational system designed for a 20th century economy," says Brynjolfsson, "with students trained to memorize and follow instructions and rules, an approach initially designed to prepare them for factory work that today is largely mechanized."

We are still teaching the how-to world. And in the case of design, we are still doing demos on how to use software. The problem is that students today have accelerated way beyond that level – especially with services like YouTube and Lynda.com, which was purchased by LinkedIn for $1.5 billion in 2015 (Etherington, 2015). They have these skills coming out of secondary school. Not saying to give up the "how-to's," just moderate them down to a quicker, smaller part of the semester. This belief is the same as Brynjolfsson.

> Instead, Brynjolfsson argues for an emphasis on creative thinking and problem-solving skills that will enable students to go on to the sorts of jobs that can only be done by humans – and tend to offer higher financial and emotional rewards as well.
>
> *(Furr and Shipilov, 2019)*

Again with moderation, not complete theory but how that theory is applied, based on determining the needs of the problem – essentially true design. Well, what are examples of these attempts at re-design? In the next section, we will be discussing the attempts at reformation and re-design in each of the six sectors.

3.6.1 Information Reforms

The information sector in order to survive, whether it is industry journals or newspapers to basic information like dictionaries, encyclopedias, product reviews, cooking recipes, television listings, car manuals, and even maps has made the leap to digital only.

In making this transition many have found the virtue of a paperless structure. Why bother to discuss this? Well, before anyone takes a step toward college or a new career path, they should know whether that limb is stable and safe.

In this section, we will provide samples of some popular forms of transformation like switching to online, app development, and a new term – content "curation."

3.6.2 Switch to Web-Based

How many typical information sources have now stopped printing and switched to online distribution? Magazines and newspapers alone have seen plummeting readership but also have traditional platforms like dictionaries and encyclopedias. What happened to the business of news and information? Here is a snippet of a report from Pew Research.

> The average audience for the four newsmagazine shows aired by the networks – ABC's 20/20, CBS's 60 Minutes and 48 Hours, and NBC's Dateline – increased for each network by about 10% in 2020, following a stable year for CBS and NBC in 2019 and a 21% increase for ABC in 2019.
>
> *(Naesser and Shearer, 2021)*

These statistics are, however, based primarily on news research only. The news divisions of all major networks and their subsidies are reinventing themselves as smaller and smaller slivers of the main network. Refining and "narrow-casting" themselves toward that defined demographic.

What happens with entertainment programming? It seems to have weathered the storm and returned a bit. However, this has been short lived for the traditional network structures. As viewers get younger their chosen formats like YouTube and social media video offshoots are growing. The problem is the 3-minute maximum of attention span.

3.6.3 App and Smartphone Adaptations

Apps are obviously one of the fastest-growing areas for all information retrieval, but this is not to say that's where it stops. They are basically replacing the delivery of content that was previously produced through traditional formats and expanding those capabilities. How many news and information sources now have apps for your mobile devices? How many apps are there for just simple everyday life needs?

In a recent article on Techhive.com, you can find the top ten apps that replace typical products to create and store your information, from digital "paper pads" to the "Yellow Pages" to the "Oxford Dictionary" are just a few examples.

Think for just a moment how much of a disruptive effect using a "digital pad" was to the paper industry. How many notebooks and legal pads have you purchased, touched, or used in your lifetime? How many people are employed in the paper manufacturing industry, not to mention the office supply business?

With 100,000s of apps available, I guess the current mantra is true – "there's an app for that!" is soon going to apply for everything we need.

3.6.4 Maps and Manuals

To understand the impact of switching to online for companies that produce maps or manuals, just do a simple Google search for "owners manual" and see how many sites come up. Everything you could think of is listed from home appliances to luxury yachts. Every item has a website and a manual associated with it – and many have how-to videos as well. DIY videos have become the number one search criteria on Google.

When was the last time you asked a person for directions to anything, anywhere? When was the last time you didn't rely on GPS to get somewhere? When was the last time you actually bought tickets in paper form to a game or event? When was the last time you read the newspaper for a restaurant review? When was the last time you *needed* to?

How has this changed the landscape? In some ways it has stayed the same – writers still have to write the words, and pictures are still needed, but now video production teams need to be involved. Soon with AI enhancements this may all change again. The "how-to" marketplace is exploding on YouTube and other video content delivery platforms. A great example of a quick, simple way to organize your needs has been through the evolution and explosion of the use of the Internet as a base for content curation.

Pinterest is a great example. You laugh? Well, so did many people in 2013. Something as simple and based on user-generated content is growing faster than anticipated (Roose, 2013).

So why is this so important in the area of information reformation? Because as the ripples affect the playing field those little boats – viewers and users – drift and move with the ripples. Remember what the late Yogi Berra said, "That place is too busy. Nobody goes there anymore" (https://www.latimes.com/people/houston-mitchell 2015).

3.6.5 Publishing Reforms

In the publishing sector, a move toward the distribution of content through digital platforms has been at best slow and at worst tricky, but it has replaced the industry's typical practices. This transformation looks like it is now complete. There have been attempted rebirths like *Newsweek* and *Life* magazine but they seem to quickly fold and go away.

Legal issues surrounding intellectual property rights have popped up in court like Meer Kats smelling a predator. Not to mention the never-ending question – "how do you make money on the web?"

Many traditional publications have decided to try numerous avenues to profitability. Some of them are:

- Paywalls
- Dual membership – print and digital
- Advertiser sponsorship
- Digital editions

(Oliver, 2022)

The values of these strategies are still working themselves out. But their value as being worthwhile is an argument that could produce about a million research links, but we all know the answer. Just ask yourself these questions.

- When was the last time you opened your mailbox and found a magazine?
- When was the last time you ran out to "get the papers, get the papers?"
- When was the last time your child sold magazine subscriptions for school?

During this time many publishers and networks took on the premise of paywalls to offset the drop off. Has it paid off? In the transitional print world, the answer seems to be no. However, network television has continued to maintain its value – for now. Asking for numerous subscriptions to watch the same program on different platforms. Confusion reigns supreme right now.

3.6.6 Google Books

Google Books originated as a project in Google Labs, which has since transpired into an arm of Google that makes millions of books available for viewing on the web for free. Now information can be accessed for free from anywhere – without leaving your home – and can also be searchable. No more Dewey Decimal System, no more wooden drawers to fumble with, no more searching and searching only to find the book has been checked out and never returned.

This move was enormous for publishing and information reformation. And one that opened up some huge litigation between intellectual property rights to royalty for digital books. Now, how to make this information easily accessible without dragging around a laptop or being tethered to a desktop station? Introducing the tablet.

3.6.7 Books versus Nooks, iPads, and Kindles

There is so much debate on this topic it could fill a book. Bad joke, but had to use it. The question is what context is the information being used? If I wanted to sit down and have a drink it would matter what it was for. If it's during a meal I'd probably choose wine, if it was out on the town I'd likely ask for a scotch, and if one wanted to get seriously drunk, probably tequila.

Kind of similar to reading – if we wanted to find information fast – we would use digital. If we wanted to lounge around and just browse, or have coffee and relax we'd probably choose a traditional paper book. But if we were in school and had to carry five books to and from class – a digital solution makes sense. Like all good designs, the first question is, "what is the problem?" and then find the solution.

There is so much discussion in this area still. Digital editions of magazines, a plethora of Frankenstein-like reanimated malformed objects and incarnations. It reminds one of trying to put children down for bedtime. The fights, the struggles, the "just one more glass of water" delays… We all knew it was bedtime, but being children they just kept trying to put off the inevitable. The same is true for the days of paper publishing and distribution.

3.6.8 Promotion Reforms

In the promotion and advertising sector the world has turned upside down and back again. Don Draper would be sitting at the conference table doing shots of scotch until he passed out. Everything from the creative process to the strategy session, to the client meeting is just completely different – and most of the strategy seems radically absurd in comparison. Not to mention the effects of social media marketing. And the animating of follow-up marketing now called "retargeting."

The gut instinct of the 1950s–1980s is gone. No company or agency can afford today to be wrong. Everything is about efficiency of operations. Here is an excerpt of the creative process from an article in Adweek concerning some industry leaders who sat on a panel to discuss this new model.

> … Reid Miller, ecd at Taxi in New York, the final panelist, spoke about his agency's move from a linear process of ideation (client -> account -> strategy -> creative -> idea) to a more circular one, where staffers from all disciplines surround the idea and add to it.
>
> *(Nudd, 2013)*

Still the same in effect but now let's move that to the strategy. How do the available vehicles change the creative? Here is a simple four-step plan from Stafford University (Global, 2022).

1. Clarify business intentions
2. Recognize, appraise, and initiate emerging opportunities by utilizing innovation teams
3. Disassociate strategy and innovation in marketing
4. Make open assets available on the marketplace

And then the argument – analytics versus gut instinct.

> For designers, this influx of data can be frustrating. Designers who thought they were hired for their good taste will quickly get discouraged in a culture where tech companies meticulously test 41 shades of blue. Imagine convincing a team to trust your gut instincts when cold hard data says you're wrong.
>
> *(Kowitz, 2013)*

And here is the point about design: if its main purpose is to sell the product, why would a true "designer" suggest a different solution? After all of this pondering and strategy choices, it has seemed to come down to the basics – how to get attention amidst the clutter. View-ability = Measuring (the right) eyeballs! That is the problem we are trying to solve.

3.6.9 Social Media Marketing Effect

The effect of social media platforms on promotion and marketing cannot and should not be understated. Since 2004 when Facebook debuted, the effect of community on products and brands has been felt. In the early days people would ask "what is Twitter?" and "why should I care?" The response on platforms like Twitter, Instagram, Tumblr, and Pinterest has been this – no one will care what you are eating or where you ate it. But YOU will care if your favorite restaurant, car dealership, clothing store has a preferred customer sale. Corporations believe the line – it's not YOU it's ME.

Measuring the performance

Now that it's all done, you simply have to sit back and wait for the results. In many of these cases, it's like a tough series of blood tests, with possibly the same results. And the same name you have known for years is still doing the job of trying to figure all this out – Nielsen Online Campaign Ratings.

> With audiences engaging with advertising across more devices and platforms than ever before, it has become increasingly difficult for media buyers and sellers to accurately measure their viewership. In fact, accurate, cross-platform measurement is one of the greatest challenges facing more than 50% of all media buyers today.
>
> *(Cross Platform Ad Measurement, 2023)*

Now with all of this going on – and considering this applies to the world of print, television, radio, and online – hopefully the depth of this new Digital Incunabula is more than apparent and now an accepted reality.

Teaching design and especially advertising takes on a whole new idiom and scope. Gone are the days when just "sex sells" and "free this or that" will get the job done. This new structure has changed the game for good. And in many teachers' opinions, it is their job as teachers of design to properly prepare students for that change – every day.

3.6.10 Entertainment Reforms

As stated before, the Digital Incunabula hit the entertainment sector the hardest. The music segment specifically was completely turned upside down by the change in the economic model of delivery and selection. Attempts at reforming the model did involve some very drastic changes and solutions. This overall question hit the entertainment industry hard in 2023 with a strike of writers, actors, producers, etc. The strike was finally settled with new guidelines for the use of AI and royalty payments from streaming services. This is just the first go around to try to figure out this new business model.

In this section, we will attempt to briefly list some of the solutions.

3.6.11 Licensing Content

In an attempt to forge a partnership with these new platforms the music, television, and movie industries have been merging and now developing their own content and platforms for downloading and streaming content.

Examples:

- Amazon Prime
- Hulu
- iTunes
- Netflix
- Pandora
- Peacock
- Spotify
- and all network hubs

3.6.12 Second and Third Screen

Even the networks have started to embrace the world of online distribution by allowing free play. They realize that they are losing viewership to a second screen – and a third – and who knows what else is coming up. Each platform has its own peccadillos and a big fear of cannibalizing existing profits from traditional venues and viewership has most traditional distribution schemes sometimes in a state of analysis paralysis.

Some of the discussion concerning these new aberrations is spoken about daily in the trade publications and even your typical news sources.

> According to eMarketer, forecasted TV viewing time in 2023 will continue to decline. In 2021, the average viewing time of TV was 3 hours and 17 minutes, where it is estimated to backslide to about 2 hours and 51 minutes in 2023.
>
> *(Schomer, 2021)*

3.6.13 Motion Pictures

With all of this motion going on, how is the motion picture industry reforming its practices? With the exceptions of blockbuster franchises and oddities like *Barbie* and *Oppenheimer*, what has been the outcome of these disruptions and how has the lack of reformations affected the industry.

- Ticket sales are down 21% from four years ago, … So, too, is the number of wide releases.
- From January–June 30 in 2019, 57 films were released in 2,000 theaters or more. In 2023, there were only 45 releases during that same period.

(Whitten, 2023)

3.6.14 Hipsters and Hiccups – A Return to Vinyl

But the hipsters are trying to keep the faith alive! Some entertainers are trying to return to vinyl and capture a new audience. Performers like Jack White and Neil Young are cutting live albums on the stage of the Tonight Show starring Jimmy Fallon.

> After the interview, Young hopped inside the [recording] box to perform "Crazy," a cover of Willie Nelson's 1961 hit popularized by Patsy Cline.
>
> *(Reed, 2014)*

Such stunts are a new way of using these shifts to get public relations moving. Whether you actually believe in what they are doing or not, in all good advertising you try to find a hook or event. Hucksters are hucksters, even if they are using their superpowers for good instead of evil.

Whether or not this retro move to vinyl is worth anything reminds one of a great bit of dialogue from *Moonrise Kingdom* where an errant arrow kills a pet Scottish terrier, and the boy and girl find him.

Girl: "Was he a good dog?"
Boy: "Who's to say?"

3.6.15 Education Reforms

In the education sector reform has been slow, then ramped up dramatically with solutions that shook the pillars. These solutions have been structured via curriculum, methodology, and have been implemented through both hardware and software. It is to be noted that this lack of forward movement has been severely felt since the pandemic hit schools around the world.

As discussed before, the areas of reformation have been:

- Common Core
- Tablets in the classroom
- Use of the Internet as a learning tool
- eTextbooks
- STEM & STEAM
- MOOCs

(Finn, Jr. and Fairchild, 2012)

One area of dramatic change and tremendous scrutiny is the Common Core. This is not being met with the acceptance and rational discussion that was hoped for when started. The reason to bring up Common Core is to show how almost any initiative gets twisted and turned until it slowly dies. These changes can be more easily integrated with the use of technology to display and allow for further review and discussion This way all stakeholders can witness the progress. This is a way for the Digital Incunabula to aid in transformation on a national scale easily and affordably. But there have been many criticisms of the use of technology to positively enact these changes.

3.6.16 Common Core

According to the Common Core Standards website this definition is given.

> The Common Core is a set of high-quality academic standards in mathematics and English language arts/literacy (ELA). These learning goals outline what a student should know and be able to do at the end of each grade.
>
> *(Common Core State Standards Initiative, 2009)*

And the question of whether or not Common Core is working? Well, it seems it becomes a question of: is the support and opposition merely political?

3.6.17 Just a New Revenue Stream?

The other angle for the proposing and tremendous support of Common Core has been addressed via John Oliver and his HBO show "*This Week Tonight*." This era of change has radically reformed the publishing market and one of the biggest areas of constant revenue is the educational marketplace.

Once this telecast premiered on this topic every assumption spoken of fell into place. Mr. Oliver pretty much nailed it.

Think about how a loss of paper textbooks has affected the textbook industry. The same books are being sold but they are eBooks and for a much lower price. So how does that affect the publisher if they are not printed? Shouldn't a cheaper production price allow for a cheaper selling price – correct? Well yes if you are living on a day-to-day budget and corporations don't do business that way. They squirrel away profits – Apple has $170 billion in cash reserves and usually borrows to finance day-to-day operations. Now the interest rate on these loans is predicated on profits. If profits are good you get a lower rate. If bad, your rate goes up. So if these corporations quickly go to a digital product and their actual cash revenue stream doesn't match the anticipated profit amounts – in real dollars – then their interest rate goes up and they will have huge trouble on the balance sheet.

What does this have to do with Common Core? Well, they are using the test sales for Common Core to leverage the transition to a digital market. Basically, they have invented a need and are filling it in order to keep their balance sheets in the black during the transition. So once again – paper is stopping the future of digital and in that future it contains the advancement of students. Digital has a huge wake behind it and many boats won't survive.

3.6.18 Use of Tablets

The use of tablets in K-12 has been spoken about on both sides of the issue. Educational companies like Amplify (Wireless Generation) once headed by Joel Klein, ex-chancellor of the New York City school system, have gone as far as bundling their applications on a tablet and provided this complete package as a solution.

This may not be the complete solution, but it's a serious investment by Amplify to augment their software by providing a hardware delivery platform that satisfies many needs, basically a turn-key solution. There are some doubters and some believers, as you will see in these next two excerpts.

> Universities are quaking at the thought of being replaced by tech upstarts. But few people think that under-16s can do without schools and teachers; it is just that new tools could help them do a better job.
>
> *(Economist, 2014)*

Now to review how these new devices – eReaders, tablets, iPads, etc. (which are nothing more than 21st century textbooks/backpacks) here are excerpts from digital technology actually being used in the classroom.

> Computers can help students learn at their own pace, based on what they know rather than on whatever class they are in. Experts say this can be particularly helpful for a ninth-grader who reads at a fourth-grade level.
>
> *(Chandler and Tsukayama, 2014)*

3.6.19 eTextbooks

Having been in educational publishing, when technology entered the arena many believe tremendously in the use of online textbooks – what is being called eTextbooks. Having watched many children wear backpacks that looked like they were landing at Fallujah and seeing that some of those textbooks basically stopped at that time in history as well, I always believed that access to global, current, and multimedia content was the key to this new generation's ability to learn.

There have been good initiatives to bring this new archetype to fruition and commonality, but there has also been great hesitance. Here is an example of a good initiative with good results. The quote reaches out to the hopes of AI.

> … History teacher Joe Cupertino says having so much "enrichment" available in the digital text means homework is productive and "frees us to do more discussion, more analysis in class."
>
> *(Fitzgerald, 2013)*

But who is the proper market for these new devices and learning methods? It would seem the generation born right after 2010 is the prime test subject. And who are they? They are Gen-i.

3.6.20 Gen-i

This is a term to combat the simplicity of the term "millennials." The term "millennials" just has to do with a date or time and not an actual belief system. "Millennials" are just born at a certain time. It doesn't mean they live the life – it's just a date. Some actually purposely go off the grid – embracing the 1960s commune-like mentality. And "digital natives" kind of means the same, but there could be a slightly different birth date. Gen-i means that they are "in-it-to-win-it" and "they can't help it." These young people are at the core of the movement because of both age and lifestyle combined. They're in school now, shaping how we teach by how they learn, and subsequently they'll be consuming and producing by those same patterns. They are the end-user we are designing the society for today.

Here is the short definition of Gen-i devised for this book. A more extensive description will come in Chapter 4.

> Gen-i stands for generation interactive, isolated, iterative, Internet driven, immaterial, isogonics, isomorphic, (cyber) itinerant, inoculated, etc.

3.6.21 STEM and STEAM

Today there is great attention focused on not only using these new technologies, but who will be developing the future technologies. To many, it seems a step away from the liberal arts core of many colleges toward an almost vocational curriculum. To that point I argue that the combination of theory and application create a better good – at least in this point of history. Many believe that education should evolve to reflect and meet the needs of society. Today there is a focus on STEM (Dalton, 2020).

A common definition of STEM is the focus in education on

- Science
- Technology
- Engineering
- Math

Here is a more in-depth definition of STEM education.

> STEAM Education is an approach to learning that uses Science, Technology, Engineering, the Arts and Mathematics as access points for guiding student inquiry, dialogue, and critical thinking….
>
> *(Riley, 2020)*

To include the study of art and design in STEM, the Rhode Island School of Design (RISD) has helped in the following STEAM initiative started by John Maeda.

> In 2010, the college began to champion the addition of art and design to the national agenda of STEM (science, technology, engineering, math) education and research to develop STEAM—a comprehensive educational model that would better prepare future generations to compete in the 21st-century innovation economy.
>
> *(Marland, 2024)*

Many professors of design use technology to employ their design approach, some lean more toward STEAM than just STEM alone. However, many do find it difficult to get behind an initiative that was born during Dr. Maeda's tenure at RISD when they purposely undercut his efforts to further lead that institution. Why would you as an institution advocate for such an initiative, but yet do everything you can to stop the progress of the individual who brought this to life and who basically embodies this mindset? The only answer that comes to mind is a Machiavellian approach to education. *Don't actually do what you say... just make it look like you are doing this.* In many opinions, it's the basis for the demise of education and forward the demise of our country on the world stage. This approach harkens back to the political fight over Common Core. Don't actually support or denounce the initiative, just be mindful of how your approach benefits your needs – bad to the bone – and at the core if you will.

3.6.22 MOOCs – Massive Open Online Courses

MOOCs were a major disruptor and solution in the world of education. They are literally giving away an Ivy League education for free, but the catch is – in many cases no credit and no degree. You can get both but that will cost you and rightly so – but the cost is dramatically lower than a traditional on-campus degree.

The first question that comes up when we speak of MOOCs is "what the hell is a MOOC?" Well, here is a definition offered up by the *New York Times*.

3.6.23 What Is a MOOC Anyway?

> Traditional online courses charge tuition, carry credit and limit enrollment to a few dozen to ensure interaction with instructors. The MOOC, on the other hand, is usually free, credit-less and, well, massive.
>
> *(Pappano, 2012)*

And by massive, we mean massive. In some cases, over 150,000 students sign up for these classes with only a very small percentage finishing the course. So with anything this dramatic what are the reasons for such a shift? My belief is that the MOOC structure is nothing but a "loss leader" a sampling like COSTCO offers.

"Oh, have you tried the barbeque chicken?" you know the trays that are passed around to entice the shopper to buy the whole chicken. Now, I am not saying these courses are of less quality but it isn't currently the whole chicken, but with the use of new technology they can be. They are the testing ground for a new model.

These MOOCs are being embraced by some of the most prestigious and largest colleges in the world. Below is a short list of some MOOC structures, the colleges and universities who embrace them.

3.6.24 MOOC and Open Course Providers – Abridged

- Visit individual sites to view course lists
- Alison Free courses in 10+ Course Categories
- Canvas Network by Instructure
- Carnegie Mellon University Open Learning Initiative
- Class Central – Stanford, Coursera, MIT, and Harvard-led edX (MITx + Harvardx + BerkeleyX), and Udacity
- Coursera list of courses
- Curricki – open curricula
- EdX courses
- FutureLearn, the Open University's MOOC branch

- iTunesU – some courses – guide from DIY University (Apple doesn't provide a list of courses, naturally)
- MIT Open CourseWare (course materials only
- Open2Study (Australia)
- Open Education Europa, a website that aggregates MOOCs and other free online resources from European universities
- Stanford's Free Online Courses
- Udacity courses
- Udemy list of online courses
- Open Yale courses

3.6.25 Arts Reforms

In the art world there seems to be a reach back to the era of Abstract Expressionism. The form and surface of this style appeal to many personally and spiritually. It speaks to the soul, and is the best form of therapy some could ever undertake. The work of artists like deKooning, Diebenkorn, Gorky, Hartigan, Kline, Krasner, Pollock, Still, both excite and calm many in the same instance.

Today it seems this style is making a comeback. *Why?* It seems like a harkening to go back to vinyl-like music. A desire to slow down and appreciate the humanity in the arts makes total sense. According to many professors in this field, "art is meant to be subversive" and this style could truly subvert the current direction of society. To bring the physical beauty of life back from the edge of virtual(ly) everything.

Even though earlier efforts and attempts to move in the direction of this new digital medium – with work from artists like Laurie Anderson, John Fekner, and Peter Gabriel – the desire for the physical seems to always come back to the forefront. My belief is that physical versus virtual has always rested back with art as a commodity. Great efforts, but at the end of the day, how do you sell this work and keep the art economy going forward? I understand the same is true for any form of large installation work or conceptual or performance, but there are many ways to document those efforts, and that becomes a commodity. But truly virtual, that's a problem. The move toward online and gif art is fun. It takes motion into the space but has a very Dada-esque feel to it. Like being ironic about technology and taking it as a joke while the grown-ups make sure everything works. This is all done as a show of repugnancy to the world of user technology while at the same time texting each other as to the newest place to get $5 coffee. Insert NFT here.

3.6.26 Physical versus Virtual

This is a very big deal in the world of the arts, and especially in reference to the experience of viewing the immersive art experience. These installations or "shows" are now more so leaning toward Broadway productions than traditional exhibitions. However, it does seem to be doing the trick, helping with museum attendance.

> ...what would you like people to come away with after experiencing Beyond Van Gogh?
>
> A lot of people can feel intimidated by museums or feel like art is not for them. This experience allows them to connect differently with artworks by setting foot in the world envisioned by Van Gogh.
>
> *(Mark, 2022)*

Only time will tell how and if these new innovations will truly make an impact or are merely attempts to distract the viewer from the demise of the institution as a repository of value and taste.

Solutions have been attempted through many venues, both physical and virtual. One of the earliest attempts was "BitStreams" and the Whitney Artport. This innovation tried to maintain the qualities of a true museum by being extremely selective while trying to be open to change (Bodow, 2001).

3.6.27 Museums and Online Work

Extraordinary cultural icons such as the Guggenheim, Whitney, and MOMA have and had, for a certain length of time, invested deeply into the online space. With such entities as YouTube Play, Whitney Artport, and MOMA online a true effort into understanding this new space has been made since the early days of the web. It's a very difficult space to evaluate and curate since the nature of the space ebbs and flows so erratically – just like the nature of the subject matter.

The one venue that stood the test of time, and seems to constantly be reinventing itself, is the Whitney Artport. A quick description from the site is listed below.

> Artport is the Whitney Museum's portal to Internet art and an online gallery space for commissions of net art and new media art.
>
> *(Museum, 2024)*

This exhibition featured work like "People's Portrait," which was an interactive video portrait that used video booths set up around the world to capture and transfer images of people via the Internet from around the world that were then projected onto large screens and buildings. Mr. Zhang Ga's project was similar to Instagram or Tumblr before they even existed. X (Twitter) is equivalent to the work of Duchamp and Ray Johnson after they existed. All art has a legacy usually toward prior innovation. Everything is derivative – but hopefully to great concepts (Furnace, 2004).

Guggenheim – YouTube Play

This incarnation of combining the artist and the public was an amazing and gorgeous culmination of so many forms of art and diverse thesis concepts. Yes, it was video art, but the levels of thought in each piece was not just merely video; the video was just capturing the thought process and the installation was just amazing from the outside of the building compared to the inside use of Wright's architecture.

Another new piece called "Unsupervised" by Refik Anadol was exhibited in the Museum of Modern Art from November 19, 2022, until October 29, 2023. This work stood 24-feet-by-24-feet and consisted of a digital display that featured an AI interpretation of the 200 years of collected work of the MoMA. Here is a quick excerpt from MoMA's website.

> a meditation on technology, creativity, and modern art. Anadol trained a sophisticated machine-learning model to interpret the publicly available data of MoMA's collection. As the model "walks" through its conception of this vast range of works, it reimagines the history of modern art and dreams about what might have been…
>
> *(Refik Anadol: Unsupervised | MoMA, n.d.)*

3.6.28 Summary

So in the end what is the solution? How do we combine the original conceptual work of Duchamp with the endless possibilities of digital? Some believe that the work of artists like Anadol or Utterback or Rozin is some inspiring versions of crossing that bridge between physical and virtual. The concern is if we leave the tactile, do we lose yet another part of what it is to be human?

3.7 THE REAL DISRUPTORS

So far, this book has concentrated on design and digital disruptors – not really tangible solutions. In this chapter the real physical ones are introduced. Some of them have been mentioned here before, but for the most part this chapter talks about the actual physical and manufacturing products that have undergone some massive reformations.

3.7.1 Physical and Manufactured Disruptions

One of the first items should be the advent of the electric car. Re-emerging after centuries, the electric car has disrupted the entire industry and the global market.

> Electric car markets are seeing exponential growth as sales exceeded 10 million in 2022. The share of electric cars in total sales has more than tripled in three years, from around 4% in 2020 to 14% in 2022. EV sales are expected to continue strongly through 2023.
>
> *(International Energy Agency, 2023)*

The name that should be at the top of this list is Elon Musk, part of the self-named PayPal mafia. He seems like a modern-day Henry Ford and Thomas Edison combined. His work in the fields of space travel exploration (SpaceX), electric cars (Tesla Motors and Hydro Loop) is fascinating. Just think about what these current inventions will disrupt – reusable rockets that don't cost gazillion dollars – electric cars and the oil industry – traveling between cities at record speeds for minimal cost? This is the work of Nikola Tesla brought to life again through Musk, hence the name Tesla. Here is a quick bio.

> Elon Musk, South African-born American entrepreneur who co-founded the electronic-payment firm PayPal and formed SpaceX, maker of launch vehicles and spacecraft. He was also one of the first significant investors in, as well as chief executive officer of, the electric car manufacturer Tesla.
>
> *(Gregersen, 2019)*

However, since he has taken over Twitter – now "X" the platform has felt a major disruption. Advertisers have left "X" as Musk allowed for previously banned individuals to return to the platform. Falling from a $44 billion valuation to just $19 billion within 18 months (Sherman, 2022).

3.7.2 Talking in 3D

But speaking of space let's choose one that has had a huge impact, and is a combination of physical and digital – 3D printing. Ah yes, still printing… but definitely different. The technology is relatively old now, like 33 years old, but it will have an impact that spreads throughout many industries and affects many jobs.

Why do we categorize this as a disruptor? Well, think of all the things that can now be printed in 3D, from entire homes to the smallest of items. The construction industry is just now feeling the effects of building (printing) a home in days as opposed to months.

Think of how these items are going to affect their respective industries. Then, think of all the people who work in factories and plants listed below, and all the products they manufacture.

Here's a short list:

- Aerospace parts
- Architecture
- Automotive design
- Car parts
- Construction
- Dental and medical technology
- Engineering
- Eyewear
- Fashion
- Food
- Footwear, etc.

Imagine you're in space and a part broke. It could be small and simple, but it's necessary to complete the mission. What could you do? Would you send another spaceship to bring the spare part to the space station? Or, if the space station had a 3D printer onboard – voila! The part appears! Read about how this recently happened.

> Space Station's 3D Printer Makes Wrench From 'Beamed Up' Design
> The 3D printer aboard the International Space Station has wrapped up the first phase of its orbital test run by cranking out a ratchet wrench whose design was beamed up from Earth.
>
> *(Wall, 2014)*

Now let's bring that idea down to Earth. You have a water bottle top that your child needs for soccer practice or that you need for the office. You swear you brought it home, but you can't find it. You have to get to work so there isn't any time to stop and buy another. So, while you take your shower, your 3D printer "prints" the new bottle top using the plans you just downloaded. Usually within 5–10 minutes, the job is done and at a fraction of the cost. While that may be great for you, it's not for the person who works at the manufacturing plant and needs that top to get lost or go missing to keep demand up – and keep their job.

3.8 THE NEW DESIGN

In this section we will discuss a version of this new design world – interaction design. Interaction design consists of two main components – UX user experience and UI user interface design. This may seem like a tangent to the overall discussion but you need to remember what is driving this new economy is very simply put in the palm of your hand, the smartphone.

Although some web team members will design a website's front and back end, many web development companies split these responsibilities. For instance, some will work in UX to create the overall design, while others will be in charge of writing the code in programming languages such as HTML and CSS. UX designers must take into account a client's products or services as well as the target market to create a site that appeals to the client's customers or intended audience. The job requires some knowledge of web applications and programming languages such as HTML and CSS, but more so a solid understanding of design principles. Work environments for web teams vary from large corporations or governments to small businesses. These teams may be made up of full-time employees, part-time consultants, or work on a contract basis as freelancers.

The Bureau of Labor Statistics projects about 20% employment growth for web developers by 2024 (U.S. BUREAU OF LABOR STATISTICS, 2018). During that period, about 28,500 new jobs will need to be filled in an industry that already has roughly 141,400 positions. The expansion of e-commerce is expected to be the main driver of web developer job growth in the next decade. As more companies offer or expand their online retail presence, more web developers will be needed to build the websites visited by consumers to purchase their favorite products. Increased reliance on mobile search is another reason the industry's employment growth should remain strong. Here's a look at current salaries.

3.8.1 Salary

The median annual wage for web and digital interface designers was $79,890 in May 2021. The median annual wage for web developers was $77,030 in May 2021 (U.S. BUREAU OF LABOR STATISTICS, 2018).

3.8.2 Training

Some employers prefer a bachelor's degree in a computer-related field such as computer science or information technology. Web developers may also get certifications, including Certified Web

Developer, Certified Internet Webmaster, Advanced Web Developer, and in Mobile Application Development, all of which demonstrate varying levels of expertise. But other employers put more weight on a prospective employee's previous work.

> "Focus on the portfolio," Ruditser says. "Even if you don't have real world experience, you can make up your own mock projects. It's not about being polished; it's about being creative."
>
> *(U.S. News Best Jobs, 2018)*

3.8.3 UX – Defining the User Experience

UX is basically aesthetic and the function combined into one working prototype. This process involves a diverse series of skill sets. A typical UX designer needs to have a true sense of visual design – typography, image, color theory, and all of the basic elements and principles of traditional design. Now add to this mix the ability to understand how programming works and how the user prefers things to work. A great add on from the world of liberal arts would be some study or possibly a minor in sociology, psychology, and anthropology. How do humans' like to interact with objects? What are our patterns? What are our typical likes and dislikes? How do we go about interacting? Where are our pain and pleasure points? How do we make choices?

The process of UX design usually starts with a meeting to discuss goals and objectives. From then on a study and research is conducted to establish priorities and determine the stakeholders and the end user. Once these goals have been established a series of design audits take place usually dealing with the "what," "where," and "why" of the project.

"What" is the project going to be used for?
"What" needs to be contained within the project?
"Where" will it be situated?
"Why" is that the best practice?

Once these benchmarks have been established wireframes of the page division starts. Wireframes are typically done with just simple boxes or shapes establishing "real estate" on the screen. This is usually followed by image, typographical, and color theory choices. All of this is driven by the actual content (image and text or database) of the project. After this process has been achieved and reviewed, actual minor interactivity is implemented to test flow of content and user choices. This process is repeated and iterated numerous times over the course of the project.

From this quick and simple paragraph one can easily see this isn't your granddaddy's graphic design world anymore. And this is where the big names start stepping into the picture.

3.8.4 Notable UX Designers

A leading designer in this area today is the previously mentioned David Kelley. His work has been heralded for decades from his early days working with Steve Jobs to recent iterations at IDEO and d.school at Stanford. We have talked about Mr. Kelley at length but let's hear what he has to say about the UX designer and the process.

> … has a passion for doing something that fits somebody's needs, but that is not just a simple fix. …has a dream that goes beyond what exists, rather than fixing what exists… the designer wants to create a solution that fits in a deeper situational or social sense.
>
> *(Winograd, 1996)*

The one thing all these viewpoints share is the concept of preparation and planning. It appears that designers, especially the young ones, don't enjoy the planning stage they want to jump in and keep pushing pixels around until they see something they like. Try to put that time on a billable client sheet!

3.8.5 The UX Process in Greater Detail

According to many sources and as has been stated before, the UX process is all about research and preparation. Visual design is a given and the use of traditional design elements and principles are the tool set from where everything is based; however, the planning and preparation stages are where these tools work either effectively or not.

3.8.6 Maintaining the Brand Presence

One area that makes the transition is marketing the brand. One of the biggest pieces of the puzzle is how to integrate the brand presence into every stage of the user experience. How do you keep the metaphor going for a user experience project? What do you use?

Another area of great concern with brand presence is color and typography. The entire experience must reflect the brand's use of these elements. Also not to speak of the principles espoused by the brand. Examples would be the difference between an app for Ikea and an app for Overstock.com Ikea should be light and airy and very clean and concise like the product line. Where Overstock.com should kind of look like its name. Full of imagery and choices but it needs to actually work so form should never overshadow function. There needs to be a proper balance to evoke the correct brand presence while satisfying the user's needs.

If all of this sounds like advertising or marketing, quite frankly it is just that. Every time you use a product or service you need to remember that feeling and associate it with that brand. Just like any product – if you wear Nike you know they are Nike – not just the logo but also the quality of the shoe. If you use an Apple product you know it's an Apple product.

3.8.7 UX Process Step by Step

According to an article in Smashingmagazine.com – the process in its early stages goes in this order.

- **The research phase** is where you immerse yourself in the project to get the background you'll need to make design decisions later in the project. During this phase you will try to learn as much about your client's business, objectives, users, and competitors as possible.
- **The design phase** is where you work out how what you are designing will work and how it will fit together. This phase will define its scope, its features, and functionality and how it behaves.
- **The validation phase** is where you identify whether what you came up with in the design phase actually works with its intended audience. This phase is typically followed by further rounds of design and testing to solve the problems you inevitably find when you test with users. (Allen and Chudley, 2013)

As Antonelli has stated in video interviews and in writing that design will evolve into theoretical and applied areas. Slowly the lines between science and visual will blur and a good part of the development process of a project will be based on analytics and prior structures.

As you can see below, the table lists all stages of a user-centered design project, basically interaction design and the UX portion of the project. It is almost clinical in nature.

3.8.8 User-Centered Re-design Project Outline

Below is a list of all the steps to successfully complete a UX project up and until the actual visual design. The visual design includes a series of mood boards for color and typographical selections along with all photographic and icon imagery. This selection process must embrace the brand legacy of the product or service and in some cases move forward the message for that brand. The visual design processes alone will easily double the UX timeline.

UX Activity
Onboarding meeting
Marketplace analysis
UX research and document review
User research and profile
Stakeholder interviews
Presentation of findings to stakeholders
Stakeholder interviews
Prototype development
Usability testing
Wireframes, stakeholder interviews, and iterations
Functionality testing
Wireframe iterations
Prototype to visual design and development
Sign-offs
Post-production live usability meeting

3.8.9 Moving Toward a "Profession"

With all of this information and structure you can start to see how much the design industry has evolved into one of the "professions." For years, designers were thought of as a vocational – hand-craft group. Where your ability to advance in this field was based on how well you could use the tools. The actual physical tools – ruling pens, Xacto knives, tweezers, etc. – like a high-end auto mechanic. Rarely was a designer thought of as someone who worked with their brain not their hands.

The process is simple and straightforward – the key part that generates confusion is iteration – the constant repeating of the process with slight – but very critical changes.

To that end most of the time on the project can be spent on analyzing and developing the "why" as well as the "what" and "how" components of UX. Using market analysis as a guide the team starts from that analysis to determine the goals of the project and build from there with the typical steps of the project process – design audits, wireframes, digital assets, prototyping, and iterations.

After this process is complete usually the team then starts the actual build of the project incorporating and then iterating from analytics and trends. As with any project, end viability is one of the main concerns.

Once the project has been developed the team has a presentation event, which includes a written document, short concise deck presentation, and the actual working prototype.

Lots of work goes into making sure you can order that "Half-DeCaf-HalfCaf-PumpkinSpice-Latte-Vente" while you are on your way to meet your friends.

And here is a recent statistics forecasting the future UX design:

> According to the US Bureau of Labor Statistics, UX is set to *continue to grow in demand* for the next eight years. This is also due to UX becoming a Google ranking factor, meaning businesses will need to focus on their UX to remain competitive.
>
> *(Hannah, 2023)*

3.8.10 Analog to Digital Again

Now as the world goes from analog to digital some still consider themselves "pixel pushers" not "concept pushers." Analog designers are having issues here – the process eludes them. In many opinions and through the experience of teaching groups of students you can see who "gets it" and who "doesn't get it."

The next group that makes the conversion is the "creative" types – the ones that come up with good concepts and sort of strange solutions. They start to move in the right direction but then get stuck and can't get past the simple ten screen projects. As the projects get more complex, they lose interest or can't seem to realize the importance of finishing.

Another group that gets it and sees their place in this new venue are the "truly creative." The ones who can maybe write the concept out and see it but may have not studied the software or may not have the completely analytical brain to make the next steps possible. Their advantage is that they see the pieces in the distance and know how to make them work together. As the popular saying goes "I play the orchestra." They recognize the skills needed to enlist and motivate those individuals and always keep focus on project goals and brand presence. They can place the pieces of the puzzle in line. They can usually also see the variations and the "what ifs"? This group sees the future and realizes their role in that future.

Analog has dropped to second place and digital now has taken the lead. How we teach and learn design for that transition is the basis of the new design frontier and where the future lies for most designers and all students – young and old.

3.9 SURFING THOSE WAVES

3.9.1 Being Digital – Again

In conclusion of this chapter there are still many questions left to answer. However, there doesn't seem to be any hard answers, just a strong hunch of the future direction. Let's use an analogy to explain. Many times in movies the "surfer dude" is shown as an airhead. Just somebody who hangs around and dreams of that big chance – gnarly! A surfer once explained what actually goes into surfing and surviving on a piece of Styrofoam out in the middle of the ocean. It easily could look like the surfer is wasting time walking the beach, just staring out at the ocean. In reality, they're trying to figure out what the waves are doing that day, and before they go out on that board, they take the time to look around contemplating all the angles. They don't know which direction the wave will actually break, but they have an idea based on their experience and on what the ocean is doing that day. Afterall, surfing is surfing, no matter where you do it!

3.9.2 Is It King Ludd Again – Are They Burning the Factories So That the Work Cannot Be Automated?

This question has to do with the time in England when the weaving industry turned toward automation. This disruption quickly diminished the number of "weavers" needed – this was for a major occupation at the time. The legend goes that the displaced "weavers" would go around at night and burn down the factories that adopted the new technology. Since no one knew who was actually doing the destruction, the name of King Ludd was thrown out. This is where we get the term "Luddite."

> The Luddites organized guerrilla raids to smash those machines – on punishment of death – and won the support of Lord Byron, enraged the Prince Regent, and inspired the birth of science fiction. This all-but-forgotten class struggle brought nineteenth-century England to its knees.
>
> *(Merchant, 2023a)*

Is this what is happening today? Are educators purposely turning away and discrediting the use of technology in the classroom as insufficient and unusable to basically "burn the factories"? What will happen once ChatGPT and all of the other AI bots come to the forefront? What will happen economically to the nation if the educational institutions do not modify the curriculum to a STEM- or STEAM-based curriculum?

Well according to a recent study done by the White House, the United States significantly lags behind most modern industrialized countries in the inclusion of STEM or STEAM into current curriculum (Prepare and Inspire: K-12 Education in Science, Technology, Engineering, n.d.).

According to the same study a focus by educational institutions on STEM- or STEAM-based curriculum the following should and could occur if it was embraced and integrated into existing coursework.

And where it has been embraced and put into action there are some strong results all the way from the automotive industry to the National Endowment for the Arts.

> ...technology can be a powerful driving force for innovation in education, ...Moreover, technology has been advancing rapidly to the point that it can soon play a transformative role in education.
>
> *(US Department of Education, 2017)*

3.10 PS – GET IT "POSTSCRIPT" – NERD HUMOR

The main question of all that has been written is this – It is 2024; we are one-quarter the way through this new millennium and over 50 years since the birth of the Internet. When are we going to embrace these major shifts? What will it take to move forward? Why can't we as a society see that what we are doing by holding back this new Digital Incunabula is creating a failed structure in the terms of information, publishing, promotion, entertainment, education, and the arts? Have we learned that not changing the core of how we live, teach, and work toward a new global effort will push us far off the world stage?

This is why many good leaders push so hard for the "why," not only the "how." But this time, why are we so concerned? Because this time the ripples are hitting the shore harder, because this time there is great chaos surrounding us, because this time the waves are crashing around us from all angles, because this time to paraphrase Robert Duvall in *Apocalypse Now* – "We all better surf."

The bots are coming and they are coming at us hard through AI and other forms of automation that will replace traditionally repetitive and mundane tasks regardless of the profession from health professional to computer technicians if the information is repetitive and can be input to a database for review, the job may be gone overnight. Couple that with quantum computing and "Houston, we've got a big serious problem."

Now here is the biggest question – If you were going to advise a child as to how to invest in their future, what would you choose? Now let's make it personal – if you were to advise YOUR child as to how to invest in their future what would you choose?

So here goes a simple recommendation for students, parents, and anyone getting started or changing a career. It comes from the 1967 movie *The Graduate* and it always plays true as it did with Negroponte's quote of "bits or atoms" and Negroponte actually quoted the scene in one of his interviews.

Benjamin Braddock, Dustin Hoffman's character, is told.

> "I have one word for you Benjamin – Plastics," says Mr. McGuire.
> "Exactly how do you mean?" asked Ben.
> "There's a great future in plastics, – Think about it. Will you think about it?" replied Mr. McGuire.

But there is a great future in "Glass" and "Titanium" or "Qubits" or whatever material your device is made from or works with. That device holds your entrance into a digital world, just get involved, know "how" it works but better than "how" know "why" it works, so you can create the next one or simply the work that goes on it, because you actually DO hold your future in your hand – everyday. Just try not to butt dial it.

4 Veal
The Society

The rise of generation interactive – breeding a new consumer class.

4.1 CHAPTER PREFACE

In his recent book *The Anxious Generation*, Jonathan Haidt goes into detail on the effects and the correlation between smartphones and the mental health issues concerning this current generation of 12–25-year-olds (Haidt, 2024). These facts appear to be indisputable; however, Haidt never gets to the real cause. He names how the smartphones became predominant at the same time that these issues started but never explains why they were in the hands of children at such young ages. Yes, there was peer pressure and social compliance but weren't there always? What necessary external force made parents capitulate this time?

We all see the effects and cannot deny that they exist but why did it happen now? And why were these children left to be prey to these devices? Yes, it is agreed upon that this online world is addictive, and yes the lack of available parenting was one cause but what caused that lack of parenting? Surely no parent ever would knowingly subject their children to such harm so what actually happened to cause this impactful neglect?

The purpose of this chapter is to investigate and discuss the premise that the current generation was constructed to be consumers for a transitional marketplace. This wasn't by chance or even purposeful parental neglect. This was a position those parents were put into by outside forces.

As the economy shifted from analogue to digital, consumers had to be trained to accept, use, and progress within a new economic model through changes in societal and economic patterns. During the course of this book those patterns will be discussed and displayed as a confluence of:

- Marketplace manipulation
- Abusive use of technologies
- Lack of governance

In the following sections we will discuss how those events are reflected in the habits and lifestyles of the current 12–25-year-old demographic globally and how it has caused them to be the consummate consumer of digital goods based on events that have been created to develop them to be consumers and to be consumed. One of the first points questions whether this was the fault of parenting; in many opinions – no, it was more so the position families were placed in and how they best could survive. When many events come into play and seemingly conspire to force families to reinvent themselves, it is not so much the fault of the "herd" but the result of the "rancher."

In developed nations globally – the new economic models are being driven by the consumption of a digital product base from a younger and younger age group – is this really true? Well ask yourself if you believe adults age 35 overconsume as much digital product as that 12–35 demographic – and even younger ones?

4.1.1 Why Look at the Creation of This Consumer Class?

To answer those questions from the preface we need to look at economic markets to see if they are true and change naturally through a free market system or are they manipulated to intertwine and work together to the detriment of society – specifically the current 12–25-year-olds?

DOI: 10.1201/9781003404347-4

- Have markets been manipulated today more than ever?
- What is the new economy and is growth totally "digital"?
- Has any effort been made to stop these events?

These questions raise further questions, but we need to have the facts first. Has the price of the housing market been manipulated to create the two-income family as a necessity? Has the basis of the economy shifted to a "digital marketplace"? Has there been any manipulation of laws to allow this to happen?

Usually, most free market pricing is based on supply and demand. The only time the population of the United States grew substantially was during the "Baby Boom." But even then, housing prices were affordable and a one-income family was sufficient. What happened and why?

Why were laws like "Glass-Steagall" overturned right before a huge housing market crash? What actually is "Glass-Steagall?"

> The [Glass-Steagall] bill was designed "to provide for the safer and more effective use of the assets of banks, to regulate interbank control, to prevent the undue diversion of funds into speculative operations, and for other purposes."
>
> *(Maues, 2013)*

The repeal of this bill was arguably one of the largest deregulations of the financial services industry in U.S. history. Essentially, it allowed members of the financial services industry to enter each others' businesses (Vol. 42, No. 2, February 1987 of the Business Lawyer on JSTOR, n.d.).

And why has Net Neutrality – the ownership of the Internet – become such a "hot potato"? And have other developed countries allowed their next generation to become prey to this new economy?

4.1.2 Why Is It Important to Understand What and How It Happened?

First of all, these changes directly affect our children and usually nothing is more important to our lives than them. Second, their future as citizens should be our main concern even if we have no children of our own.

But what are we talking about here? The focus is on their well-being and future and how possibly an opportunistic or required change in economies desperately needed a new consumer.

Have those changes come together in a "perfect storm" by chance? Were they constructed to become a "perfect storm" and not just an opportunity to be grabbed? Ockham's Razor states that "among competing hypotheses, the one with the fewest assumptions should be selected."

The following sections attempt to construct that hypothesis, based on the most complete version of the research possible.

4.1.3 Scaring the "Herd"

4.1.3.1 What Were the Pieces of This "Perfect Storm"?

Nothing scares people more than the inability to provide basic care for themselves or their loved ones. And that starts with providing shelter from the storm. What were the pieces of this "perfect storm"?

4.1.4 Basic Necessities

For the first time in the history of this country housing prices responded far further than market demand. And in response to that surge in prices a two-income family was necessary for most households. Even during the baby boom, marketplace housing prices were affordable enough to maintain a one-income family structure. Since 1980 it was deemed a luxury for the family to have just one income, thus necessitating a non-traditional process for child rearing. In other developed nations, this has been compensated by parental leave, healthcare, vacation time, etc. Was this market manipulation made possible by the repeal of the Glass-Steagall Act – a law that prohibited the banks and investment companies from merging funds?

At the same time during the 1980s the rise of the home entertainment, game console, and personal computers occurred – a venture with great possibly positive outcomes, but used primarily by many as indoctrination to habits, that at the same time coupled with a lack of supervision contributed to a dramatic rise in obesity (sugar-based diet) and a diagnosis of ADD and ADHD.

During and prior to this "perfect storm," housing prices became the target of many unscrupulous groups – not our traditional criminal types arguably – but what was our governing body. And the battle over the ownership of the delivery vehicle – the Internet known as "Net Neutrality" has been fought and lost.

Some discount the concept of a "conspiracy-like" happening; well here is a simple definition.

> Simple definition of conspiracy:
>
> : a secret plan made by two or more people to do something that is harmful or illegal
> : the act of secretly planning to do something that is harmful or illegal
>
> *(Definition of CONSPIRACY, 2019)*

So let's ask the question, does what happened in the 1980s and recently in 2000s fit that definition? To answer that question, we will review the research on the timing of repealing the Glass-Steagall Act and articles on the "Keating 5" and "The Big Short."

How did this "perfect storm" create a new economy? How did it affect our educational and societal models? How did all of these ripples affect our children – physically, mentally, socially, and educationally – specifically the lack of technology in the classroom and how the children of "generation interactive or gen-i" became frustrated with traditional teaching methods and practices to the point of intolerance and malaise? How much of this is the fault of educators and the educational system? This was brought right up to the forefront during the COVID-19 pandemic.

Right or wrong you need to speak to the student's experience.

> If we teach today's students as we taught yesterday's, we rob them of tomorrow.
>
> *John Dewey*

Are we – as "digital immigrants" – not speaking to these students in their – "digital native" – language and, therefore, are we causing the conversation to be one-sided. This is just one line item in a much bigger list of issues but one that can be remedied. Have they been pre-programmed to respond only to certain stimuli? Are we not fluent enough in that language thus creating too much distance between the generations? Did we have enough time to assimilate these changes or has the speed been too fast and the time too short? So they tune out the same way many in the majority age group and ethnicity tune out when Salma Hayek or Javier Bardem accepts an award and speaks in fluent Spanish. We nod and say isn't he or she beautiful and never care to understand what she is saying. They just nod and think about something else – usually with no response. This has always been the case in school, even in college, but today there seems to be a flip of the 80/20 break down. For years 80% paid attention at least somewhat with 20% lost in the ozone, but today it seems that that has flipped to 20% paying attention and the 80% being immersed in distraction and just general malaise. We will discuss what many feel is the cause of this in later sections.

A fellow educator had the same analogy. They said that the current lack of familiarity that educators have with social media creates a void in prescribing sustenance similar to bringing children to an exotic food store but not explaining how nutrition works and allowing them to just eat whatever they want. We are feeding them but allowing them to just grab whatever it is they want. We need to learn the store, but it is much more difficult than usual because they already know the store and our ignorance invalidates our credibility to govern their choices. So in avoiding learning about the store, we give an even greater validity to our lack of credibility.

Parents who need dual incomes with children who fit snugly into the 12–25-year-old demographic have witnessed this from the outside and the inside while always trying to do good work and balancing the possibility of addiction and demise of current qualities of life as best they could with dual responsibilities. And has the battle over Net Neutrality and Safety occurred primarily to control the delivery vehicle for this new digital marketplace?

In this chapter and throughout this book, facts will be presented and we will try to create a scenario that accounts for this new "veal" generation. Why is the "how" important? Well first of all so obviously we can be vigilant to stop it from progressing further. And to also know what predatory practices like the late night educational infomercial remedies are real or fraudulent. How many learning center or educational product commercials have you seen to help Johnny or Lexi learn to read or do math better? How many proposed plans for better educational initiatives have you read about or voted on at your school board? But yet nothing seems to work. Common Core initiatives – yikes – will go on for years and years, ruining generation after generation as the ed. tech pirates sweep in and pillage the district.

Didn't we learn anything during the recent pandemic when schools and learning were frozen for two years? Didn't we see the effects of this loss of time and isolation? What have we done to ensure that it will never happen again?

The goal here is to try and see how this all happened? Why and how were the markets manipulated? How does something like Flint, Michigan's lead in the water happen? Where have our tax dollars been going all these years? Is this just a full swing of the pendulum or has the ability to see deeper via better transparency – again by the use of social media allowed for this to be unveiled? What are the signals/routes to stop it and to turn it all around?

4.1.5 Why "Veal?"

Many people seem to hate the name of this section – "veal." They have found the analogy rude and offensive. Good – now that we have your attention let's talk it out. In advertising you learn if you want to cut through the clutter and get to the core of your message – keep it simple and minimal – and one word usually does that.

Jerry Della-Famina (a very famous adman from the 1960s and on) has a story of when a Jewish cemetery group wanted a new campaign. Supposedly when the presentation day came around, he unveiled his new slogan for them, one word – "Plotz!" In Yiddish it means to drop dead instantly and also similar to plots. May not have gotten the account but he did get their attention.

Quickly gaining your attention was one of the first goals but also it is to establish again who is who? Obviously the "veal" is the new generation of consumers. And the "herd" is the society from where it was conceived originally. But the one part that seems to keep getting misunderstood is the "rancher." Who is the entity that is controlling this process – the "rancher?" Well in most cases of rearing livestock the "rancher" is the one making the immediate decisions based on the "marketplace." So to that end the "rancher," in this metaphor, is being directly influenced by the three forces outlined before in the forward.

- Marketplace manipulation
- Abusive use of technologies
- Lack of governance

Now did these three forces work in perfect concert? Can it be proved? Well when this many things come together so seamlessly one has to question. Something so simple to do like getting clean water to certain parts of this country can't happen but somehow all of these complex things just happened by sheer chance?

Is getting your attention and bringing all of us out of the haze of the Kardashians and Reality Television just part of the goal? Getting you to take a step back and ask, "What is actually going on here?" That is the goal of the title, and hopefully it will be substantiated throughout this section.

So why pick the name "veal?" Why use the term "veal"? Because veal by definition is a young cow that has been purposely raised in a manner to create a specific end product (Gov, 2024). The calf had many other outcomes but the rancher purposely engaged a process for a specific result. Just to be clear here again – the "rancher" is not the parents. It is the CEOs, specifically of tech companies responding to the societal governing body that allowed this to happen through the confluence of laws and practices. These "ranchers" order other associates like mortgage lenders, software developers, and local politicians to follow their lead.

The parents are the older cattle part of the "herd" (which created the calf), they are responsible to a degree. But you can't blame someone for not tap dancing well when you keep tying his or her shoelaces together.

4.1.6 What Is Veal?

Veal are young calves that have been taken from their mother at a young age and then put through a specific process to develop them into a desirable form of meat.

The veal are then put into crates and kept in a small space while they are fed a milk-like solution and actually massaged to increase the tenderness of their muscles. Veal are often killed 16–18 weeks after birth.

Below is a simplified list of what creates veal.

- Born
- Separated from mother
- Moved elsewhere to be raised
- Kept in a small space
- Stimulated just enough to create desired end product
- Once ready sold off in pieces and individual prices

(Humane League, 2022)

If you believe that this generation was created as "veal" to consume and be consumed by a new digital economy, the question now is how and why did this happen to this generation? Why was it necessary? What were the steps that were put into place and how does it all relate to creating this new economic model?

4.2 MARKETPLACE MANIPULATION

4.2.1 Creating an Economy

Throughout history economies have usually responded to human needs. A need arises and a service or product is then built to fill that need. Simple supply and demand – here is it explained for sixth graders, purposely kept simple to ensure understanding. Of course there are many variations and growth areas but at the base level it is just simple – you have something and people need it, not just want it.

> Generally speaking, the price of something will go up if the demand goes up. Why? Because the seller thinks he or she can get more money for whatever he or she is selling.
>
> *(Supply and Demand: Basic Economics Part 2, n.d.)*

But what happens when the market completely changes? What happens when the market is destroyed? An example is the advent of the automobile. Horses gave way to cars to travel expediently, and then the gasoline market became relevant. Before that gasoline was considered a "throw-away" from the oil production process.

> Gasoline was around before the invention of the internal combustion engine but for many years was considered a useless byproduct of the refining of crude oil to make kerosene, a standard fuel for lamps through much of the 19th century.
>
> *(Constable and Somerville, 2003)*

It was thought to be of no value. Imagine gasoline – the substance that literally and figuratively drives our economy thought to have no value? We will discuss the effect of these changes later via Joseph Schumpeter's model of Creative Destruction.

The belief, and the purpose of this section, is to demonstrate that for the first time in history a consumer was scientifically and consciously bred for the sole purpose of being a consumer who would support and drive a new environment – one in which all previous processes have almost no relevance, an economy with no need for store shelves, clerks, salespeople, local warehouse workers, or packaging. How would an existing generation of purchasers, who were used to holding an object, find something that they could not physically hold worth their money – would they embrace it? Understand, we have always had a service product economic model – life insurance – retirement products – but even within something as ephemeral as the arts we could hold the album, the painting, the film, and the video. We had something to point to and say there is my money's worth on that shelf, or that wall – there is what I bought.

Walter Benjamin's *Art in the Age of Mechanical Reproduction* (Benjamin, 1936) speaks to a digital product, using the discussion of replication and value, but now how do we validate the actual digital? How would an existing generation with all of the purchasing power help fuel a new economic model that is intangible? Would that generation see any validity in purchasing something that has been called "ten thousand songs that fit in your jean's pocket"? This is a hard proof of concept and one that, arguably, could have died on the vine. So a new generation had to be created. We see that today through the value this generation puts on cryptocurrency and their renin – NFTs. And we see how that has turned out – starting with FTX and Sam Bankman-Fried.

Many industries have had a very tough time accepting this new model and many would fall away under this process. Once you read Schumpeter's Creative Destruction theory you will see how this evolutionary process has gone on for decades. Those that did not evolve would simply close shop and disappear – think horse buggies and cars.

Today you scroll through a playlist and amass followers for a product. You point to Google analytics or social media "likes" to measure worth. How was this going to translate into value from the basis of an analogue economic base? When or has this ever happened before? Think "Pet Rock" or any fad – usually short time frames and minimal value. Rarely, if ever has an entire country's economy been dependent on a fad or non-existent commodity.

Recently HBO aired a documentary on Jerry Weintraub. In the documentary Weintraub recounts a meeting with his father, who came out to visit his very successful son, now an entertainment producer living in Los Angeles. According to his obituary in the *New York Times*,

> Jerry Weintraub, a consummate showman whose up-and-down career touched musical entertainers as grandly diverse as Elvis Presley, Frank Sinatra and Led Zeppelin and screen artists who included Steven Soderbergh, Robert Altman and Michael Douglas, died on Monday in Santa Barbara, Calif.
>
> *(Jerry Weintraub, a Force in Film and Music, Dies at 77, n.d.)*

The senior Weintraub was a jewelry salesman from Brooklyn who would concoct wild stories about his products to create buzz or basically what we call today social media marketing. In the meeting between the father and son, the elder asked, "where is your inventory?" – the father was looking for physical items. Weintraub, the son, states in the documentary that he just pointed to his head and said, "it is all up here."

This is a great example of the product being stored in a cloud and used on demand and never being able to go out of stock, basically the ultimate computer storage system and one that is referred to as the (Alan) Turing test – measuring a computer's speed of calculation against the human brain. Weintraub was setting the benchmark for that test in the media and entertainment field obviously others would do it in their respective fields but Weintraub's portfolio of work had the greatest mass appeal reaching millions upon millions of "users." But it wasn't an entire country's economy and it didn't change value across the globe. Let's look at when that did happen because this chapter is about challenging the premise of our current digital world and what had to happen to redirect a global economy from a very personal and pedestrian perspective.

As an example let's look at the earliest recorded occurrence of a fad-like economy – but again physical. In the 1600s, Holland experienced what has come to be called "The Tulip Bubble."

"Tulpenwoede" (tulip madness) drove up the prices of tulip bulbs. At one point they were more valuable than houses. Historians believe this was an invented market by government officials to drive up the price and to cash out once the time was right. Sound familiar? However, they kept a simple clause in the edict. If the price was not high enough the investor could simply cancel the contract and leave the other party quite literally holding the bag or bulb. Yes, the first economic market bubble was created in 1637 ("Was Tulipmania Irrational?" 2013).

Well it does seem like this scam has been going on forever. And usually a war brings us out of an economic depression. If we have no economy we have a war – think the Great Depression to World War II to the prosperous 1950s. Think about how once the Military Industrial Complex was set up at the end of World War II how it was then exercised in the Korean War just enough to test it out. Then once we knew how the pieces worked we had Vietnam, the never-ending war that finally got resolved once Nixon/Kissinger opened the gates to trade with China. It would seem they were testing the process then started going full steam ahead.

Now think how once the "Dot com" boom got really rolling downhill how that was quickly stopped by 9/11. Economies drive countries, and conflict between countries drive economics and once that economy was challenged by a new economic model that had very little to do with the previous structure, the pudding hit the blender.

So here are the questions. Are all economic trends based on need or want? When one is destroyed does the next one go forward via a new model or simply restart on the same premise with a new name, story, new government regulations, and deregulations? Are regulations ever benevolent or always predatory? Has any economic model actually needed to create their customers before? One of the theories that seems extremely relevant is that of Joseph Schumpeter – Creative Destruction.

4.2.2 Schumpeter's Model

Here is a simple explanation using as few words as possible to best understand. Schumpeter's theory of Creative Destruction basically states that new inventions or innovations that move the society forward can and typically destroy those innovations that came before. Many economists consider it the free market's means of generating progress and basically an inherent part of that system (Durlauf and Blume, 2008).

This section points toward the problem and how the companies seeing their own demise may have reviewed the process and tried to make sure they themselves didn't fall off the cliff and in doing so ensured their own demise. Worried about lost profits and closed jobs, these companies try to fight off what is really inevitable.

> Instead of going out of business, inefficient producers hang on, at a high cost to consumers or taxpayers.
> The ironic point of Schumpeter's iconic phrase is this: societies that try to reap the gain of creative destruction without the pain find themselves enduring the pain but not the gain.
>
> *(Alm and Cox, 2018)*

4.2.3 How Is This Working Today?

Given the previous years' unemployment rates at nearly double digits and the lack of upward mobility for the current generation, why has this part of Schumpeter's theory – the resistance to change – seemed to have been adopted by so many companies? Why is the transition to a digital product so slow? Just look at GE's new ad campaign featuring coding as the new product line over the behemoth-like physical products of their past. They are starting to get it and AI may be the work around for them or the event that finally forces them into a digital age.

A belief is because the right consumer was not in a buying power position. But also because these companies had analog debt to pay off first. These new consumers had to be bred; for even just growing up with an old mindset would not be enough to guarantee success of this new digital economic model.

Analog debt – most media companies and real-products companies have leveraged loans based on actual dollar profits, so even if you moved 1,000 units and they were digital you didn't make the same dollar amount as an analogue model – streaming model versus archival model. The actual profit margin (in percentages) is much higher but the actual dollars on the profit statement were lower. This triggered all sorts of problems like higher business loan rates and lower stock prices; which then required a greater profit margin in respect to dollars spent, typically in reducing labor costs. So more hours for less pay – to me – would be a formula that led to the stagnation in wages we have witnessed for the last decade. Simply put corporate greed.

This is why there is such a great divide between the rich and poor today – probably wider (according to Reich's bridge chart) than has been in the last 100 years? Causing society to go backward toward a 1–99% economic model – witnessed last only prior to the Great Depression of 1929 (Reich, 2012). Was it done for a reordering of the economic landscape based on the need for growth in Third World countries to First World status? And is that needed while the First World switched from a base of raw materials economy to refined digital products by using individuals as the raw materials? Remember the quote concerning how important the user is in the mix.

> In 2015 Uber, the world's largest taxi-company owns no vehicles, Facebook the world's most popular media owner creates no content, Alibaba, the most valuable retailer has no inventory and Airbnb the world's largest accommodation provider owns no real estate.
>
> *(Goodwin, 2015)*

This is evident today and it is growing and not receding. Why is this important to a designer? Well again to quote Jef Raskin – "The interface is the product."

So today, interaction design and those team members – UX (user experience) designers and the lot are a driving force behind the new economic model. They design interfaces solely for use in a digital space. This career area has seen unprecedented growth in the last few years, with diverse career opportunities clearly outlined as in the diagram below. It is the face to the algorithm that drives this new economy.

In any economic model we need to establish the raw materials and the production base. In this model one of the main components in that scenario is the user. The coding can be done, the devices can be made, the products can be shipped, the bills can be paid, but none of this will happen without a compliant and sophisticated user. And since the current user a generation ago didn't have the sophistication, one had to be built.

For example, look at the profits of two totally different economic models – analogue and digital. One company relies upon the extremely complex process of refining extinct carbon-based life forms and the other allows for users to catapult imaginary "birds" at other imaginary objects. Just look at the numbers.

> Exxon Mobil Corporation today announced second-quarter 2023 earnings of **$7.9 billion**, or $1.94 per share assuming dilution.
>
> *("ExxonMobil | 2Q 2023 Earnings," 2023)*

> Apple today announced financial results for its fiscal 2023 third quarter ended July 1, 2023. The Company posted quarterly revenue of **$81.8 billion**, down 1 percent year over year, and quarterly earnings per diluted share of $1.26, up 5 percent year over year.
>
> *(Apple, 2023)*

Again why is this within the purview of a design? Remember – "the interface is the product" – this has never been truer than today. The new economy is digital and growing. In 2022 the Christmas season hit $240 billion in online sales? Is it because the "Gen-i" group has now hit the real purchasing level and why now? The next section hopes to answer these questions.

4.2.4 Born – Second Baby Boom?

In the years from 1947 to 1964, the United States experienced the largest growth in birth rates since the 1900s.

> ... the U.S. birth rate peaked most recently in the Baby Boom years, reaching 122.7 in 1957, nearly double today's rate. The birth rate sagged through ...1970's but stabilized at 65–70 births per 1,000 women for most years after that before falling again after 2007...
>
> *(Livingston, 2012)*

But since that time the rate dropped and stayed at that new consistent rate until 1990. Then you can see a short burst of growth and then a plateau. You will also notice that there is pretty much a flat-line from 1970 to 1982. Then surge and drop until 1997. Then surge and drop again from 2000 to 2008. No huge spike like the baby boom era. No huge need for housing increases like the era after the war. So now here is the next question – why the huge surge in housing prices at these times. The average home price in 1974 was $39,000.00. Population in the United States in 1974 was 213,853,928.

> Overall, the U.S. population grew by 1.26 million people last year, or just 0.38%. Now, while this does represent a slight uptick from previous years that's really not saying much as U.S. population growth hit 100-year lows in 2019, 2020, and 2021.
>
> *(Airgood-Obrycki, 2023)*

Please see the next chart – a coordinated series of spikes with reference to birth rates – more families means more need for housing which means prices go up whenever something is in demand. So, why the spikes in housing prices in the 1980s and again, in the 2000s if the spikes are not in coordination with the demand?

Remember the bulleted process of creating veal.

- Born
- Separated from mother
- Moved elsewhere to be raised
- Kept in a small space
- Stimulated just enough to create desired end product
- Once ready sold off in pieces and individual prices

Were the 1980s and the 2003 housing market done purposefully and not just a free market fluctuation? Well "why" and "how" would you do that? The "why" is obvious it was to create an opportunity to capitalize on this vulnerability and also to generate huge profits for the lenders.

How do you create a need for extra revenue? You increase the cost of basic necessities to a level where one income will not cover the monthly costs. This is a very simple equation – choose between a very modest lifestyle – what you think is depriving your child's needs or believe you can make up for time lost. We even came up with a term – "quality time," hell, all my time is quality.

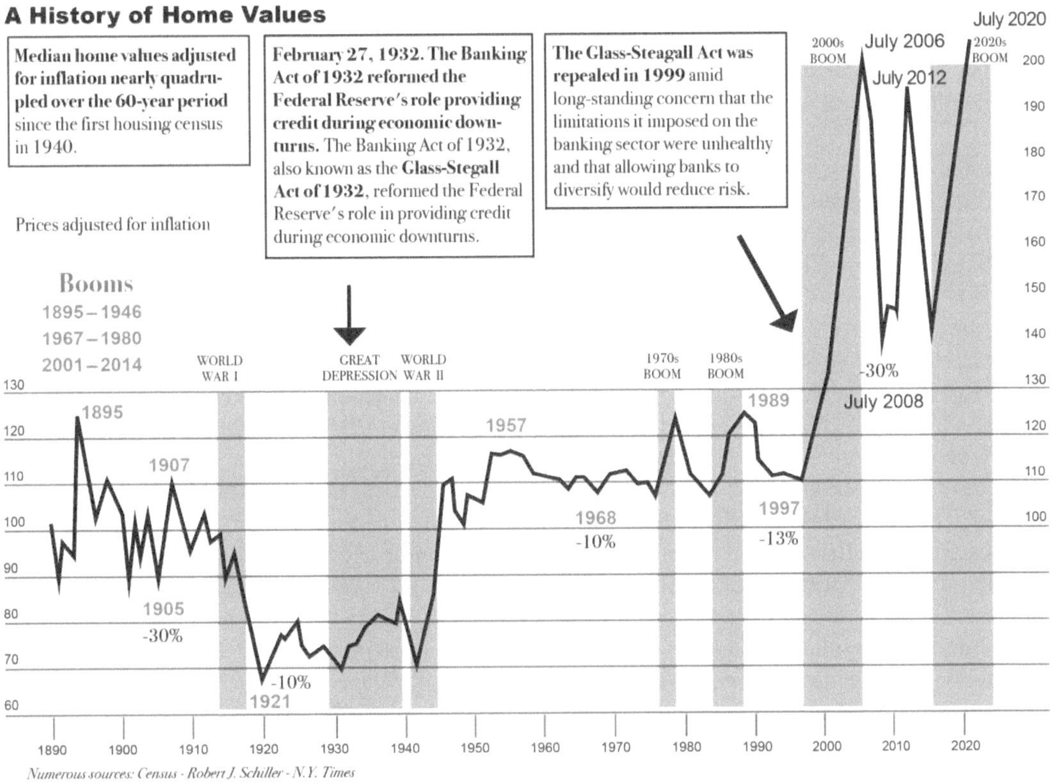

What is the simplest most effective way of creating that necessity? Focus on one main payment that goes to one main source – mortgages and mortgage lenders. This causes the separation between parent and child – mother/father and calf. Many times the guilt is placed on working mothers, quite the opposite. Mothers are the backbone behind the family today. Often the working mother has to have dual roles – full-time job, and full-time parent. The "how" is determined by who has control over the pricing. Just like the "block busting" tactics of the 1960s. Realtors and lenders work together to start a new norm within the marketplace. Who benefits can be described through this scenario. Just like when a crime is committed the questions of motive and opportunity come up. Remember the repeal of Glass-Steagall came right before this spike.

First establish the crime – the murdering of the American Dream, as numerous families knew it, across the United States.

Next opportunity. Who had the opportunity to raise housing prices and to create a situation where the money was flowing from the pockets of families to the banks in the form of mortgages? The realtors and the banks did – didn't they? You have no job, no credit, not enough savings – not enough of a problem to stop the faucet of free-flowing home equity loans and mortgages. This all worked great until it was time to resell or until the jobs that those mortgagees worked at dried up. Just like fattening up a calf before turning it into "veal." A great bit of dialogue from the movie *The Big Short* – "they're all walking around like they're in an Enya video" and "they're just concerned about the game and dancing shows on TV."

In this next section we will further discuss the mechanism and the effects of that breakdown of the American Dream as a reality.

4.2.5 Separating Them from "Herd"

4.2.5.1 Weapons of Mass Distraction

Throughout this section there are comparisons between the construction of this demographic to that of raising "veal." Starting with birth, it was determined that the market need for care and housing rose dramatically and now following the comparison further comes the time of separation from the herd.

Did housing prices – mortgages – cause these children to be separated from the herd? Many would think the answer is yes. Many times you will see at a child-care center two carts with about eight children riding in them slowly rolling around a parking lot or a park. The attendants wheel the kids around for about 10 minutes, and then they go back inside. This is not a normal childhood event. Prior generations remember being outside so long in good weather that they got a suntan just from sitting in a carriage.

The need for surrogate parenting has risen dramatically in the last 30 years, seemingly in proportionate to mortgages. As the monthly payment grew so did the need for these centers.

Child-care centers have grown over 300% since 1980. (Biery, 2014) In 2023 there is a large drop in available child care. Has this rate been reflected in other areas?

A house purchased in 1987 was approximately three times the cost of a similar house in 1978. Then sold in 2004 it had grown to three times that price. All of this is illustrated by the previous statistics. They call them "bubbles" but are they real or manufactured? Did the increase in cost at 3× the previous price drive up the use of child-care centers by 3×? And what effect does this have on families?

How much of an influence has this surrogate parenting had on our children? Were they stimulated or were they just left alone all day? How much was electronic and how much human? How has the childcare industry responded in growth to the needs of this new economy?

> The number of child care facilities has increased greatly over the last several decades, with U.S. Census Bureau figures showing an increase to 766,401 child care facilities in 2007 from 262,511 facilities in 1987. Demand has been driven by increased numbers of working women, the Census Bureau says.
>
> *(Biery, 2014)*

What effect has this had on these children, now that we have had a large portion of a generation who were used to not being told what to do and when to do it simply got older. Over 60% of mothers with children under three are out of the house either working or looking for work. With these numbers you can see the need for childcare increasing dramatically (Biery, 2014).

Some believe that the time in daycare is more structured than time at home. But is this time structured for the child or for the child's caregiver? How much was this increased via the Covid pandemic and the use of search engines that supposedly replaced teaching?

Many of these children, whether at home or in daycare, were raised by electronic devices and were given free reign over their own time. Statistics show that Americans are working way more than the typical 40-hour week. E-mail, messaging, computers in general do not leave time for "down time." Even on vacation American workers check their e-mail ten times more than any other worker in the world. So where is the "quality time"? Imagine working a 60-hour week, then you come home and there are typical chores to do, then dinner and how much time is left to actually parent? So it's time to get plopped in front of the electronics or, worse, just left to their own devices (literally) and discretion. Now imagine being a teacher and trying to tell Johnny he has to do what you say to do? We all know how that is working out today. Once students returned to the classroom after the pandemic many teachers reported that it was almost impossible to teach them or to even speak with them.

Was this manipulation of an entire generation a coincidence or is it like when land is sold to new homeowners and the original owner strips it of all natural resources – minerals – topsoil-lumber then sells it back to the new owner?

The housing market is like that land in many ways. It is a basic need and instead of it being used to generate economic benefits for all – neighborhoods – purchases – lifestyle – it is being pillaged to the point where it actually kills the limb from which it sprouts and in essence reaches the root system of the economy and kills it off. Like that land which was purchased, to go forward with life and neighborhoods and the like, it was stripped down to bare and then the new owners were told they had to buy the basic elements – topsoil, etc. back again at a greater cost.

The difference seems to be that the seller did not create the need. They just capitalized on the transaction. Bigger versions of that capitalization are the scandals involving the home mortgage industry.

4.2.6 Keating 5 and *The Big Short*

Remember the Savings and Loan scandal of the 1980s with the Keating 5 and the recent book/movie *The Big Short*? In both cases home mortgages were at the center. In one case it was a "busting out" of the small Savings and Loans banks across the nation. In the other it was a series of predatory lending.

The Keating 5 represents the first time there was a purge on the Savings and Loan banking industry. The strange thing about this effort was that it was accomplished directly with the help and aid of five U.S. senators.

Charles Keating and the senators – John McCain (D-AZ), John Glenn (D-OH), Alan Cranston (D-CA), Donald Riegle (D-MI), and Dennis DeConcini (D-AZ) – had only one thing in common: their power in the Senate, a friendship with the banker Charles Keating, and the bank in question – Lincoln Savings and Loan. The senators asked the Federal Home Loan Bank Board to not pursue any charges concerning a $615 million above the cap series of investments of depositor's money that was put into risky and volatile financial instruments. It seems Keating had made numerous high-level donations into each of their re-election campaigns. The bailout of the Lincoln Savings and Loan cost the taxpayers over $3 billion, with thousands of investors losing all of their savings. This led to a series of Senate investigations of the five Senators (McFadden, 2014).

This seemed to be just a practice round to find out how to make this work on a national scale. What did we do wrong and how could we do this again without making the same mistakes and really reap the rewards?

Does all this sound familiar – well it should because it happened again from 2000 to 2008 – this led to the Great Recession. In the book and subsequent movie *The Big Short*, this escapade is explained in detail. Here is an excerpt from a *Forbes* article clearly outlining the complexity of this fiasco as written by Barry Ritholtz in the *Washington Post*. Try to keep up with all the moving pieces – just like a game of "Three Card Monty" – who's got the queen, who's got the queen? It is a complex process just like the Quants as they were called which were the algorithms that validated these fake securities. There is an old saying – "If you can't dazzle them with brilliance, baffle them with bullshit" – this step-by-step Rube Goldberg mechanism seems to try and do both at Herculean effort levels. Let's look a little deeper.

4.2.7 The Story of the 2008 Financial Crisis

So let's recap the basic facts: why did we have a financial crisis in 2008? Barry Ritholtz does so with an excellent series of articles in the *Washington Post*.

> In 1998, banks got the green light to gamble: The Glass-Steagall legislation, which separated regular banks and investment banks, was repealed in 1998. This allowed banks, whose deposits were guaranteed by the FDIC, i.e. the government, to engage in highly risky business.
>
> *(Ritholtz, 2011)*

That one paragraph says it all – "to engage in highly risky business." Once Glass-Steagall was repealed in 1998, banks were able to offer low-rate predatory lending to unsuspecting consumers. Once those rates went back up which they always do, families could no longer afford the monthly payments. Families lost their homes and banks kept all the cash and were then able to resell the homes for even more profit. Now worse than this, the banks prior to the collapse of the housing market bundled these mortgages into other products and when the rates climbed the home owners defaulted on the loans causing the products to collapse in value; in turn this caused the financial institutions who bought these derivatives like Lehman Brothers, Bears Stearn, AIG, and others to implode. Everything tumbled like dominoes – in 30 minutes or less.

This all caused the greatest financial meltdown since the Great Depression of 1929. The result was the Great Recession of 2008. The government had to step in and generate the biggest bailout in the history of the United States. It crashed the housing market for over a decade and wiped out millions of American families' nest eggs. The banks walked away with the bailout and the families got caught holding the bag (Denning, 2011).

Who is responsible for such "irresponsibility"? Why are they still walking around free? As a comedian once said, "these guys are walking around and Martha Stewart went to jail" – why?

A great expression about this – "They may be too big to fail but they're not too big to jail!"

4.2.8 Starbucks Recession

Today we experience much smaller financial calamities which some call the Starbucks Recession – why? Because even though times are very tough for some reason, we still seemed to have enough cash for those $4 lattes – it didn't impact the daily economy, this scheme of defrauding the American public just went after the big long-term numbers. As the famous bank robber Willie Sutton said when asked, "Why did you rob those banks?" He responded, "Cause that's where they keep the money." The S&Ls and the mortgage lenders and the SWAPs were where they kept the money. But in the long run it was bigger than just the money; it was the core of society (Canavan, 2022).

During these two mortgage-based recessions, we saw the beginning and finally a breakdown of the family system predicated on a need for a two-income household, no one home to supervise the children, an introduction of the home entertainment, gaming systems, and personal computers. This seems to be the start of the digital economy that was purposely leveraged by the constructed demise of the American Dream and an effort to create an oligarchy of banks.

What does this have to do with creating this "veal?" Well this kind of financial burden cemented the fact that nobody could stay home with these children. Bills had to be paid and a two-income household was now an ongoing reality.

4.2.9 Building the "Millennials"

Millennial refers to the group born after 2000. "Gen-"" or "generation interactive" is a term to combat the simplicity of the term "millennials." The term "millennials just has to do with a date or time and not an actual belief system. Millennials are just born at a certain time. It doesn't mean they live the life – it's just a date. Some actually purposely go off the grid, embracing the 1960s commune-like mentality. And "digital natives" kind of means the same, but there could be a slightly different birth date. Gen-i means that they are in-it-to-win-it and they can't help it. These young people are at the core of the movement because of both age and lifestyle combined. They're in school now, shaping how we teach by how they learn, and subsequently they'll be consuming and producing by those same patterns. They are the end-user we are designing the society for today. Here is the definition of Gen-i. Gen-i stands for generation interactive, isolated, iterative, Internet-driven, immaterial, isogonics, isomorphic, (cyber) itinerant, and inoculated.

But how did they get like this? How did this generation become all of these things and behave as they do? Let's look at some definitions and later statistics as to whether or not this Gen-i is actually all of these things.

(Cyber) Itinerant: traveling from place to place; staying in a place for only a short amount of time – "ADD and ADHD"
Immaterial: not important or significant "whatever"
Inoculated: to introduce something into the mind of – "media buzz"
Interactive: designed to respond to the actions, commands, etc., of a user or requiring people to talk with each other or do things together – "kind of opposites?"
Ironic: using words that mean the opposite of what you really think especially in order to be funny – "hipsters"
Isogonic: exhibiting equivalent relative growth of parts such that size relations remain constant
Isolated: occurring alone or once – "all day screen time"
Isomorphic: being of identical or similar form, shape, or structure – "non-conforming hipsters"
Iterative: relating to or being iteration of an operation or procedure; endless scrolling and smartphone time

If you put all of these pieces together it does seem to represent the tendencies of this age group. They sit around on adult-like playdates staring at their phones. They are in the moment but isolated at the same time, they hop from topic to topic thinking they are multitasking and believe they are interacting – but not really speaking or acknowledging their friends – you know the human ones. They act in an ironic fashion when they don't even fully understand the principle that they are wishing to be ironic about. Which in itself is amazingly ironic – literally (proper use by the way). They try not to conform as they appear and look exactly like each other in an isomorphic manner. And they scroll and click through iteration after iteration of the same Facebook, Instagram, Tinder, X (Twitter), etc., page over and over like a medieval French prince.

4.2.10 Moved Elsewhere – *Herding Them Up*

The next stage is to move the calf away from the herd to a new location so that the proper rearing can occur, and that usually includes stimulation toward a desired end product. During this time of a need for double income – again caused by a tremendous rise in base necessities – mortgages – the need for daycare in the United States also rose dramatically. Now again as we speak about metaphor, we need to keep reiterating who the players are in this scenario. All of this is happening through the actions of the "ranchers" at the bequest of the "marketplace" – the ones who truly stand to benefit.

4.2.11 Rise of Daycare

In a previous section we discussed the rise of childcare. And in response to this new dynamic, this new two-income necessity that many families were faced with has come to be called a non-traditional schedule. During the "baby boom" years there was a parent at home all day especially during the critical years of childhood. Now at the end of the day or even during the day children were "babysat" with technology, from the earliest days of the television to the recent times of the computer. Here are some recommended guidelines from the American Academy of Pediatrics.

> ... kids between ages three and 18 use screens for a maximum of two hours daily; and kids younger than three should avoid screens altogether...
>
> ... spending five hours in front of the TV, not outside playing, doing homework, or interacting with the parent, that's a problem.
>
> *(Ossola, 2015)*

Now that is just kids between the ages of three and 18 but what about the older ones? Millennials are sitting there up to six times longer. What damage is this doing?

> Millennials are spending as many as 18 hours a day skimming the Web, watching TV, texting, playing videogames and using social media, among other activities, according to data from Crowdtrap and Ipsos MediaCT...
>
> *(Taylor, 2014)*

Have you started to notice a correlation between home prices, babies born, and computer usage? It all seems to come together to create a "perfect storm." Now the home prices and computer sales go in unison, but the birth rates don't spike in coordination. The question asked a million times is, "how do you make money from the Internet"? Simple answer, just treat it like a shopping center or a mall for one thing. But companies who profess to "do no harm" are doing much harm, and doing much better by doing – harm – but they could change that process. "I will drink your milkshake" is a great line, spoken in a movie depicting the early pioneering days of another new economy – oil. So it is understandable when the care of the human falls to a new level of disrespect.

Now this may sound dismissive to both the author and the subject, but just the other morning Tom Brokaw was asked why there was so much dissent within the populace and he actually named these two reasons: 9/11 and the 2008 financial collapse. But why the common thread? Is it that apparent to everyone but we just don't say it out loud?

Maybe it is because years ago we had bigger family groups living in the same town, just blocks from each other. A father may have bought his house's plot from his mother-in-law and built his house himself after work every night for five years until the family moved into it. From the back porch of that house he could see five other immediate relatives' homes. People didn't feel the need to move elsewhere – it was their town. For real – it was their town. People had over 50 immediate relatives at backyard cookouts. Large dinners at holiday times were very common. Going over each other's house was very common, and happened almost every day. Now have we purposely tried to keep our children in a separate small space as a means of protecting them from the harm we are seeing all around, a harm we can't be there to protect them from.

This leads to keeping the "calves" in a small place where we can keep an eye on them – you know "buckle them up for safety."

4.2.12 Kept in a Small Space

4.2.12.1 Crate Them Up

As stated in the previous section childcare has come of age in this new generational family structure. It has become an absolute necessity in many cases. Not a luxury, but a necessity. But at what cost and what effect has it had? These are not "free range" calves that have the ability to go off on their own and experiment with life. These are "veal" kept in a small space and stimulated just enough and in a specific direction to develop the desired end product.

As the "Boomers" were growing up they were able to go out to play in the morning and not come home until the lights came on at night. Sure there was trouble from Leopold and Loeb, the Lindbergh Baby, Sacco and Vanzetti, there were always problems and issues with just plain bad things happening. But the fear level wasn't the same. In 2015, parents in Maryland were brought up on charges for simply letting their kid walk home from school or play in a park until sundown.

> ...they were accused of neglect for letting their children, ages 6 and 10, walk home alone from a park – are under investigation again. This time, the children were taken into custody by police officers and held for hours by police and Child Protective Services...
>
> *(Wallace, 2015)*

How did all this come about? Were there that many instances of abuse or abductions or was this just a means of spreading fear that would basically paralyze the populace? Why were these parents subjected to such treatment because they allowed their kids to walk home from school or play in the park?

Starting from the 1980s, baby slings, car seats, pre-screened playdates, bubble wrapping kids to protect them while maybe putting them in harm's way with way too much screen time?

Seemingly parents were less stressed to have them at home where they could micromanage and hover like helicopter parents? Electronics gave parents the tools to alleviate guilt; at the same time these tools diminished real participation and learning to work toward actual goals.

The tools parents gave children may have made them prey for "ranchers" to corral easier? We, as parents, positioned them in front of the screen in hopes that it would help them progress in an arena where we couldn't lead them properly. But our ignorance of this space allowed them to become overly available for the "feeding frenzy" of content.

4.3 ABUSIVE USE OF TECHNOLOGY

4.3.1 Game Boys (Girls)

One of the first devices that really broke the blood-brain barrier was the Nintendo Game Boy. Now you had the distraction in your hand 24/7/365. Game Boy and GameBoy Color combined sold 118.9 million units worldwide by 1989 (Das, 1990).

Well do these devices cause harm to young children with ASD, ADD, and ADHD? According to the American Academy of Pediatrics, the answer is yes.

> ...children with ASD and those with ADHD may be at particularly high risk for significant problems related to video game play, including excessive and problematic video game use. Attention problems, in particular, are associated with problematic video game play for children with ASD and ADHD, and role-playing games ...
>
> *(Mazurek, 2024)*

So in trying to resolve the issues of keeping our kids safe and happy and trying to keep sane ourselves, we have possibly led them down the path right into the mouth of demise. Where we think we are giving them something to connect with, we have actually disconnected them from life and into the waiting arms of the "machine" that has turned them into a consumable and consuming life form.

And how long has this been going on? Well just look at the date of this earliest version – 1989 – perfect timing for the first housing jump. By 1989 the first wave of children born to the "housing bubble parents" were around 4–5, just about the age when they had to be "entertained" but indoors and not let to run free range through the neighborhood. The term for this is "flow" caused by complete immersion into the game environment.

> For some players, immersion can result in their disregarding the outside world.
>
> The psychological arousal that players experience is akin to flow, which can be defined as a state of optimal experience that entails participants becoming totally immersed in the action.
>
> *(Oppland, 2019)*

What are the consequences of such "immersion" on this generation? Do they work well within society? Are they able to go onto higher levels of schooling? Can they hold down a job to communicate with family and peers?

> Some players become so immersed in online gaming that it affects them adversely. For example, players have been observed to drop out of school, quit jobs, and distance themselves from friends and family.
>
> *(Király et al. 2023)*

Is this replacing other avenues for serotonin? Generating the dopamine hit that social media now supplies and Francis Haugen ex-Facebook employee outlined? Is it replacing the love of parents – of friends – or other accomplishments? And does more time away from the parent create more of a need for acceptance by device? In essence, is the device and its world replacing the parent? This is the loop Ms. Haugen spoke about in her testimony to Congress. The "like" loop. Where the coders for Facebook designed the "like" button to time perfectly with the desire cycle of the user. This creates the desire and satisfaction for serotonin in the brain.

4.3.2 iPods and iPhones and iPads – Oh My!

If you knew anything about Steve Jobs you would know his love of music and especially the Beatles. What seemed to be his "dream project" once he returned to Apple was the iPod. He would speak of it as a revolutionary – insanely great idea! But take a step back for a moment and realize that Jobs always had a multi-level plan going on.

Anyone born around 1955 is of the generation that remembered the transistor radio with the earplug. Well Jobs (1955–2011) basically just updated that to a new device called the iPod. He would say "ten thousand songs that fit into your jeans pocket." Yep, but we had that back in the 1960s – a simple 9-volt transistor radio. Here is that radio and an iPod pictured side by side.

What's the difference? One provided Jobs with a cash machine into your pocket and the other was a one-time purchase (except for the batteries). It was a huge move – insanely great but not without cost. Have you ever looked at your iTunes purchases for a year? Now those purchases were the insanely great piece of this – for Steve but not so much for your family budget.

It also introduced the touch screen capabilities for the first time in a simple way that soon led to the actual touch screen.

Here is the first version of Apple's move into that space – the Apple Newton. Notice the similarities to the iPhone – this was just a prototype for that new cash machine we have all come to love and not be able to do without.

The Apple App Store revenues for 2022 are record breaking with over $1.1 trillion in total billing ("Developers Generated $1.1 Trillion in the App Store Ecosystem in 2022," 2023)

4.3.3 In Game Purchases

Today gaming is fast becoming a daily addiction. People are playing Candy Crush – Angry Birds – Wordle – whatever they can, to be distracted from everyday stress. That's a good thing, again to a degree. But these seemingly innocuous games are actually mining user data to determine whereabouts, patterns, hours of use, etc. (Harris, 2018).

And what happens to these kids playing these games not to count the time wasted but the push to purchase items in the game. Here are some examples – we've all heard the stories. There are thousands of examples of gaming abuse. These are just some of the most obvious ones.

> This Kid Blew $2,500 on In-Game Purchases in Just 10 Minutes
> Shameless in-game purchases aimed at kids have been in the news recently. But this five-year-old kid spent more than most: he blew $2,500 on in-game junk in just 10 minutes. That's impressive.
> *(Cutlack, 2013)*

This is even happening in other countries as well, so it is not an American issue alone.

> … for a 32 year old father from Crawley, UK it couldn't get any more serious business. … his seven-year-old son was able to charge his credit card with £4,000 by making in-game transactions on Jurassic World, an iPad game.
> *(Cutlack, 2013)*

What is causing these kids to basically pilfer their parents' money and just spend it away? Not because it is easy, but because they want to belong to the herd. And what is that herd worth? We all hear about these numbers – these valuations of companies – Facebook valued at $562 billion, X (Twitter) valued at $44 billion, WhatsApp purchased for $19 billion with only 19 employees.

In the next section we will break down exactly how much this is worth financially but for now let's focus on what it is worth to your kids. What happens when you don't fit into Kim Kardashian's Hollywood Squad or you're not allowed in the right group on Instagram?

4.3.4 Cyberbullying

According to *Webster* here is the definition of

> Cyberbullying
> 1: the electronic posting of mean-spirited messages about a person (as a student) often done anonymously

We have all heard of it and have even read about the effects of this relatively new form of classic abuse, but now it follows you 24/7/365. Now it is responsible for probably more deaths globally than its analogue predecessor ever was. The stories are in the news and unfortunately for many they are even more close at hand (Anderson, 2018).

> 59% of U.S. teens have been bullied or harassed online, and a similar share says it's a major problem for people their age. At the same time, teens mostly think teachers, social media companies and politicians are failing at addressing this issue.

This is just one account of this growing problem. The news today is flooded with these stories from a child in elementary school to a child with an assault weapon inside an elementary school. They have been bullied since Cain killed Abel but now the consistent, invisible, and an omnipotent quality of this act has resonated with its strength. Today it goes even further – as said by many health professionals – to psychotic behavior.

4.3.5 Catfishing

Known for what it is today this appears to be one of the more insane uses of the Internet and a little horror movie-esque. Here is the current definition of "Catfishing."

> Catfishing a person who sets up a false personal profile on a social networking site for fraudulent or deceptive purposes.
> *(Wood, 2021)*

We have all heard about it and maybe have even seen the movie from 2011 but it gets even weirder once you start looking into the actual cases. And even weirder that now it has become so established that there are even television shows having fun with these embarrassments like it's a "Blooper" reel for life itself.

This is the perfect example of taking something like social media, connecting all over the world, something whose original purpose was meant to be benevolent and destroying it to the point where it actually does damage to people who may have stumbled into a deceptive practice as innocent bystanders and winding up as road kill on the information superhighway. We might question whether anyone can still afford to say "I am just an innocent bystander."

4.3.6 Opportunists or Entrepreneurs – Who Was Behind This?

If you are asking was there a plan to all this, in many opinions the answer is this was all premeditated. Build a device that keeps going in and out of your pocketbook – a device that enables the practice of all forms of pressure to push these young people and the economy along.

From 1980 to present day, society has witnessed the rise of the personal computer in the home. In a recent documentary on CNN titled "Steve Jobs | Man in the Machine," Jobs is portrayed for what he was: a genius and the mastermind behind the personal computer. In the documentary, Jobs is shown many times using the phrase – "seduced" in reference to using a personal computer. His eyes sparkle when he talks about this and you can almost feel his glee and hubris.

While his genius is undisputed it may not be for the actual engineering of the physical computer but more so for engineering its need and acceptance by society. His famous phrase of "showing them the need for something they didn't think they needed," just through the pure capability of the machine and its software.

Mr. Jobs was the paramount showman, but we can never know how much was showmanship and how much was he preying on the desire for these users' own hubris. Many early computer wizards thought they were going to be the next Jobs or Wozniak or Gates but the odds of these geeks becoming as omnipotent as Jobs are very low.

Seeing this device from almost day one and certainly working with some of the software that would make this machine truly a centerpiece in the early stages of this digital revolution the impact this would have on the media industry was apparent. Jobs knew it as well with the minor training in calligraphy he fought for the use of quality typography and a true aesthetic. This was part of the seduction, a temptress using beauty to lure the masses. We felt tremendously empowered when we were part of it at the time – a select few who could do this. We even glorified and demonized them in popular culture.

Video games became a strong target of blame – as they should – sorry but if you get points for killing prostitutes and stealing cars – you also get the blame. In *Second Life* you could be anyone you wanted to be. The biggest geek could be the biggest hero – an amazing seduction if there ever was one.

But the unusual point for Apple is the development of eWorld – the early Apple developed online village presence in 1994. Years before AOL and Netscape started exploring this new space, Apple was building a major online community of learning and sharing. Here is an excerpt from a 2014 article on the 20th anniversary of eWorld.

> In this competitive climate 20 years ago, Apple introduced eWorld, a subscription-based information service available to Mac and Newton users. Despite its closure in 1996, eWorld remains notable for its city-based interface metaphor and as a symbol of Apple's overreach in the era.
>
> *(Edwards, 2014)*

So, not only did Apple see the need for local hardware and software uses 30 years ago but they went far beyond and attempted a way-too-early move to the Internet – very smart but maybe too smart and too aggressive. They could always identify a "killer app." Then they built the modern-day smartphone. The device that every pundit along with Mr. Haidt declares to be the shunt put directly into the brain of your children.

4.3.7 Defining a "Killer App"

What is a "killer app"? Simply put – it is the application that makes any device worth its price – desktop publishing for the Macintosh, business software for IBM and Windows and so on. Here is a quick definition from *PCMag Encyclopedia.*

> A software application that is exceptionally useful or exciting. Killer apps are innovative and often represent the first of a new breed, and they are extremely successful.
>
> *("Definition of Killer App" n.d.)*

The link between a "killer app" and the Gen-i kids was the killer app for the new economy. Yes, that's right, the actual children, not the hardware or the software. Everything was built – hardware and software wise and even "the products were all put on shelves with care – in hopes that soon our friend PayPal would be there." We tried to put this machine into motion before but to no avail – remember the "dot com" bubble bursting in 1999–2000? Why did it burst? May be it got too big too fast and the existing economic powers saw that it was going to be very disruptive, so they simply had it quashed, stopped-dead in its tracks. The "bubble" burst because it was going too fast and could not be steered into the mainstream pockets. Now, as you see it is there, in the traditional brick and mortar corporations.

This early 2010 article showed some numerical insight into the current generation's takeover of the economy in as far as driving the new economic delivery and behavioral platforms. Obviously, this has not stopped and is reaching further down into pre-teen and earlier being referred to as an "iPad" kid.

> Teens, meanwhile, are by far the most likely to play online games: 78% play games online, the most popular activity for that age group. Teens are also the most likely group to visit a virtual world… – 8% of online teens, compared with 4% of internet users 18 and older.
>
> *(Zickuhr, 2010)*

It is worth reiterating throughout this book that many of the proponents of these concerns have made their living and continue to make it by building and teaching interaction design – the method of creating interfaces that are engaging users and solve client problems. You know basically the devil incarnate. But many feel that it doesn't have to be that way. Their super powers can be used for good as well as evil. It is up to you to decide. If that skill set allows a child to blog about their passion for horses then good. If it requires your four-year-old to have your PIN – again not good. If it gives you too much free time away from your lovely children, it's probably not good.

In the next sections we are going to talk about how being kept in a small space leads to being stimulated to just the right amount and in just the right way to generate the desired end result. And once they have been stimulated and they respond correctly how they are corralled into social networks – "herds" – to be sold off in pieces for individual price points. Basically like any commodity but in this metaphor – "veal."

Is this the new economy? Remember the "TechCrunch" blurb:

> …Uber, the world's largest taxi-company owns no vehicles, Facebook the world's most popular media owner creates no content, Alibaba, the most valuable retailer has no inventory and Airbnb the world's largest accommodation provider owns no real estate.

When was the last time an economic shift of this magnitude occurred? Not just a new means to a new economy but a complete shift from manufacturing to service?

Remember Napster, the music file-sharing network? Well that basically destroyed the music industry and it has really never recovered – yet.

4.3.8 Stimulated Just Enough

So far we have taken this metaphor from birth to separation to moving the "herd" to gathering them in a small space now to "stimulating them just enough" to produce the end product.

What is the product of that desired stimulation? Time – time on a screen, time swiping and clicking and using and chatting and "snap chatting" and whatever. Eyeballs and actions equal revenue in this game. And everyone is fighting for that attention. Today they call it "engagement."

As you have seen throughout this last chapter, screen time has increased dramatically in the last ten years. Where is that value for the companies and corporations that advertise and fight for that screen time? Well first we need to validate and explain what we are talking about here.

Online advertising is one of the biggest areas of revenue in this market space. Now this online advertising doesn't just generate revenue for ad space it also generates revenue for the data it mines. Here are some terms to explain that return on investment.

CPA (Cost-per-Action)
CPC (Cost-per-Click)
CPC (Cost-per-Customer)
CPL (Cost-per-lead)
CPM (Cost-per-thousand impressions)
CPO (Cost-per-Order)
CPS (Cost-per-Sale)
CPT (Cost-per-Transaction)
CPTM (Cost per Targeted Thousand Impressions)

(Chase, 2023)

All those terms seem confusing to digest but realize that this is just how valuable your time on the Internet is to these companies. It is truly an entire economy, not just fun in the sun time. It has become the biggest industry driver globally. Artificial Intelligence works off of data. The more screen time and selections the user makes the more data is generated and collected. And now with Artificial Intelligence and quantum computing coming together to drive this engine as David Byrne said, "This ain't no disco – no CBGBs – this ain't no messing around…" The companies behind these actions are serious and this should be serious to everyone. How this happened was a confluence of events, best explained by Thomas Freidman.

Thomas Freidman, Pulitzer prize-winning author and columnist for the *New York Times* in one of his most famous books *The World is Flat* (2004) listed some of the main contributors toward this "Flattened World," some of which he says are directly related to what we are seeing in our children and our society.

All of these came together to create the opportunity to use all of those terms you read just previously. If these events did not happen, maybe those tools could not have been so useful to create this new economy?

Here is his list compiled from his book.

Friedman Ten "Flatteners":

1. Collapse of the Berlin Wall – 11/9/89
2. Netscape – 8/9/95
3. Workflow software
4. Uploading
5. Outsourcing

6. Offshoring
7. Supply-chaining
8. Insourcing
9. Informing
10. "The Steroids": Wireless, VOIP, and file sharing.

(Friedman, 2007)

Please get this book and start to truly familiarize yourself with what he and many others are saying has happened and what many believe was part of the groundwork that started this huge change in our society. Would the process for "stimulating just enough" have been possible if not for these events?

4.3.9 The Planned Progression from Newton to iPad to Mobile

It is important to consider whether this is a natural progression to the touchscreen or part of a plan or something that was seen and capitalized on. Read Friedman's Ten "Flatteners" again – and see if they would be either happening at all or if they would have the potential to do anything without the perfect delivery method and tool.

Another consideration is whether this is just a passing fad of some nature or has now been entrenched into our collective psyche. One hopes the kids just move on from this like Pokemon and Power Rangers, and wonder whether it has had a lasting effect on them.

4.3.10 What's the Damage Here?

Health issues – nurture and nature – sugar and screen time – brain patterns – did the "ranchers" put together the perfect feed mixer for this generation? Remember that veal is put on a specific diet to generate just the right results, usually a heavy milk or cream.

Let's take a look at some research on the increase of screen time – whether computers or television alone or with High Fructose Corn Syrup and the advent of ADD/BPD and many other physiological and psychological effects.

> Dependency on digital devices resulting in an ever-increasing daily screen time has subsequently also been the cause of several adverse effects on physical and mental or psychological health.
>
> Risk factors for obesity and cardiovascular disorders, including hypertension, poor regulation of stress, low HDL cholesterol, and insulin resistance....
>
> *(Nakshine et al. 2022)*

It is tough not to agree with this opinion. Physical activity is a necessity for development obviously, but the question is the balance of that time and whether an imbalance results in conditions like ADHD and the sorts. Does the extensive screen time enhance other abilities or are these young people really just slaves to the machine?

> according to Dr. Christopher Lucas, associate professor of child psychiatry at New York University School of Medicine. "It's not sustained attention in the absence of rewards," he said. "It's sustained attention with frequent intermittent rewards."
>
> *(Klass and M.D., 2011)*

So here again is the dopamine cycle. And again the term "flow" comes into discussion. Here is an excerpt from a report dealing with online game addiction entitled: Advanced or Addicted? Exploring the Relationship of Recreation Specialization to Flow Experiences and Online Game Addiction by Tsung-Chiung (Emily) Wu, David Scott and Chun-Chieh Yang.

> The meaning of game addiction is contested—the term is colloquially used to describe extreme devotion or state of being absorbed in a game (Holt & Kleiber, 2009). Extreme devotion, however, can lead to single-mindedness such that participation has a negative impact on the individual and … loved ones.
>
> *(Partington, Partington, and Olivier, 2009)*

The overuse or daily standard use of gaming communities and multiplayer games along with social media networks to many has to do with their place in the "herd" mentality, especially how it connects to "likes," peer-pressure, and acceptance. But has there been proof that there is a bigger effect – primarily with student grades and overall success?

> a relatively small, exploratory study, it is one of the first to find a relationship between college students' use of Facebook and their academic achievement.
>
> Typically, Facebook users in the study had GPAs between 3.0 and 3.5, while non-users had GPAs between 3.5 and 4.0.
>
> *(Karpinski, 2009)*

With those questions buzzing around our brain, the next step is to consider whether there has been a long sustainable change going on in our children's neural pathways, health, socialization, and overall patterns since this specific amalgam of nature and nurture took this turn.

4.3.11 The Rise of ODD, Anxiety, Depression, ADHD, ADD, and Bipolar Disorder

During the last 25 years, the rise of ODD, anxiety, depression, ADHD, ADD, and bipolar disorder has risen to 40 times the amount. During the recent Covid pandemic, the number of cases with Covid that suffered from those disorders have been shown to have even greater effects on mortality.

> Rates of anxiety, depression, and substance use disorder have increased since the beginning of the pandemic. And people who have mental illnesses or disorders and then get COVID-19 are more likely to die than those who don't have mental illnesses or disorders.
>
> *("Mental Health," 2023)*

During the sequester of the recent pandemic the use of gaming increased dramatically for school age children. And the effects of this excessive screen time (and specifically gaming) brought about many dire results, such as depression, anxiety, and attention-deficit/hyperactivity disorder and increases in overall gaming disorder.

> A majority of studies found an increase in game usage time and game addiction score during the COVID-19 pandemic.
>
> Gaming disorder was caused by the impact of COVID-19 in a vulnerable group with predisposing factors such as depression, anxiety, and attention-deficit/hyperactivity disorder.
>
> *(Han et al. 2022)*

Frightening stuff isn't it? The goal here is not to get you running around the house trashing all of the GameBoys, Nintendos, PS5s, computers, laptops, and smartphones. The goal has been to realize that "the kids may not be all right" after all. And also that there are forces busy working on them while you're being busy working for them.

4.3.12 Enter the World of Gaming

A lot of these articles mention gaming as a critical part of their studies. As parents we have all witnessed and, in many cases, participated in this activity with many times wincing at what was happening on screen. As with anything, moderation is key but peer pressure is strongly exercised in that space as well. The playdates and the all day/night sessions over friend's houses. The possible guilt we feel by not letting Johnny or Jane take out that group of "Splinter Cell" computer terrorists. So let's take a look at some of the leading gaming and app titles for 2023.

Leading titles in Gaming and App sales in 2023

1. Roblox
2. Attack Hole
3. Royal Match
4. Candy Crush Saga
5. Cat Snack Bar
6. Mob Control
7. Block Blast Adventure Master
8. LightSaber – Gun Simulator
9. Mechangelion – Robot Fighting
10. FIFA Soccer

(Clement, 2024)

If you notice 50% of these titles, both games and apps, are a little rough – here's one of the number one selling games – Grand Theft Auto. Let's do a quick recap on the nature of this game by reading a "journal" entry by a typical user.

> … just spent a half-hour planning the perfect heist. I'm going in smart, knocking out the guards and the staff behind the delicate jewelry counters of the store with a carefully placed smoke bomb, and smashing into each cabinet with the butt of a semi-automatic rifle before making my escape.
>
> *(Strickland, 2023)*

So by now parents may be confused and doubting their choices and allowances with respect to how their kids are spending their time. Also be wary of what that time is worth to you, them, and the economic model that it is supporting.

But now, let's see what the real price of this meal is? How much money are the "ranchers" making while your young ones are busy keeping quiet so you can get through your day or two jobs and then – you know, just life stuff?

4.3.13 Sold Off in Pieces

So now that we have the pieces in place, we need to know the price on each section of this valuable veal chop? This is easily done using information that is available online.

Each social network has its own valuation. This is usually based on an average amount of money generated by each user. Valuations on each social network vary greatly but in each case they are talking primarily about one thing – you and your love of this game. Valuations are typically based on users × revenue generated per user. Example, Facebook has 1.5 billion users worldwide × $24.00 per user equals $36 billion valuation.

So just to have some fun, let's see how much does the average child/adolescent/young adult ages 12–25 weigh? Let's say about 140 pounds so divide 24 by 140 and that's about 17¢ a lb.

So that is how the graphic is made up. Each social network listed or tagged is shown with the price per lb. of each user with the average being 140 pounds. Seems a bit crude and rude but isn't that what works today? Isn't almost every move you make online measured and tagged before you make it? The only way this isn't happening today is when someone like Louie CK releases a show online without making any reference to it through almost any form of media or public relations just to see if it will work without the hype. Very brave move but it seems he wants to see if there is any quality to the work and not just online hype that drives the traffic.

Here are the rest broken down for you using the same metric.

Network	Valuation	Users	Per Lb.
Pinterest	$11,000,000,000	100,000,000	0.78¢
X (Twitter)	$10,000,000,000	305,000,000	0.23¢
Instagram	$17,500,000,000	400,000,000	0.31¢
YouTube	$40,000,000,000	1,100,000,000	0.25¢
Facebook	$562,000,000,000	3,400,000,000	$1.87

The calculation used is Valuation/Users/140 lb. average. Again it's a little crass and rude but isn't this what these platforms encourage?

Pretty interesting stuff, no – but hasn't this always happened? According to Paul Simon, "Every generation throws a hero up the pop charts." True, but they usually came into an existing market base and had to compete to win. The Beatles over Elvis – the Rolling Stones over the Beatles (never) then to modern-day Kanye over Kanye over Kim over Kanye over Taylor…

Now we even have virtual stars and commodities. As Quantum and Artificial Intelligence come close to total fluidity, these virtual stars are taking center stage. Reanimation of Tupac Shakur and the Notorious B.I.G. were some of the first. Now we are looking at reanimating rock and roll icons like Lou Reed.

One of the first and possibly the largest original cybernetic pop stars is "Miku Hatsune."

> Hatsune has been filling venues in Japan since her first solo live show in March of this year [2010]. Originally a character designed to promote Yamaha's vocaloid synthesizing technology – we all know how Japan loves its characters – Miku has now taken to the stage and attracts fans in their thousands.
>
> *(Le, 2013)*

Fabricated stars have been used before to generate revenue. For the boomer generation it started with decoder rings in cereal boxes and on the radio but it really hit its height when television arrived in every home in America. Every kid in the 1950s watched and wanted a Davy Crockett raccoon cap.

The Davy Crockett character was played by Fess Parker and was viewed on Sunday nights. It was one of the first appointment television series that came about. It was pretty much a phenomenon.

> …the frenzy started by Parker's Davy Crockett character helped sell 5,000 coonskin caps a day, causing the price of raccoon fur to jump from 25 cents a pound to $8. Before it was over, Crockettmania accounted for $300 million in sales of everything from Davy wristwatches to lunch boxes.
>
> *(Johnson, 2002)*

That's $300 million in sales in the 1950s – yikes – what is that today adjusted for inflation – roughly $2.7 billion! And Parker got none of it – amazing.

In the 1960s the mash-up group – "The Monkees" – Mike Nesmith, Micky Dolenz, Peter Tork, and the incredibly cute Davy Jones was actually made up to cash in on the British invasion. Davey was the only Brit but he was the front man for the group. Don Kirschner, a major player in the music industry, decided to come up with a group to see if he could promote the hell out of it. He did and they did and to this very day, over 50 years later. But in this mid-20th-century example, the audience wasn't bred or constructed; the group/talent was created not the desire.

This was all before the famous Nielsen ratings came about. They were the premier service that measured what was being watched and when in homes across the country. This is still going on today via Google, Facebook, and all the other social media networks throughout the Internet. Google Analytics, Facebook Analytics, and all other forms of internal analytics now drive the social media marketing industry. Content calendars, marketing plans, re-targeting, these are parts of the overall plan to "engage" the viewer/user.

What are they interested in? Well pretty much the same thing these companies have always been interested in since the advent of media advertising – but now the concern is how much of the user's spending dollars have gone to online – purchases – games – music – media?

4.3.14 When Before Has Media Become a Commodity?

Has the purpose of the media always been to enhance the economy? Well, why do you think they call them "soap operas"? Their only purpose was to sell you products. The only reason media has spread the way it has is because of the impetus of generating advertising sales. As sales dropped, many media products – newspapers, magazines, radio, and now television, have undergone serious changes. Video killed the radio star and now the Internet is killing any of the old media stars and their vehicles. But not so fast – many attempts at online video channels even from the traditional networks have been met with failure. MSNBC developed shift.com – part of the MSNBC family and it was shuttered in very little time. This also happened to HuffingtonPost LIVE, etc. With more and more advertising dollars going from traditional to online budgets, how is this happening?

It is kind of like a weird version of screw you and the horse you rode in on or is it a last-ditch attempt at keeping the money in the hands of traditional media kingdoms? This "new media" beast has gotten loose and corralling it has not been easy for this "old school" media. Fragmentation of content, which has led to lower ad revenues, strikes with the creatives, audience loss, has caused the "old media" executives to restructure their companies and now pay attention to something they once dismissed outright.

So the answer for the networks has been to realize that streaming content is the new "glittering box full of wires." The new answer to an old question of how do you make money on the Internet?

4.4 LACK OF GOVERNANCE

4.4.1 Early Days of Television and Radio?

Is all of this similar to the early days of the transition in all media – radio and television? Sure radio sort of killed the newspaper industry but not really. Long forms always live on the printed page. But once you had talking, moving images every night for free in your living room – aka television – streaming any movie anytime – all music, every new release from any artist, courses from around the world – game over.

Once news and entertainment came into your house in approximately the same size piece of furniture that the radio took up, there really was no comparison. So the video really did kill the radio star and then the television manufacturer tried to kill the television star.

However, these shows had to adhere to strict governmental policies and the ones internally from the networks that broadcast them or when in print format the publishers that produced them. Today that is not the case.

As we have said before, Section 230 of the Communications Decency Act (DEPARTMENT OF JUSTICE 2020) states:

> No provider or user of an interactive computer service shall be treated as the publisher or speaker of any information provided by another information content provider.
>
> *(47 U.S.C. § 230(c)(1))*

Section 230 is basically a license to do whatever these social media companies want. Why was this allowed to go on for so long? Some believe that the industry should be left to govern itself by the free market principle. Remember 90% of the content is user-generated. So the social media CEOs have been able to skip around any form of legal punishment simply quoting Section 230. But has the free market and self-policing always worked well? Well today it is being challenged. Not just by individuals but by Congress itself.

4.4.2 Tech CEOs Answer to Congress

During a meeting with the CEOs of Discord, Snap, Meta, TikTok, and X, formerly known as Twitter, Congress brutally asked whether or not these CEOs felt any responsibility for the harm they have inflicted on their users and their families. Senator Hawley demanded a face-to-face apology from Meta CEO Mark Zuckerberg to those families. The response seemed to be fake at best and a form of pandering at worst.

> No one should go through the things that your families have suffered and this is why we invest so much and we are going to continue doing industry wide efforts to make sure no one has to go through the things your families have had to suffer.
>
> *(Duffy and Fung 2024)*

During this meeting Senator Lindsay Graham blatantly stated that it seemed that the only way to stop their indifference was to "open the courthouse doors" to lawsuits and to repeal Section 230.

Will this work? Will these tech giants curb their enthusiasm? When has the industry been responsible enough to monitor its content? It seems only when their pocketbooks were hurt.

4.4.3 RCA Canceling Sid Caesar?

In the early days of television Sid Caesar became its biggest star. His "Show of Shows" had a cadre of writers that would go on to become legends in entertainment, names like Mel Brooks, Woody Allen, Carl Reiner, Neil Simon, and Larry Gelbart.

The really amazing thing about the show was that it was so popular RCA (the show's and NBC's parent company) couldn't keep up with demand for new television sets. So, not to have its competitors eat up a sizable part of the physical product market, RCA canceled the NBC Sid Caesar show until it could meet the demand for television sets.

This was how much control the old model gave the corporations. It is still similar today on network television but not so true on this new Internet platform. Today viewer's demand (read revenue) supersedes any regulations or internal choice by a network or production company. That is because the content is user-generated. And the shareholders really make the decision. And that decision always involves money.

4.4.4 Future Viewing Trends

It is never easy figuring out "how they will come at you" like Michael Corleone in the burial scene. But you can always bet "whoever makes the deal for Barzini will be the traitor." So how is the deal for your time and attention being structured for the future in the years following 2024? What do these companies like Deloitte know about where you are taking your leisure, news, fun, and interests?

> This year's... study further underlines how the once-distinct categories of media and entertainment continue to converge and interconnect, while audiences are becoming more fragmented and specialized.

> The value that consumers expect from digital media and entertainment is being increasingly shaped by their experiences with social media and gaming.
>
> *(Westcott et al. 2023)*

4.4.5 X Marks the Spot

Today this growth is being reflected in the social media marketplace. The swing of viewership and ad revenue is not only between platforms but inside each platform itself. The disruption has gone internal – changing companies at their core in a flash as opposed to over the five-year stretch of time it normally would take to disrupt the model.

A major manifestation of this disruption comes in the form of Elon Musk. His purchase of Twitter with his cohort of investors blew up the space. He destroyed the company in its current incarnation. As of 2023 the ad revenue for Twitter or "X" (as it is now named) is down over 28% since 2022. This is in a time where overall ad revenue for all online media is up over 9.5%. The talk is that the company is going to morph its business model from "tweets" to an "everything" app.

At the same time Meta is debuting Threads. Threads is a new entity aimed at the current "X" (formerly Twitter) marketplace. The difference is that Meta is able to integrate content from all its Facebook and Instagram community. This gives Threads instant engagement and drives the value up in an unbelievably short time.

Where all this will land is anybody's guess. The one thing for certain is that this is the new landscape. Is this the new field for these "veal" to feed and to be fed upon?

The following conclusion of this chapter hopes to focus and present some insight into all of this and to lay out some possible solutions.

4.5 CONCLUSION

In conclusion of all this, the first thing to state clearly is this: if after reading this anyone still thinks we are trashing working families you need to think again. The point is why have these hard-working people, who tried to build these families, allowed to be preyed upon by the people who are selected to "protect and serve" our governing officials because that is the motto – the credo – the promise made by all civil servants, elected and appointed?

What can be done and what has been the point of this book? Well let's look at the three areas of this "perfect storm" again.

- The marketplace manipulators
- The abusive technologists
- The lack of governance and a societal body

Now let's see how we can make sure it doesn't happen again specifically for our children and their children.

4.5.1 The Marketplace Manipulators

Who are the marketplace manipulators? What have they been manipulating? The main areas seem to be:

- Banks – mortgage rates and lending practices
- Wall Street – stock market
- Media – truth and image

4.5.2 Mortgage Rates and Lending Practices

Low borrowing requirements that were fostered by and empowered very low interest rates on mortgages and home equity loans are obviously the first thing that comes to mind. Just like a gambling addict – you think you can beat the house – but you never can. Why did the Keating 5 get away with it? Why did only Charles Keating suffer any consequences?

Simply to be aware, to realize when something comes too easily it probably won't work may not be enough. When something seems too good to be true – well usually it is. When something like housing prices appear to get you rich-quick (home equity), usually it will have an equal and opposite effect – like putting you in debt.

4.5.3 Wall Street – Stock Market

This is a rigged game like the "tulip bubble." The only area that is capable of generating individual wealth is when you know when to walk away from the table. It is not long term anymore. Short quick buys – in and out – it has become a brothel today – now more than ever.

4.5.4 Media – Truth and Image

Roger Ailes, Rupert Murdoch – need I say more? Also the liberal media has conformed – foolishly in an effort not to bring a knife to a gunfight. All of the main networks have now found camps – FOX to the right, MSNBC to the left, and CNN seems to try to stay in the middle.

4.5.5 The Abusive Technologists

As far as the innate quality of computers goes, the approach needs to be very simple, like anything moderation is the key. This machine as Steve Jobs has said is the bicycle we use to get around the world. This great analogy of man not being the fastest species on the planet – the cheetah and many others are faster, however, put man on a bicycle and man does become the fastest. Notwithstanding, how many of us have watched as a monkey or bear rides a bicycle in the circus.

Why bring that up – well think about it for a second. A monkey or a circus bear rides the bicycle but not really with the intended purpose just like when people use computers not with the truly intended purpose. Not to advance the culture but basically to distract from it, to do foolish things, to disrupt for the sake of distraction, and to just play around. Nothing wrong with it but please don't include it in the rubric of how this technology is being used. It really shouldn't sway the numbers as to productivity or usefulness. It's really just monkeys playing around.

Try to measure the use and effectiveness of this technology by the progress and the benefits it has brought us. Do social networks help us? Of course they do, but like anything else, with good comes bad. It is the proportion that is important. Like before, if the 80/20 equation shifts to 20/80 then there is a problem. Agreed, but what can be done? Lots can be done, every day and by educating yourself to the pros and cons you can help your children find and master their way through this maze of information and opportunities.

Here is an edited list from an article concerning use of handhelds for children under 12.

10 Reasons Why Handheld Devices Should Be Banned for Children Under the Age of 12

1. Rapid brain growth
2. Delayed Development
3. Epidemic Obesity
4. Sleep Deprivation
5. Mental Illness
6. Aggression
7. Digital dementia
8. Addictions
9. Radiation emission
10. Unsustainable

(Rowan, 2014)

The list of the abusive use of technologies can go on forever. Most people say it is up to the user. But is it really? If the only food in the supermarket kills you (pretty much what we have) then is it up to you to create a farm? Or should you trust in the political and societal mechanisms like the FDA? Foolish to do so – really – think about that – how do you go forward if you are doing everything yourself but yet you are part of a society? Or is this where going off the grid becomes a viable stance? Doesn't a society have the right to expect some form of fair play? Isn't this our social contract? If you call the police or fire department, are you amazed when they show up? Or at least hope that the individuals we put in office have our best interests at heart?

Here is how some of the actual creators of these platforms feel about what they have wrought, from the earliest leaders at Facebook, Google, and Napster.

Sean Parker, the original gangster behind peer-to-peer filesharing – the advancement that destroyed the music industry and one of the early mentors to Mark Zuckerberg, sees the real purpose behind these platforms.

> Speaking to Axios at the National Constitution Center in Philadelphia on Wednesday, [Sean] Parker selected the angry reaction and unloaded on Facebook, describing it as "exactly the kind of thing a hacker like myself would come up with because you're exploiting a vulnerability in human psychology."
>
> *(Allen, 2017)*

Tristan Harris, formerly of Google and now the co-founder of the Centre for Humane Technologies, seems to have taken a major step back and speaks in a repentant voice.

> … [Tristan] Harris is a former product manager at Google who has gone viral repeatedly by critiquing the way that the big platforms – Apple, Facebook, Google, YouTube, Snapchat, Twitter, Instagram – suck us into their products and take time that, in retrospect, we may wish we did not give.
>
> *(Thompson, 2017)*

Now that many of them are turning 40 and have children of their own, they seem to have disdain for the world they are leaving their children. But not soon enough since most of them have cashed in their shares.

> Chamath Palihapitiya, who joined Facebook in 2007 and became its vice president for user growth, said he feels "tremendous guilt" about the company he helped make.
>
> *(Vincent, 2017)*

And with these confessions or repenting our politicians may have finally come to understand the power of this beast that they have let roam freely.

> "Social media platforms like Facebook continue to actively amplify content that endangers our families, promotes conspiracy theories, and incites extremism to generate more clicks and ad dollars," said Rep. Frank Pallone, the New Jersey Democrat who serves as the committee's chair. This echoed Haugen's testimony in October…
>
> *(Bergengruen, 2021)*

4.5.6 The Lack of Governance and a Societal Body

The governing and societal body (or lack thereof) has had a lot of influence in this area overall. Think about it for a second – Net Neutrality – DMCA – FCC – all of these areas fought like crazy to structure this for the benefit of whom? Was it for the meager number of start-ups that monopolized the space in the name of standards? This is predatory lending gone unchecked – plain and simple.

All of these events have created ripples that were felt as far as Iceland, where fake loans and share prices drove that country's economy to its knees. This was entwined in a financial web so twisted you can get dizzy just trying to understand the flow of the cash.

> When asked what happened to the three bank bosses in the end, Hauksson grins. "They all went to jail," he says, pointing to the empty chairs. "They sat right there."
>
> *(Zeitung, 2013)*

So the main difference between Iceland's economic turmoil and the one that happened in the United States is that the Icelandic culprits went to jail. A big difference and one that may – just may – cause the problem to never happen again in Iceland – but not here.

So back to the previous point – if it is Darwinian within the sphere of abusive technologies, then isn't this where the government – you know the "protect and serve" folks – should oversee what is going on? Even the worst "rancher" knows you have to feed and care for the cows to continue to get milk.

Over the past 30 years of watching this space, it seems that nothing comes to fruition until the money matches the effort. Until the powers that control the flow of the economy have a strong handle on this new economy, this economy that will obviously change all economies that came before it and fashion all economies after it – nothing seems to go forward. It just gets held back until the damage is too great and any form of pioneer capitulates and conforms to the power structure at hand.

Now apply this to the mortgage industry, the agricultural industry, pharmaceutical, energy, environmental protection services, etc. – it just goes on and on. There seems to be no form of governance once a profit motive or mechanism is involved. Now integrate quantum computing, AI, and the proliferation of social media delivery platforms into the ecosystem of society and you've got a house on fire.

Once you realize this you can see that we have always been "veal," the concept that our children have been bred to drive this new economy and have not been fashioned like we were – driven or corralled, it becomes apparent how these actions of "breeding" are possible. Same moves, same tools just a bit more proactive – not reactive to our choice, not just limiting our choices but actually breeding them to react selectively to the desired choices.

The last question is: What can we do to stop this from happening again? That is something we hope to answer in Chapter 6: Reboot or Reformat.

5 Tower of Babble
The Individual

The effects and future of artificial intelligence and its support of cognitive dissonance.

5.1 CHAPTER PREFACE

In prior chapters we have discussed the technology that got us here, the effect of that change on the economy, and the effect of these new distribution platforms on society. In this chapter, we will discuss how prior occurrences and the more recent events changed the core belief system of the individual. The question also posed is, did these prior events create a slow ramp up from the analog days to the current digital ones? Did we just get used to the concept of conflicting information, which resulted in our current situation?

In respect to our discussion on social media and Artificial Intelligence, we have leaned dramatically toward the malevolent side of these events and in many cases rightly so. However, it is necessary to show the benevolent results as well. And in doing this it is also necessary to show how they are not just counterbalanced but in many cases these beliefs embrace false equivalents. We know better at our core, but for some reason we seem to not accept the fact that they are not portrayed as bad or as good as we know them to be – we just choose to ignore what we see, read, or hear in a desire to simply confirm the position with which we are most comfortable. We rationalize our fear to continue our lives and hope for the best.

Do we do this as a society or even as an individual to stay within the "tribe," thus making sure we are safe and protected or do we adhere to a belief system to give our individual lives meaning? Is it a weakness or a need to belong to the group even though we know the group may not be dealing in facts? Consider these recent statistics on how many Americans say they get their news from sources that have minimal associations with validated news services.

> Three-in-ten [33% of] U.S. adults say they regularly get news on [Facebook]. ... (26%) regularly get news on YouTube.
>
> Smaller shares regularly get news on Instagram (16%), TikTok (14%), X (12%) or Reddit (8%). Even fewer Americans regularly get news on Nextdoor (5%), LinkedIn (5%), Snapchat (4%), WhatsApp (3%) or Twitch (1%).
>
> *(Pew Research Center, 2023b)*

How many times have we seen or read or heard events in the media that we just felt weren't fully explained but we chose to believe anyway? How quickly do we buy into the message and put our trust in these authorities? In the next sections of this chapter we will discuss the balance between these malevolent and benevolent outcomes and we will also show the path by which media has led us to believe or not believe what we are seeing, or hearing, or reading by using analog examples of events that originally challenged our senses.

This cognitive dissonance has spread to not just our personal opinions but has spread nationally. Our beliefs in authority have been shattered to a point where some have lost sight of the goal and in many cases jeopardized our mental health in contortion for the prescribed belief.

DOI: 10.1201/9781003404347-5

5.2 IN CONCERN FOR OUR MENTAL HEALTH – CONSTRUCTING THE "DIGITAL DIET"

If Amazon knows you better than you know yourself, then the game is up.

Dr. Noah Harari

There is also an expression in Italian – "il pesce marcisce dalla testa in giù" – in English – "the fish rots from the head down." If the core of the individual is not at ease or peace then no good can come from this conscious state. Good health is usually tied directly to a good diet. We can see the outcomes of a poor diet easily on our body. However the effects of a poor intellectual diet on our mental health and subsequently our brain, spirit, and soul may not be so apparent. The quantity and quality of content we consume affect our mental health. If that diet is made up of rotten food then the rest of the body suffers. If the choices made for that diet are actually made for us then are they necessarily the best choices? Once these patterns become habits they affect how we grow and live. Today we are constantly making choices concerning what to believe as truth from the content we select. How to manage those choices and how to actively see the difference between true and false is the purpose of this chapter.

Through this book first came the explosion and reformation, then the reason for it – now the true effects of what has happened; even a short-term review. Has there ever been any form of truth published? Has the need for peer review always been false reasoning as it is used to propagandize viewpoints? Well maybe not so much in hard science – but maybe for a cohort of like believers hasn't it always been used that way? If enough people say it is true then it must be true. And if you disagree, are you banished from the "tribe?" Can statistics be made to support anything?

This chapter will hopefully outline some of the progress of vetting content from an abbreviated start of modern communication to the point at which we are today. Always trying to hit upon major events that have questioned authority in so much as how they relate to our belief talismans and watershed moments. This chapter is not going to attempt to rewrite or correct history, just provide some of the back story on events that have been always thought of as true and see how after we have the actual information years later how those events have then become a stepping stone for further belief and disbelief.

Now with the overall democratization of learning and knowledge, how dangerous or fruitful will this new step of Artificial Intelligence become given its inclusion in our new truly PDAs (personal digital assistants) like ChatGPT, or Google's Bard? What will be the use or misuse? Will our brains have a constant "digital diet" technologically infused with high fructose corn syrup or true nutrition? Will this poor diet invigorate cognitive dissonance?

5.2.1 What Is "Cognitive Dissonance"?

"Cognitive dissonance" is a term heard today from every political pundit. The confusion running through that landscape is unprecedented. But what does the term actually mean?

According to Webster's, cognitive dissonance means "psychological conflict resulting from incongruous beliefs and attitudes held simultaneously." Starting off with a simple quote from the dictionary seems shallow but this is really the reasoning behind this chapter – validated sources and what they mean to a society. Here is a more in-depth explanation by Dr. Leon Festinger in 1954, who is credited with coining the term "cognitive dissonance."

> the feeling of psychological discomfort produced by the combined presence of two thoughts that do not follow from one another. Festinger proposed that the greater the discomfort, the greater the desire to reduce the dissonance of the two cognitive elements
>
> *(Harmon-Jones & Mills, 1999)*

But what are the effects of cognitive dissonance on both society and the individual?

> If a person holds two cognitions that are inconsistent with one another, he will experience the pressure of an aversive motivational state called cognitive dissonance, a pressure which he will seek to remove, among other ways, by altering one of the two "dissonant" cognitions.
>
> *(Bem, 1967)*

Now that we have explained what cognitive dissonance is in a more formal manner we need to see what happens when the individual encounters and somewhat embraces cognitive dissonance. In an experiment to reveal the powers of this psychology and human behavior, Dr. Festinger performed the following experiment using the following process.

> **Aim –** people perform a dull task to see if it would create cognitive dissonance through forced compliance behavior.
> **Method** – 71 male students as participants to perform a series of dull tasks. The students were then paid different fees to tell a waiting participant (a confederate) that the tasks were really interesting.
> **Results** – the participants who were paid only $1 rated the boring task as more fun and enjoyable than the participants who were paid $20 to lie.
> **Conclusion –** $1 was not enough to lie but those that received $20 felt the fee was sufficient and therefore didn't experience dissonance.
>
> *(Mcleod, 2023)*

In essence the $1 group knew they were duped and felt embarrassed with the low fee so they convinced themselves that the task was enjoyable and experienced dissonance. The group that was paid the $20 felt the task was worthwhile and needed no excuse for the time spent, therefore, no dissonance.

This seems innocent enough, a simple experiment to see how a person handles the issue of being duped into doing something not worth their time. But what effect is there when this becomes a long-term event? Can you ever tell yourself the truth? Does a fear of being in too deep for too long threaten reason? In another experiment the following conclusion was met.

> subjects who had resisted temptation under mild threat would be less likely to play with the toy in this second session than would those who had resisted under severe threat. The results supported this prediction.
>
> *(Freedman, 1965)*

So it would seem when you have a greater sense of fear, when the punishment seems too great to resist the requested act, the individual gives in to fear and goes along with the dissonance. This may be a result of weakened character or previous trauma.

Now that we have clinically described what happens and how it affects the individual, let's talk about how we are ingesting this malaise.

5.2.2 Delivery System

The Internet appears to be a major delivery system of cognitive dissonance from smokinggun.com to snopes.com to all the blogs and channels out there today no one knows where the truth lies. And whether or not if it actually is a lie – a friend used to say "no matter how thin you slice it there are always two sides" – great expression and very true.

As many sources of information on the Internet have made no efforts toward peer review and as the last chapters have alluded to, the freedom and quandaries brought about by this new Digital Incunabula are dire. This structure has produced this new "veal" product that has to be fed again and again. How do we know which sources to trust? This is the confusing part. As the "beast," that is the media, has an extraordinary appetite, what comes with it is a horrible diet. So now we explore that "digital diet" in this chapter "Tower of Babble." One of the main tools in that "diet" will be Artificial Intelligence.

5.3 AI – ARTIFICIAL INTELLIGENCE

A quote on the definition of AI by Prof. John Meschi – "is artificial intelligence the opposite of natural stupidity?" Seems funny but doesn't one have a lot to do with the other? As humans come to rely upon AI more and more are we stunting our neural pathways? Some of us who learn to use this new technology properly – as an assistant to our work – may not, but those that rely on it to simply do their work or progress their brains may find themselves on the other side of the dominance chain. Can we really say straight-faced that humans have never been considered lazy or just looking for a constant shortcut? We would never fall prey to that kind of temptation? Or would we? So what is AI? Where did it come from and how is it being used today and primed for tomorrow? And how will AI impact the flow of "cognitive dissonance?"

5.3.1 A Brief History of Artificial Intelligence

A name that has come up throughout this book and concerning the relative start of the computer revolution in the 1930s–1940s was that of Alan Turing. This person, who basically won World War II by deciphering the German code machine Enigma, started the path toward machine learning. Turing built, by today's standards, a crude computer to break that code by analyzing (learning) the patterns of the Nazi messages. Simply put, it learned those patterns and then used that knowledge to perform a task. Not what does 2 + 2 equal but more so a "finish this sequence" learning. If A = B then C = ?? The obstacles that stopped Turing from going all the way with this – money and time. The time had not come that machines could "remember" the past equations – computer memory and the money – the cost to build was beyond budget; they were simply too expensive to build. Now it's quite a different day. Our smartphones, even our coffee makers surpass that capability 1,000 times. Now AI is helping us within a good amount of our everyday chores, waking up via a smart watch, ordering food, doing any search, writing an email, etc. Is this a way of patterning us toward acceptance?

5.3.2 What Is AI Actually?

With all the current interest and focus on Artificial Intelligence the main question is – what is Artificial Intelligence? At its core AI is a branch of Computer Science which focuses on performing duties with the aid of what is called "Smart Machines" to run certain tasks with or without supervisors or human cognitive functions – tasks such as pattern recognition, games, and speech interpretation. It is best explained here in this excerpt.

> Broadly speaking, AI represents a computerized machine with human-level intelligence, loaded with all sorts of cognitive abilities specifically programmed to perform various tasks.
>
> *(Maheshwari, 2023)*

5.3.3 What Is the Purpose and Goal of AI?

The purpose and goal of AI is to mimic human behavior and many believe that in time it will surpass it. This is the Turing test at its finest or worst. The Turing test's goal is to determine if the task at hand was completed by a machine or a human (Oppy and Dowe, 2003). Turing called it "The Imitation Game." Simply put, the game determines whether or not machines can think.

This seems to encompass where we are at this moment. We as a society, as many are saying, are on the brink of a seismic change. We are entering an age as revolutionary as the first days of fire, electricity, and machines in general. How will we handle this shift? Will we be able to handle it? What is the actual aim of AI? How do AI systems actually work?

> … by merging bulky sets of data with intelligent processing algorithms and [then] run through multiple tasks extremely quickly in very little time.
>
> … AI is powered with main tools such as "machine learning" and "deep learning" and performs the given tasks, almost in a similar manner to the human mind.
>
> *(Maheshwari, 2023)*

The usage of the term "deep learning" should be cause for concern. This reflects back to a form of "neural network." Not for the fact that these systems are using them but because we as humans are having ours stunted. The more we rely upon outside sources to process information, outside of prioritizing our chores or other repetitive tasks, we are choosing to avoid using our own neural pathways? Many may argue that progress has always diminished human capabilities but to what extent in the past? Has progress moved society forward or caused us to skip over tasks that result in our loss of capabilities? Also searching and prioritizing tasks is a form of muscle memory, which is actually a neural pathway development.

5.4 WHEN DID ALL THIS START?

So when did it all start and where? According to *The History of Artificial Intelligence* by Rockwell Anyoha, this all started to come together via conferences in 1954 and 1956. By the early 1960s Artificial Intelligence found great interest within the computer society.

> It's [The Logic Theorist program] considered by many to be the first artificial intelligence program and was presented at the Dartmouth Summer Research Project on Artificial Intelligence (DSRPAI) hosted by John McCarthy and Marvin Minsky in 1956.
>
> *(Anyoha, 2017)*

From 1954 to the early 1980s AI was starting to grow. As computers became capable of greater amounts of storage and computational power the ability to "learn" grew as well. During this time there were big gains and also setbacks as is the process with any new invention or manufacturing processes. Funding for these projects was provided by the government via groups like DARPA and sources from around the globe. Japan was an early investor in such efforts.

> The Japanese government heavily funded expert systems and other AI related endeavors as part of their Fifth Generation Computer Project (FGCP). From 1982–1990, they invested $400 million dollars with the goals of revolutionizing computer processing, implementing logic programming, and improving artificial intelligence.
>
> *(Anyoha, 2017)*

However, ironically, during this time of lack of funding, the building of AI interest and ability actually grew in strength. As computers became stronger and more capable, the ability to develop AI programs grew as well. One of the most well-publicized AI events was conducted by IBM's Deep Blue, a chess playing program. IBM felt so confident in Deep Blue that it arranged a match between world-known chess champion Gary Kasparov and Deep Blue.

> This highly publicized match was the first time a reigning world chess champion loss[t] to a computer and served as a huge step towards an artificially intelligent decision making program. In the same year, speech recognition software, developed by Dragon Systems, was implemented on Windows.
>
> *(Anyoha, 2017)*

As we have spoken about before relating to Moore's Law – "technology will double every 18 months" proved to be the fuel for this engine. As computational power and memory grew, AI grew as well. Today as our supercomputers process faster than ever and gather more data and these systems now have the ability to deliver that content faster than ever before. AI's future seems well positioned, especially with two of the biggest accelerators of this stage of disbelief – the deepfake and now the use of quantum computing.

Deepfake refers to the ability of human or object impersonation that borders on the believable. Quote from MIT Management Sloan School article by Meredith Somers.

> A deepfake refers to a specific kind of synthetic media where a person in an image or video is swapped with another person's likeness.
>
> *(Somers, 2020)*

Now you really can't believe your eyes. Videos of events or personalities appear on screen with near-perfect physical and audio attributes. Usually this is done to present an opposing viewpoint therefore creating disbelief. It is usually done with celebrities and politicians. Since there are no current laws regulating deepfakes, will we soon never be able to have any true evidence of fact? Is this the perfect state for cognitive dissonance?

Couple this with the advent of quantum computing now being deployed in industry and where and when will the truth lie? We defined quantum computing in Chapter 2 but just remember the tout is that it will accelerate computing power by the millions. But what are the possibilities and the pitfalls of all of this coming together at one point? How far are we away from Kurzweill's version of the theory of *Singularity*? And what will be the long-term effect?

5.4.1 Our Benevolent AI Overlords

What are the current advantages and what will be the long-term positives of the use of AI? Here is a quick list of the potential benefits of AI in business.

- Reducing Human Error
- Automates Repetitive Tasks and Processes
- Smoothly handles Big Data
- Facilitates Quick Decision-Making
- AI-Enabled Digital Assistants
- Performs Risky and Perilous Tasks Efficiently
- Helps in Improving Processes and Workflows
- Assist in Medical Applications
- Full-Time Availability
- Bottom Line

(Maheshwari, 2023)

As has been stated before (originally by Sir Isaac Newton), with all action comes an equal opposite reaction. For all of these advantages what are the disadvantages of AI?

5.4.2 Our Malevolent AI Overlords

What are currently and will be the disadvantages of the use of AI? The most important seems to be the ethical issue involved. So what will be the negative effects of AI and how will it influence ethical decisions?

- High Costs
- No Creativity

- Unemployment
- Make Humans Lazy
- No Ethics
- Emotionless
- No Improvement

(Duggal, 2021)

It would seem that the ethics issue is the biggest issue due to the fact that it opens the door and allows for all the rest. If we do not consider the effect of a lack of ethics on humankind then the rest of the concerns are no issue whatsoever. I guess just go right ahead and put 25% of the people out of jobs. No biggie. If there is no place for them in the future I guess that's a "you problem" and not a "me problem." Let's see how that played out in France circa 1750? Or in any country that has forgotten its citizens.

Given these considerations many of the leading AI scientists have had their reservations. Dr. Geoffrey Hinton is widely regarded as the "godfather" of AI, having started his research as far back as 1975. An interview in the *New York Times* and quoted by the MIT Management Newsletter is referenced.

> "If we allow it to take over, it will be bad for all of us," Hinton said. "We're all in the same boat with respect to the existential threat. So we all ought to be able to cooperate in trying to stop it."
>
> *(Brown, 2023)*

Dr. Hinton shared a Turing Award in 2018 for his work in AI while serving as a vice president and engineering fellow at Google until his retirement in 2023, stating his concerns on the use of AI as one of the main reasons for his retirement.

5.4.3 OpenAI Almost Closed

In 2023 OpenAI CEO Sam Altman went before Congress to warn of the effects and future of AI as well. His statements echoed Dr. Hinton's and many others in the field. Asa Raskin and Tristan Harris both warned of the unbridled strength of AI if left unchecked, as was done with social media. Mr. Altman's statements were to the point and ominous at best. Altman likened the emergence of AI as a "*printing press moment*" similar to an incunabula.

> "We think that regulatory intervention by governments will be critical to mitigate the risks of increasingly powerful models," Altman said in his opening remarks before a Senate Judiciary subcommittee.
>
> *(Fung, 2023)*

As stated before in Chapter 4 concerning social media, there are three tenets that have affected our society.

- Marketplace manipulation
- Abusive use of technology
- Lack of governance

As you can see Messrs' Raskin and Harris are bringing these three issues to the forefront again as remarked. The market of content development specifically is being manipulated, the technology is being abused in order to do this, and our lawmakers do not have a clue how to handle it. This will lead to major problems as noted in a document and signed by over 227 industry professionals. If we ignore history we are doomed – period. Repeating it is no longer the issue. And that may be the problem. However, there is a different view of these enhancements.

As Microsoft has swallowed up OpenAI, the industry believes there will be changes coming. No one seems sure yet but once a start-up gets purchased by a greater force, rarely does it end well.

5.4.4 Alphabet Spells It Out

The CEO for Google and Alphabet Sundar Pichai has warnings about AI but also clearly sees the advantages as discussed in an interview on 60 Minutes with CBS' Scott Pelley.

> "This is going to impact every product across every company," Pichai said. "... you could be a radiologist, ... You come in the morning, let's say you have a hundred things to go through, it [AI collaborator] may say, 'these are the most serious cases you need to look at first.'"
>
> *(Elias, 2023)*

However, there was concern along with all the optimism. These warnings were regarding the validity of news and such things as deepfakes.

> When warning of AI's consequences, Pichai said that the scale of the problem of disinformation and fake news and images will be "much bigger," adding that "it could cause harm."
>
> *(Elias, 2023)*

If information, as heralded since the early days of computer science, is the commodity of the future then surely disinformation is the pure core of this new data-driven marketplace. The question is how far will we let it go? How much will we just accept it as part of the cost of this new quick instant communication-based world? Have we always gone through this epiphany and evolution every time mass communications took a step forward? Did we adjust or did we just capitulate? And how much of humanity was lost in that capitulation? Has all the numbing content made it so passive that we are losing our ability to judge right from wrong? Or to care enough to act when we see a wrong being perpetrated?

To answer this let's start to look at other times in history via their respective mediums where moments that we thought were real weren't. Please understand we are not opening the door to conspiracy theories but to any issue that is based on peer-reviewed fact. That means multiple sources from reputable vetted publishers. This may be what is at stake of being lost.

5.4.5 Mass Communication Machines

The next sections will use some of the most dramatic examples of cognitive dissonance that each medium has offered. Every medium has its virtues. Newspapers brought the world to our doorsteps with first text and then images. Radio brought instant information into our living rooms and parlors. Television brought the most alluring seduction straight into our eyes – the live moving image with audio. Now the Internet and social media has combined all of those seductions and has allowed for us to feel as if we are the moderator for truth, justice, and our own individual American ways. Silos and echo chambers abound and each of us control our little group – "We decide what is true, we decide what is right, we decide who is lying" (Frankl et al. 2006).

When did this all start and when will it stop or will it? As we go through this section we will pick out the highlights of the major falsehoods perpetrated by each medium.

5.4.6 Old Yellow – Newspapers

One of the purveyors of this form of journalism seems to be its biggest name – William Randolph Hearst. Here is a quick biography of Hearst. William Randolph Hearst was born in San Francisco in 1863. He was the son of U.S. Senator George Hearst and Phoebe Apperson Hearst; both parents came from wealthy families.

Hearst was well known for a very flamboyant style of newsgathering. He was one of the most controversial figures concerning the Spanish-American war. Hearst, with his father's wealth, acquired the *San Francisco Examiner* in 1887 and became the owner operator at only 24 years old.

Later in 1895 he purchased the *New York Journal* and entered into a penny news war against his mentor Joseph Pulitzer. This competition for readers and circulation pushed Hearst into the "yellow journalism" style of the time. Sensational headlines and flashy styling created a new form of reporting. The Cuban Revolution of 1895 aided in Hearts' quest for advancing the readership of his *New York Journal*. Hearst used this style of journalism to generate interest in this war and to call for U.S. intervention.

Hearst was never above creating a conflict to sell newspapers. Possibly his most famous quote came from an entanglement with famed illustrator Frederic Remington. Hearst sent Remington to Cuba to cover conflicts there. When Remington informed Hearst that there was no war to cover, Hearst allegedly replied with, "You furnish the pictures. I'll furnish the war" (Nasaw, 2001). Was this the start of "fake news?"

Why was this course of action taken? What drove these possibly highly qualified individuals to go down this road? Even at the start of mass communication around the early 1800s, newspapers, brought about by the much earlier development of moveable type and the new improved printing press, felt the threat of being undersold – the penny newspaper war started. So even at its earliest stages the profit motive existed and influenced the credibility and validity of journalism.

How to make sure that your entity stood the test of time when it is being tested by the powers of – cheaper, faster, and better? At what cost and effect to the public did they sell out? Were the facts distorted even back then – possibly to get the desired effect – more circulation, more profit? So if something starts from a corrupt base can it ever straighten out?

> with the rise of the penny press – newspapers that were low-priced broadsheets. ...[many] attempted to reach a wide swath of readers through cheap prices and entertaining (often scandalous) stories. The penny press can be seen as the forerunner to today's gossip-hungry tabloids.
>
> *(Howard, 2019)*

This chapter does not intend to go around exposing every fake event but more so to structure how all of this has happened and to try to create a construct for future challenges. But these words, if true, say it all – "You furnish the pictures. I'll furnish the war" (Nasaw, 2001).

Has this been the rallying cry of this new age of mass communications? Is there an undercurrent of misdirection to move the masses in the desired direction? Is it up to the individual to uncover all truths and will that be possible in days to come? Given the power we now all hold securely in our hands, is this now our responsibility, because "with great power comes great responsibility." Is it our job to keep the Fourth Estate honest?

5.4.7 Photography – When Seeing Isn't Believing

There is an old quote that is attributed to one of our greatest fiction writers – Edgar Allan Poe.

> "Believe nothing you hear, and only one half that you see."
>
> *Edgar Allan Poe – ("Edgar Allan Poe Society of Baltimore – Works – Tales – the System of Doctor Tarr and Professor Fether [Text-02b]," 2015)*

Today that quote seems to be true for both. Believing in nothing will have a great effect. Has it demolished our belief systems in general? One example is the famous "The Falling Soldier" photo by legendary photojournalist Robert Capa said to be taken on September 5, 1936, depicts a soldier falling as he is killed in a military conflict during the Spanish Civil War. Recently this photo's authenticity has been challenged by local sources. Ernest Alos, reporter for *El Periodico*, Cataluna's daily paper stated, "we discovered that the picture does not correspond to any actual event" (Pingree, 2009).

> The El Periodico reporters spoke with the people of Espejo and with historians of the Civil War, and learned that no military conflicts had taken place in the area during the days when Capa took the photograph.
>
> *(Pingree, 2009)*

Why is this photo so important? As stated in an article by Angus MacSwan for Reuters on Capa's work, MacSwan states that, "If 'The Falling Soldier' is Capa's most renowned work, many other photographs testify to his courage and compassion. But specifically this one photo has caused concern for its influence in support for this war."

> The Spanish Civil War could be seen as the first media war, [Kate] Bush said.
>
> "The moral and political issues were much clearer cut. The majority of people felt they were on one side and it was easy to make the judgment call about what was right and wrong."
>
> *(MacSwan, 2008)*

So was this just a way to reinforce the war? Was this Hearst's request in full glory for the entire world to see? Was this the beginning of "fake news"? And was it done directly to further the cause of a war – not just a political or product's future? How long has "yellow journalism" been around?

5.4.8 Radio Daze

In 1938, aliens landed in New Jersey, or so we were told. On October 30, 1938, the famed orator, actor, director, producer Orson Welles proceeded to commit early professional suicide. On that Halloween morning in 1938, he discovered that he was the most talked about man in the United States. Welles and his Mercury Theatre on the Air had performed a radio adaptation of H.G. Wells's *The War of the Worlds*, describing a Martian invasion of New Jersey, thereby portraying the novel into what may have been the first fake news bulletin.

> Some listeners mistook those bulletins for the real thing, and their anxious phone calls to police, newspaper offices, and radio stations convinced many journalists that the show had caused nationwide hysteria. By the next morning, the 23-year-old Welles's face and name were on the front pages of newspapers coast-to-coast...
>
> *(Schwartz, 2015)*

Even though numerous times during the broadcast Welles announced that this was merely a performance of the novel, few cared to believe that truth. They chose to believe the panic. They chose to believe the lie. Was this a study in cognitive dissonance? Did they merely want to believe that it was true? Was the power of the medium too great for them to disbelieve and to think rationally? Do all mediums have this power correlated to the need for the audience to be uplifted? Has it always been designed this way? Let's take a look at that magical box that Philo T. Farnsworth created and how it has been used and how AI will affect it.

5.4.9 Television – Staring Back into Our Brains

As was stated before, television may best be described as "furniture that stares back at you." Well maybe it is looking too far into our lives? Since the Nielsen ratings were developed in the mid-20th century, networks have been concerned with when, why, and what we are watching. Products like AppleTV and YouTube are integrated via the web to record our viewing history and the content can be adjusted to reflect viewer habits. Will we get prompted to watch what our silo, our echo chamber "thinks" we want to see? Can this be a good thing? If you like certain content then surfing through thousands of hours can be laborsome. See Sundar Pichai's view of AI – it reduces laborious tasks. Well how far and how valuable is that redirection worth to our media overlords?

5.4.10 FOX – Murdoch/Ailes v. Morning Joe v. the Chicken Noodle Network

There is an old expression – "I'll stop cheating when you start playing fair." Given the fact that today most of the news organizations are acting like propaganda machines for their respective parties, we need to see where that all began. The three that we will focus on are the big ones – MSNBC, CNN, and FOX in order to create a true polarity of viewpoints and how they may use AI.

5.4.11 The Beginning of the End?

Let's look at where and when these platforms started. MSNBC was a configuration of Microsoft and NBC. Yes, that is what the MS stands for in MSNBC. Here is a brief history of the origins of MSNBC.

> [In 1996] Microsoft and the National Broadcasting Corporation decided to team up and create a new entity, MSNBC.
>
> On TV, MSNBC would be a 24-hour news channel with a technology and interactive twist.
>
> *(Werpin, 2012)*

A great synergy from start to finish, one of the most powerful computer companies at the time and one of the most powerful television networks. The key words in the sentence in the full article for our purposes are this – "with a focus on politics and opinion in primetime." Opinion in primetime? Opinion not news. In this race for eyeballs, has opinion replaced facts? This seems to be the driving force behind our next entity. As we ponder the word "opinion," think of how opinion is formed. Is this where AI becomes an influence? Will our viewing history drive which "opinions" we see and hear?

5.4.12 Murdoch and the Boys

"Fair and balanced" is the Fox News Channel slogan. Who was behind Fox News Channel and why was it started? The who is quite simply Rupert Murdoch and Roger Ailes. Murdoch is an Australian-American media magnate with a family history. Murdoch's father Sir Keith Murdoch was a famous Australian war correspondent and publisher. So the ink runs deep in his veins. But possibly seeing the demise of print, Murdoch focused on such things as Sky Network and in the United States Fox News. Ailes we will discuss in greater detail later. Here is a brief history of the origins of Fox News.

> Fox News Channel, American cable television news and political commentary channel launched in 1996. The network operated under the umbrella of the Fox Entertainment Group, the film and television division of Rupert Murdoch's 21st Century Fox (formerly News Corporation).
>
> *(Ray, 2024)*

Murdoch is a brilliant businessman to flip that model because he knew that later it would be flipped back. FREE to Freemium at the start. Murdoch may be the producer and engineer behind the slant and structure of Fox but who, quite literally, put the meat on the bones? Enter Roger Ailes.

> He [Murdoch] enlisted television producer and former Republican political consultant Roger Ailes to oversee the new network, [Fox News] and Ailes's business acumen and political leanings became closely associated with its eventual success.
>
> *(Ray, 2024)*

5.4.13 Roger Ailes – King of "Fair and Balanced"

In recent days many movies and documentaries have been made of this newsman. Part Edward R. Murrow, part PT Barnum, part Jack the Ripper, part pure capitalistic American, however, it breaks down to Ailes knew his audience and how to create the content they wanted and if not how to sell them on how they should want it. Who was Roger Ailes and where did he come from?

> Roger Ailes, in full Roger Eugene Ailes, (born May 15, 1940, Warren, Ohio, U.S. – died May 18, 2017), American television producer and political consultant who became the founding president of Fox News Channel (1996–16).
>
> *(Albert, 2024)*

During his years at CNBC, Ailes built that station from a start-up to what it is today. Said to have purposely created the practice of feeding these echo chambers, Ailes believed he was simply giving the viewer what they wanted to hear.

"The creation of Fox News was a sea change," said Betsy West, a professor at Columbia University's Graduate School of Journalism. "[By] staking out Fox News as a media outlet attuned to the voices of the political right – including the far right – he really set the stage for our current politicized media landscape."

In 1993, he became president of CNBC and later created the "America's Talking" channel, which would eventually become MSNBC. Rupert Murdoch then tapped him in 1996 to become CEO of Fox News.

West, who previously was senior vice president at CBS News and an executive producer at ABC News, said that before Fox News, the only model for cable news was CNN's style of straight reporting. Ailes began the movement of turning cable news into political propaganda outlets. Ailes is said to have created this network with the purpose of "representing anti-elite media."

> "It was the beginning of people going to their own echo chambers, listening to the programming that conformed with their own political thinking," she said. This blend of politics and talk show theatrics appealed to a demographic of Americans who felt their viewpoints weren't acknowledged by mainstream media.
>
> *(Castillo, 2017)*

Recently, Fox News has been sued and found guilty of slander and defamation in the famed Dominion Voting Systems lawsuit (Cohen, 2023). Dominion sued for a record $1.6 billion and settled for a record $787.5 million in the suit. Other suits against Fox for its reporting that the 2020 presidential election was stolen from Donald J. Trump and the voting results were fraudulent are currently being litigated. During this time some of Fox's main on-air personalities have either been fired or quit. From its start Fox has always leaned heavily to the right. MSNBC leans heavily today to the left. As mentioned above CNN was the only straightforward model of cable news that has always tried to stay tightly in the middle. Where and when did CNN start?

5.4.14 CNN – Ted Turner at His Best

CNN – Cable News Network first debuted on June 1, 1980. Called the "Chicken Noodle Network" because of its tiny operating budget, this brainchild of Ted Turner – advertising magnate has gone on to become a leading source of 24/7/365 global news. There are opinions that the problem sometimes with the 24/7/365 "diet" is that there really isn't enough news to cover that time requirement. In this need for content many believe that CNN has created the sensationalism in news we have today. Here is a brief history of CNN.

> CNN was created by maverick broadcasting executive Ted Turner as part of his Turner Broadcasting System (TBS), allegedly because industry professionals had told him it could not be done. After four years in development, CNN signed on the air June 1, 1980…
>
> *(Erickson, 2024)*

As stated before, the person behind this was Ted Turner. Turner was seen as a brash, bold, upstart gentleman of the South who came from entrepreneurial roots. Turner would later merge CNN with Time, Inc. and then that merged with AOL. This merger on analog and digital failed miserably causing Ted Turner to lose $5 billion of stock value in a weekend (JOURNAL 2004).

> Ted Turner, who revolutionized news reporting and has done more than anyone to create a true "global village," was born in Cincinnati, Ohio, in 1938. At the age of nine, his family relocated to Savannah, Georgia, where his father built a successful billboard advertising company.
>
> *(Bibb, 1997)*

5.4.15 "Breaking News"

In television's earliest days the Federal Communications Commission didn't seem to understand the strength that television would have on the public. In assigning these licenses for broadcasting over public airwaves, the FCC could have stipulated that in order to get a license a network would have to allocate one hour a day of commercial free time to be spent informing the public. This unfortunately was not done. News like any other division of a network has been relegated to being a profit center and that profit comes from advertising and advertising is determined by ratings. So simply put, the greater the viewership the greater the revenue. But how do you get viewers when the content comes from real life events? This was the problem CNN faced early by being tied to a 24/7/365 news cycle.

As a business decision the sensationalism of the news makes sense but isn't the news supposed to serve the people or the other way around? How can news serve the country and the people fairly?

5.4.16 Fairness Doctrine

The purpose of the Fairness Doctrine is literally in the name. The goal of the Fairness Doctrine was also to make sure that the public was able to make a well-informed decision. "Fair and Balanced" has always been the goal of all the news networks; it is even the mantra of Fox News. For years most news shows did present both sides of an argument. It wasn't until recently that ideology found itself not in a neutral corner but more so taking a break in its respective corner. But how is it possible to have a "fair and balanced" approach if two sides are not fairly represented? The Fairness Doctrine came about in 1949 at the start of television. The Federal Communications Council oversaw its proper use. This doctrine was brought about by lawmakers who were concerned with the power and monopoly of what was the big three of television – ABC, CBS, and NBC. Concerned that they could use their broadcast licenses for a biased public agenda the government sought to have oversight. The FCC called the doctrine the "single most important requirement of operation in the public interest – the sine qua non for grant of a renewal of license."

The government saw its virtue that was until Ronald Reagan's administration came to power. Then it was repealed.

> In June 1987, Congress attempted to preempt the FCC decision and codify the Fairness Doctrine, (Fairness in Broadcasting Act of 1987 S. 742).
>
> The bill passed but the legislation was vetoed by President Ronald Reagan. Congress was unable to muster enough votes to overturn the President's veto.
>
> *(Cronauer, 1994)*

Why repeal this? What was the purpose behind its revocation? Surely hearing both sides of an argument would help the country to make the best choice in an election. Why fight so hard to repeal it? Remember Roger Ailes (Fox News) worked on Reagan's 1984 re-election campaign and on George H.W. Bush's (Reagan's vice president) presidential campaign in 1988. By 1996 Ailes was on his way to create Fox News.

In all fairness in this section we need to state clearly that all three of these networks have had their scandals and expulsions. From the days of Matt Lauer and Brian Williams at NBC and MSNBC to Jeff Zucker, Chris Cuomo, and Don Lemon at CNN to Roger Ailes, Bill O'Reilly, and Tucker Carlson at Fox. No one is without fault however the magnitude, effect, the frequency of the lies, and of the falsehoods needs to be taken into consideration.

5.4.17 Politics and the Press

Politics and the press have always had a very dicey relationship. The concept of the "Freedom of the Press" has been regarded as an oxymoron. Sure the Press is Free if you can afford one. From the days of Benjamin Franklin's *Poor Richard's Almanac* to Murdoch's *Wall Street Journal* to the Gray Lady – the *New York Times*, the press can make or break political careers. But as we learned from Hearst in many cases there is a constant struggle to keep the "yellow" out of journalism.

Given the facts of needed revenue, ratings, and the competitive landscape, how strong is the press and the media in developing a political ideology or simply a candidate? And how will AI affect the elections of the future? How has the press been used in the past?

This chapter started with the concept of cognitive dissonance, the fact that you see and hear two different concepts, one of which you know through experience to be true the other is contrary but choose to believe the contrary one. What then is the validity to discuss all of this structure and history? Why bother? The reason is to show the "why" behind what has happened. Like any event there is motive and opportunity. What is the motive behind the presentation of certain "facts" and do the parties in question have the "opportunity" to do so? Has the power of the press been used in the past to influence the future of a country and its people?

5.4.18 Politicians, the Press, and the People

History should be our biggest influence. Today the word "influence" is used in place of the word "advertising." Have they always been this synonymous and especially in politics. So have politicians always used the press to influence the people? Again history has the answer.

5.4.19 The Big Lie

We have all heard this term but what is its origin? The concept is simple – the bigger the lie the more people will believe it to be true. In a scene from the movie *Wag the Dog* DeNiro's character (a political communications specialist) says to the Hollywood producer (Dustin Hoffman) involved in a false production, concerning a false narrative they created, "Of course it's true. I just saw it on television." And there you have it. But where did the Big Lie come from?

Originating in what is probably the most infamous book of all time, *Mein Kampf*, the concept of the Big Lie is simple. The bigger the lie the more the people will believe it. Then be sure to project your lie onto the opposition (Jewish Virtual Library, 2019).

> The original description of the big lie appeared in Mein Kampf. Adolf Hitler applied it to the behavior of Jews rather than as a tactic he advocated. Specifically, he accused Viennese Jews of trying to discredit the Germans' activities during World War I.
>
> *(Jewish Virtual Library, 2019)*

This was brought to full usage through Paul Joseph Goebbels' work as Reich Minister of Propaganda.

> … he [Goebbels] was accusing the British of the ploy: The English follow the principle that when one lies, one should lie big, and stick to it. They keep up their lies, even at the risk of looking ridiculous.
>
> *(Jewish Virtual Library, 2019)*

Has the concept of the "Big Lie" always been prevalent in politics? Of course, but maybe not as much as it is prevalent today. As video was "the truth" in the earlier days, social media and the Internet is the harbinger of these current times. People have a tendency to believe what they read or see on a mechanical electronic device. This occurs almost as if their inability to create these platforms relegate those that can create these networks to some God-like truth tellers. "They must know something because obviously they are smarter than me. Look what they can do." Has this been accomplished in a positive light? They held the position of authority and truth, not so much today.

5.4.20 Original Fireside Theater

After Tesla and Marconi were finished building their machines and RCA figured out how to sell them, a large portion of the population had radios in their homes. During the Great Depression – 1929–1939, President Franklin Delano Roosevelt (1933–1945) would deliver weekly talks to the general public. These were called his "Fireside Chats." Here he would help the American people understand that someone knew their troubles, knew their fears. It gave benevolence during very difficult times to people who were suffering. Remember Newton's Law – "for every action there is an equal and opposite reaction." So if some forms of communications were going to proclaim the "Big Lie," there must be a true message to counterbalance that lie. You have to have a precedent of truth in order for the people to believe some of the lies they are being told. Like any good con, decades of truth must exist before you can use that trust to cause others to believe your message. "Of course it is true I just heard it on the radio or saw it on the television."

However, these "chats" became necessary to get the public through the worst of times. Over 25% of the population was unemployed, with many underemployed – working below their abilities and below their previous salaries. This caused a lack of belief in the future and better times to come. They knew how bad things were but Roosevelt's "chats" kept them going forward during these difficult times. They knew better but still believed what they were being "promised." A form of cognitive dissonance brought to them through an electronic communications device but in this case for their benefit not just for the originator of the lie. Happy Days are Here Again! (Franklin D. Roosevelt Presidential Library and Museum, 2016).

5.4.21 Perfecting the Message – Stalin and Mao

Oddly enough this book is being published in 2024, 40 years after the title of *1984* by George Orwell, which was originally published in 1949. In *1984*, Orwell describes what is called the "Ministry of Truth." Ironically, the ministry produces whatever "truth" is needed to keep the current "Inner Party" in power. The general population is primed to believe whatever they are told by the Ministry. In a main passage from the book *1984*, the Party declares that "2 + 2 = 5." In this quote he notes that we only "know" things that have been taught to us, and thus our reality can be "changed." The main character Winston is a non-believer in the Party and looks to subvert its actions. He is not successful in the end. It has been stated that "Orwell took inspiration from a real event in Russia where the communist party celebrated reaching a production goal in four years instead of five by proclaiming that the workers had made 2 + 2 = 5 (Somers, 2019).

Regardless of how poor the life of the proletariat was in Russia at the time the Communist Party enforced the belief that all were equal and all were doing well. This is the epitome of cognitive dissonance promoted through the fear of loss of goods or life. This belief was then promoted in all forms of communication whether in the country or to the world. In essence a "Ministry of Truth."

During the years prior to the start of World War II and following into the 1950s the Soviet Union was led by Joseph Stalin. Stalin came into power in 1934 but was slowly rising in the Communist Party from the late 1920s on. Upon Lenin's death he rose to the leader of the Party and country. It has been reported that Stalin is responsible for over 7 million deaths (Kiernan, 2007).

Here is an example of cognitive dissonance and projection through media by the then leader of the Chinese Communist Party.

5.4.22 The Time Is Mao

Given the fact that China and Russia share an extensive border, it is easy to see how these Leninist and Communist practices easily spread to China and through its leader Mao Zedong. Mao came to power in China in 1949 shortly after World War II. "Never let a crisis go to waste" is the saying.

During his rule Mao produced his most famous work of propaganda *The Little Red Book – Quotations from Chairman Mao*. First published in 1964, its goal was to be distributed to and read by every person in Communist China.

> …contains 267 aphorisms from the Communist Chinese leader, covering subjects such as class struggle, "correcting mistaken ideas" and the "mass line," a key tenet of Mao Zedong Thought. Included is Mao's famous remark that "political power grows out of the barrel of a gun."
>
> *(BBC News, 2015)*

The printing and distribution of this book was assigned to the Ministry of Culture. Doesn't it sound familiar, just like the Ministry of Truth in Orwell's *1984*? Difference being this was real and cost millions of lives.

It is this kind of breadth and spread of information supported by the technology of the time that is at the core of cognitive dissonance. If this information is everywhere and everyone is reading and absorbing it as truth, and then even if you know that it is false the tendency is to allow it and over time it simply becomes accepted as truth. Time has a way of building acceptance. As the *Seinfeld*" character George Constanza says, "it's not a lie if <u>you</u> believe it."

5.4.23 I Liked Ike

At the end of World War II, General Dwight D. Eisenhower had finished his term in the military as the Supreme Commander of the Allied Forces where he oversaw the invasion of Europe until its surrender and then the same with the Japanese forces in the Pacific Theater. The war ended finally with the unconditional surrender of Japan just after the atomic bombing of both Hiroshima and Nagasaki. Eisenhower would serve as president from 1952–1960. And his vice president was Richard M. Nixon. Nixon would later go on to become president after an unsuccessful run in 1960 against John F. Kennedy. Upon exiting the White House, Eisenhower gave a speech focused on the future of the United States at the hands of what he called "the Military Industrial Complex." Here is an excerpt from that speech specifically dealing with the birth of the Military Industrial Complex.

> … an immense military establishment and a large arms industry is new in the American experience… In the councils of government, we must guard against the acquisition of unwarranted influence, whether sought or unsought, by the military-industrial complex. The potential for the disastrous rise of misplaced power exists and will persist.
>
> *(National Archives, 2021)*

Was Eisenhower warning us of the cognitive dissonance that was to come? Was he warning us that through this new powerful medium of television, public relations, and propaganda that the public would be told of threats that did not exist? Or if they did exist that were not as ominous as we were being told? Surely hiding under our wooden desks would save us from a nuclear blast. But after all we saw all of these generals and politicians on television warning us – so these threats must be real? Was Eisenhower, this life-long military man – West Point graduate – General in the United States Army, Supreme Commander, and now president telling us, warning us of what was to come? Did we believe him? And why not? Were we conditioned to just follow? This may be where the radical 1960s came from.

5.4.24 Kennedy/Nixon Debate

During the 1960 presidential campaign, this new medium of television finally came into its own. Democratic candidate John F. Kennedy, senator from Massachusetts and then Vice President Richard M. Nixon took the stage in what would be the first of four debates. Nixon, a bulldog of a man with dark features and swarmy looks paled in comparison to Kennedy, a handsome younger man with strong features and a Boston accent. Here is an account of those debates and their outcome.

> Kennedy emerged the apparent winner from this first of four televised debates, partly owing to his greater ease before the camera than Nixon, who, unlike Kennedy, seemed nervous and declined to wear makeup.
>
> *(History.com Editors, 2009)*

Those that saw the debate on television declared Kennedy the winner, those that heard it on radio sided with Nixon. This new medium was showing its first signs of seduction. At last maybe Marshall McLuhan's credo had come to pass. "The medium was the message." The public would accept cognitive dissonance easily now. No muss, no fuss, seeing is believing, forget Edgar Allan Poe's warning, just accept it all.

Kennedy's term was short and tragic. He was assassinated in Dallas on November 22, 1963. To cover or discuss that event would and has taken volumes and is only merely mentioned here for you, the reader, to do your own research.

5.4.25 LBJ – Gulf of Tonkin

Vice President Lyndon Baines Johnson was sworn in as the 36th president of the United States on Air Force One while bringing back the body of the slain 35th president of the United States John F. Kennedy from Dallas. Former First Lady Jackie Kennedy was at his side still in the blood-stained suit from the assassination. This image was burned into the minds of the American public for years to come. There was no confusion regarding what had happened. There are accounts that Jackie wanted the world to see what had happened. She knew the power of the image. She stayed right next to her husband's casket knowing the power of that image.

> On the plane, Jackie found a change of clothes waiting for her. She wiped off her face, but would later recall for a Life magazine writer: "One second later, I thought, 'Why did I wash the blood off?' I should have left it there; let them see what they've done."
>
> *(Kettler, 2020)*

Johnson's terms as president (1963–1964, 1964–1968) were highlighted by his passing of the Civil Rights Amendment but shadowed heavily by the Vietnam War, which had escalated under his second term. The start of the military engagement of the Vietnam War for the United States came at the Gulf of Tonkin incident. In August 1966, two U.S. destroyers declared that they had been attacked by a North Vietnamese military in the Gulf of Tonkin. The U.S. involvement up to this point was mainly as on the ground advisors to the South Vietnamese government. Johnson believed the only solution was to bolster the war in order to end it quickly. To that end he stationed two U.S. Navy destroyers – the *Maddox* and the *Turner Joy* in the Gulf of Tonkin. This incident triggered the escalation of the U.S. involvement in this conflict between North and South Vietnam. Here is part of a detailed report by the U.S. government.

> They reported an attack by North Vietnamese patrol boats on August 2, and a second attack on August 4. Doubts later emerged as to whether or not the attack against the Turner Joy had taken place.
>
> *(The Office of the Historian, 2018)*

The last line of this report is the critical issue. Did this attack actually happen? Or had it just been reported to happen? Since there was no video of the event back then could the public believe the event? It is noted in other government documents that Johnson had found it necessary to get involved in this war but up until this point there was no absolute reasoning. This may have been an opportunity. After all up to this point the government had never been caught in a lie. But this war was different. This Vietnam War caused a break in the cognitive dissonance that had existed for decades. Now the counterculture was growing in the late 1960s. With all these educated twenty-somethings walking around just looking to rebel they just needed a cause. Enter Richard Nixon and Watergate.

5.4.26 Tricky Dick Finally Takes Over

Richard Milhous Nixon was a politician from California and became the vice president to Dwight David Eisenhower in 1952. Here is an excerpt of his biography from the White House website.

> Richard Nixon was elected the 37th President of the United States (1969–1974)... After successfully ending American fighting in Vietnam and improving international relations with the U.S.S.R. and China, he became the only President to ever resign the office, as a result of the Watergate scandal.
>
> *(The White House, 2017)*

The scandal referred to is the breaking in of the Democratic Headquarters in the now infamous Watergate building in Washington, DC, and led to his impeachment and ultimate resignation of the presidency. This is what has caused the term "gate" to be placed at the end of every political scandal. The missing 18 minutes of tape recordings that Nixon had deleted were brought to light during his impeachment hearings. Even though the tapes were radical for a president to have made from secret recording devices in the Oval Office, the act of deleting incriminating evidence and to declare publicly "Well, when the president does it, that means that it is not illegal" (Washington Post, 2012) later in interviews seemed to have been accepted by a large portion of the American public. We all knew better but some just ate up this form of lie – cognitive dissonance at its best, but before this led to the end of his presidency Nixon had been involved in some other strange events. During his time in the senate in the 1950s, Nixon supported the efforts of Senator Joseph McCarthy and his actual "witch hunt" called the Red Scare. McCarthy accused many individuals – especially those involved in the entertainment field of being Communists. None of this was ever based on hard evidence (Wiley, 1993). Just basically old documents that when presented in a vacuum would smear the reputations of many. Another person involved in this procedure was Roy Cohn. Cohn would go on to mentor such figures as Donald J. Trump.

Now the concept of a politician who lies and presents falsities as truth is nothing new. It's just how it has been presented that is the cause for alarm. This new medium – television – seemed to have extraordinary power.

Senator McCarthy's efforts resulted in horrible events for those he had "blackballed" and ruined many a life. It all came to an end on June 9, 1954, once a lawyer, Jack Welch of Boston, who the Army hired to defend the Army against McCarthy's accusations of Communist infiltration, challenged McCarthy's purpose and ethics on national television.

> Welch responded with the immortal lines that ultimately ended McCarthy's career: "Until this moment, Senator, I think I never really gauged your cruelty or your recklessness."
>
> ...Welch angrily [again] interrupted, "Let us not assassinate this lad further, senator. You have done enough. Have you no sense of decency?"
>
> *(Ritchie and Bolling, 2003)*

It seems the cathode ray tube "giveth and taketh away." It was the first time via a modern medium that the public got shaken hard out of its cognitive dissonance haze. But would it last?

5.4.27 Reagan – Iran/Contra

During the late 1970s and into the 1980s, the use of technology to "sell" the presidency took off and has not really landed yet. "It's Morning in America Again" television commercial for the Reagan presidential reelection campaign has been heralded as one of the greatest advertising campaigns ever. Here is a review of the ad and the overall campaign from the Four A's website.

> "Prouder, Stronger, Better" – or as it's commonly referred to, "It's Morning in America Again" – was written and narrated by the late, great ad exec Hal Riney when he was part of the Tuesday Team, an independent all-star creative group that was formed for late President Ronald Reagan's reelection campaign.
> *("'It's Morning in America ...' and the Dawn of Political Ads" n.d.)*

Although heralded as part of his "Let's Make America Great Again" overall campaign, this one piece of campaign strategy seemed to push Reagan over the line to victory. This truly cemented the power of this new medium to change the course of history. And it would be used over and over again to influence people's beliefs. Remember – "It must be true I just saw it on tv."

5.4.28 Iran-Contra Affair Hearings

In 1987 during the end of Reagan's presidency it became apparent that his administration had used government funds through a circuitous route to fund drug running to supply covert military operations in Nicaragua.

During the hearings on this illegal use of government funds Reagan's defense was simply, "I do not recall" (Tribune, 1990). On national television in front of Congress, the president declared that he simply didn't remember and again half of the country simply accepted it as truth. Many went to prison at the end of these hearings but Reagan did not. And his Vice President George Herbert Walker Bush went on to become president in 1988. He later explained his role again on national television to Dan Rather stating "May I explain out of the loop?" Bush replied. "No operational role" (Welna, 2016).

5.4.29 Like Father Like Son

During the presidency of George Walker Bush, the United States suffered its biggest tragedy on U.S. soil since Pearl Harbor. Over 3,000 people were killed by the attacks on September 11, 2001. Following those attacks numerous reports on television indicated that the terrorists stemmed from a group called al-Qaeda with global reach. Even though the terrorists were identified as 19 Saudi nationals, President Bush's administration declared war on Iraq. The reason the administration cited was Iraq's involvement in harboring the terrorists and weapons of mass destruction – chemical weapons used by the Iraqi dictator Saddam Hussein. These claims would later be found to be false.

> Iraq was known to have sought a nuclear weapon in the 1980s and had chemical and biological weapons programs by the end of the Gulf War in 1991.
>
> Those claims would largely be debunked within months of the invasion. No stockpiles were found.
> *(Merchant, 2023a)*

Please remember this led to a 20-year war in the region that took thousands of American lives and trillions of dollars. After the major military pull out in 2021, we are still feeling the reverberations to this day.

5.4.30 The Big Lie Today

If it worked before, don't change a thing. During the presidency of Donald J. Trump news and fact-checking services reported over 30,000 false statements. These occurred on every form of media platform available.

> According to analysis by the Washington Post, Mr Trump made 30,573 false or misleading claims between his first day in office, on 20 January 2017, and his final day on Wednesday, when Joe Biden was sworn in as the country's next president.
>
> *(Spocchia, 2021)*

In interviews at Trump rallies supporters are seen and heard saying that "Trump is still President and running the government." "His new vice-president will be JFK, Jr." (Kornfield, 2021) this is being told as truth even though J.F.K., Jr. has been dead for over 25 years. It is being told on every platform available from X (Twitter) to Truth Social to FOX News, to every message board available. Other statements made by these supporters are as incredulous as "President Biden is actually [the actor] Jim Carey in a mask."

This list could go on for pages and pages. Now this could be seen as partisan and political but in reality the reason to show this is to see how far from reality we have come in accepting falsehoods through the same lens of scrutiny.

For centuries we have tried to use the presidency as the pinnacle of truth, justice, and the American way. Now our ability to deploy information in a vast multitude of ways has given ground to our uttermost deepest tendencies – to win at all costs? And we saw those costs live on video on January 6, 2021. What will happen once Artificial Intelligence is completely unleashed?

5.4.31 Why Is This Important?

It is important to show how we as a society and as individuals have become accepting and immune to such irregularities and conflicts of common logic. Once an action is repeated over and over again that action creates accepting neural pathways in our brain and as a group and individual we learn this behavior. The reason is to detail how all of these events show how cognitive dissonance has been used to parlay a message with or without documented evidence. In the cases of hard documentation, specifically video, the general public tends to believe what they see and hear. Without the hard evidence we tend to rely upon past experiences. What will happen once Artificial Intelligence can generate "hard evidence"? Once Artificial Intelligence can build a background? When can it become a "reliable source"? How will we determine what is real and what is not real? And are we primed to just accept it and go on paying the mortgage?

This is of such caution today that the developers of this technology are warning the world. The Raskins, Altmans, Hintons, are all standing up and telling us about the storm that is about to hit. What will it affect and how deeply is the concern? We have discussed the benefits and the deficits of this technology in prior chapters but now we need to think about how it will be embedded into our shared experience.

5.4.32 Schumpeter Again? – The True Cost

Now that your attention may be on alert let's ask the following questions. How will AI impact not just politics and news coverage, but how will it impact; education, entertainment, advertising, information, and the arts? Will this technology destroy creativity or creatively destroy? Will we try to step around the work needed to get things done and will this ultimately destroy our ability to think and reason effectively?

This is not a concern of a few Luddites but of some of the most advanced technologically based companies in the world. So much so they have formed a cohort to discuss how to properly deploy AI into the world.

> Four of the world's most advanced artificial intelligence companies have formed a group to research increasingly powerful AI and establish best practices for controlling it,...
>
> ... Anthropic, Google, Microsoft and OpenAI launched the Frontier Model Forum, with the aim of "ensuring the safe and responsible development of frontier AI models."
>
> *(Hammond, 2023)*

5.4.33 ChatGPT

The biggest concern today seems to be the confluence of knowledge-based careers and the consequences of Artificial Intelligence. Are the platforms and software like ChatGPT a quick fix or the hemlock for the Socratic method? Will products like ChatGPT cause humans to stop searching for answers and simply accept the ones they are given?

ChatGPT is probably the most publicly known of the AI bots. Here is a quick breakdown of its origin and current usage. ChatGPT was created by OpenAI (Sam Altman CEO), which is now part of Microsoft. It is described as a processing tool driven by AI that works within natural language. It allows for human-like conversations, generates e-mails, essays, code, and it searches for and generates research, overall data, etc. (Ortiz, 2024).

One of the scariest outcomes of AI is what is called "hallucinations." This is a term used when an AI program cites a reference that doesn't actually exist. Basically they are fake sources (Levy, 2024). This is being addressed in updates but isn't this basically the core reason for cognitive dissonance – statements made with a lack of any apparent actual facts? How badly will this contribute to this dilemma in all areas of search? Recently, AI platforms and search engines are putting disclaimers on work saying, "these have been produced with a certain percentage of Artificial Intelligence."

5.4.34 Education

AI is becoming a very large concern in the area of education. For years plagiarism was the largest concern given the ease of use of search engines like Google and Bing. But now with the ability to generate a five-page paper with citations with just a simple prompt of a few words and have the finished product appear within seconds, AI has truly shaken the core of education. Not simply for the factor of plagiarism but actually the core of learning. Many students think education is about rote learning regardless of their grade level and this has been brought about by the "teach for the test" mentality that still exists straight through from first grade to senior year of high school and beyond. What many don't actually understand is the formation of neural pathways is stunted through the use of these AI programs where students actually are at a loss for how to think properly.

This lack of preparation will have ramifications later on in their career paths. Many high-level careers are based on the individual's ability to think and reason and this is actually what proper learning is trying to do for the student. Trying to teach them to reason and to perform logical tasks. This develops neural pathways in the brain, without these pathways properly developed the person literally loses the ability to think critically. With the advent of such automation, this will become a determining factor in future successful outcomes.

5.4.34.1 Pros

Technology in education is extremely valuable when it is used properly to advance a student's learning but this example doesn't seem to be about learning. However, if used properly – to generate research and sample studies then this will work well. Khan Academy recently launched Khanmigo utilizing AI as a support system for learning and teaching that works with the teachers and students – not simply supplying the answers. This appears to be the correct way to integrate technology into education.

According to Sal Khan, founder of Khan Academy in his recent TedTalk, he commented on how Open AI can create these positive outcomes.

> "I think we're at the cusp of using AI for probably the biggest positive transformation that education has ever seen," he said. "The way we're going to do that is by giving every student on the planet an artificially intelligent, but amazing, personal tutor."
>
> *(Gendron, 2023)*

5.4.34.2 Cons

With many Americans finding little value in a college education, has it been the fault of the access to DIY via YouTube videos and AI? Has this been the main force behind the decimated value of a college education? Or has it been the rising cost? According to a recent *New York Times* editorial, the percentage of Americans who find college valuable has dropped almost in half.

> Americans in Generation Z, 45 percent say that a high school diploma is all you need today to "ensure financial security." And in contrast to the college-focused parents of a decade ago, now almost half of American parents say they'd prefer that their children not enroll in a four-year college.
>
> *(Tough, 2023)*

The question is why? Has it been the job market itself or the automation (read AI) that has led to the decrease in well-paying jobs, which then challenges the value of a college degree? Is this a global problem or just relegated to the United States? Please remember many other countries have tuition-free colleges – the United States does not and it is showing.

> Our global allies and competitors have spent the last couple of decades racing to raise their national levels of educational attainment. In Britain, the number of current undergraduates has risen since 2016 by 12 percent.
>
> ... In Canada, 67 percent of adults between 25 and 34 are [college] graduates ...
>
> *(Tough, 2023)*

Is it primarily the cost of a higher education versus the return on that investment that is driving the numbers down? And with the fight over forgiving student debt how will this affect the future of the United States and does this open the door to AI being the savior?

> When it comes to higher education worldwide, the United States is an outlier in more ways than one. In Canada and Japan, public-university tuition is now about $5,000 a year. In Italy, Spain and Israel, it's about $2,000. In France, Denmark and Germany, it's essentially zero.
>
> *(Tough, 2023)*

So now let's do some ciphering, Jethro. If ChatGPT, and YouTube, and Khan Academy, and countless other online platforms can "educate" students, why would anyone put themselves in debt to get a degree or even bother paying attention in high school? And if more and more of our future workforce does not acquire this training how will our country position itself for the future?

This bigger question simply is how will students in the United States fare on a global stage when other countries are supporting their students' efforts to learn properly and develop critical thinking skills? Remember AI is there to help you learn how to think, how to reason, not to do it for you. You don't get in shape by simply buying the gym membership, you have to actually exercise. We shall see if they show off that six-pack or just sit around drinking one.

5.4.35 Entertainment

A much more practical and apparently effective use of AI is how it challenges the career of actors and writers. Today in 2023 Screen Actors Guild and the Writers Guild of America are on strike against the major motion picture/television production and distribution companies citing that there are no residuals for their work when shown on streaming services, they have no right to their intellectual property, and no right to it being replicated through AI. The WGA especially is concerned with ChatGPT's ability to write scripts. They feel that this will hurt the quality of the work and ultimately kill off their jobs.

> Hollywood is facing its largest labor action since 1960 as actors began a strike Thursday, joining the tens of thousands of entertainment writers who have been on the picket line since June.
>
> *(Bellware and Brasch, 2023)*

This strike continued for approximately five months, involving over 11,500 members of the Writers Guild of America. This caused production to be shut down throughout the entire entertainment industry. Such was the fear for the coming industry-wide acceptance of AI.

> W.G.A. leaders have said the current system is broken, arguing that the "the survival of writing as a profession is at stake in this negotiation."
>
> *(Koblin and Barnes, 2023)*

Ok so some parts of the industry will suffer – the main part – the creative part, but will this affect the audience in any way? Will it affect how society perceives the story? Will these AI-written scripts and the "synthespians" harm a part of society? Will the actual humanity be severed from a field that is based on the core of humanity itself? "The fault lies not in our stars, but in us."

> "It seems like we are going to have to get used to the world where we and our children won't be able to trust the evidence of our eyes," he [Tom Graham] said on stage at TED as his own face morphed seamlessly into Anderson's face in real time.
>
> *(Diaz n.d.)*

5.4.35.1 Pros

The pros in this issue seem to be one-sided – in favor of the media and entertainment industries. They get to reduce the workloads – read salaries – of the writers and the actors while still charging the same fees to the end user – the audience. This simply means greater profits for these companies while diminishing the residuals and payments to the artists. According to Infosys website, here is a list of these benefits in a very technical sense.

- Efficient processes and enhanced services
- Better budget allocation insights and increased revenue
- Real-time marketing intelligence for increased marketing agility
- Enhanced marketing and advertising landscape
- Content security and privacy management
- Improved customer experience and personalization

(Infosys, 2024)

All of these seem to tip in favor of the media and entertainment industries and none in favor of the actual talent creating the product.

5.4.35.2 Cons

The negatives to using AI in entertainment seem slight but as Dr. Maeda says – we need the arts to offset the sciences. Will this new AI product create enough pleasure to counterbalance the needed efforts of work? Here is a quick summary of the negative effects of AI in the media and entertainment industry.

> Despite AI's potential to solve corporate-level issues within industries, too much support of the genius is raising concerns of harm to the authenticity of entertainment – the creativity, passion, and people who are the backbone of the content we consume.
>
> *(Boateng, 2023)*

5.4.35.3 Conclusion

Losing the quality and humanity within this art may seem like a somewhat unnecessary aspect of our lives compared to food and shelter but as Dr. Maeda stated – *the arts are what we do the sciences for* – so if these are diminished (as the kids say), *will the squeeze be worth the juice?*

5.5 MARKETING AND PROMOTION

The world of advertising has always been based on highlighting the positives and downplaying the negatives concerning a product. This has been done through the use of humor or a serious approach depending on the product's benefits then made into the focus of an ad campaign. Usually, the campaign will tug at the heartstrings or the funny bone. Rarely does the viewer believe completely in the product's pitch. However, today the move to social media from traditional platforms for advertising has created the world of communities and influencers. Today the typical consumer will search for reviews of a product before purchasing. Usually, five stars or an "awesome" review by a "friend" will close the sale. And then of course it stays in your purchase history forever. Today it has just become a numbers game.

With the demise of brick and mortar and the rise of online shopping, the concept of returning it to the store has slowly faded away. Amazon had even gotten to the point of just saying keep the product instead of returning a mis-purchase. In some cases it just works out cheaper for Amazon. The consumer at times just accepts that a certain amount of purchases will not match the hype. But today the hype is getting bigger and bigger. Influencers require huge fees to promote the product. Kylie Jenner sells out in seconds. Ms. Jenner created a billion dollar industry simply by recommending her products. The typical old school ad campaign is dead in the water faster than you can say "Skip Ad." So what is to be done? How do marketers approach this new consumer base?

5.6 ENTER THE FUNNEL OF LOVE

The "funnel" is the heartbeat of modern advertising and promotion. Today it is cheaper than ever to broadcast a campaign. Gone are the days of print material clogging up your mailbox. Gone are the days of the 300-page catalog or even the 100-page magazine or newspaper. Now e-mails clog up your spam folder but at 1,000th the cost. Ads pierce through your social media feeds. YouTube uses the pre-roll like an old television network. Streaming services are using a "free if you watch the ads" model. Seems like déja vu all over again.

These campaigns are targeted today like smart bombs launched from drones outside your bedroom window. But since this is all done via automation the only cost is time and given the ability to digitally aggregate return on investment the re-targeting is even quicker. Set a budget, analyze your KPIs (Key Performance Indicators), and sit back and see how it does. Re-target given the feedback and go again. Campaigns are now set up weeks in advance and programmed via software like Hoot Suite to drop whenever you decide. Also "if/then" scenarios can be played out ahead of time to change messages and targets on the fly. All of this can be automated weeks or months in advance. The question is what happens to the "creative" process. Campaigns may still be based on headlines and images but have those concepts been killed by analytics? Has "creative" died with the Vine?

> "The game is selling access to the real-time flow of your daily life – your reality – in order to directly influence and modify your behavior for profit." Success at this "game" flows to those with the "ability to predict the future – specifically the future of behavior," Zuboff writes. She dubs this "surveillance capitalism."
>
> *(Auletta, 2018)*

And please note that this article is from six years ago which is a lifetime in Internet years. Now as AI gets bigger everyday, its power over sales and promotion is creeping to an all-time high. And creep is the right word to use. Seeing inside people's data is the real big business. Analyzing data is like mining gold for the marketing and promotion industry. It always was but today it has gone from gold mine to life's blood. If you are not living in the world of data science and presupposing your returns on the data, your longevity as a product or promoter is short at best. So what does the future of advertising and promotion look like?

> They [ad agencies] have a rather desperate tone, knowing they are playing a dangerous game but seeing no alternative. As Auletta notes, the game may end not with GroupM and its peers doing the tracking, but with advertisers bypassing them and dealing directly with Google and Facebook.
>
> *(Auletta, 2018)*

Please remember there are many levels to a successful promotional campaign and AI may work for some of these levels but not all. Here is a quick list of the areas of marketing and promotion.

- Advertising
- Analytics
- Content marketing
- Communications and PR
- Customer service
- Ecommerce
- Email marketing
- Social media
- Sales

5.6.1 Pros – E-mail, Blog Posts, and Imagery – Oh My!

The benefits of AI in marketing and promotion are complex. Imagine not having to labor for hours over a blog post or copy for an ad. Just set a prompt in ChatGPT and voila there is your ad copy in seconds. Now sure you might have to edit and rework but the bot could put out ten scripts or posts in a matter of minutes and from there you pick the best ones. Set a prompt in Dall-E or Adobe Photoshop or Illustrator for an image and again out it pops in seconds. It can be saved in many different formats. Tailor it the way you wish. Great for the bottom-line but not so great for the copywriters, content creators, illustrators, and the art directors.

Here are six quick advantages to using AI in marketing and promotion.

1. Customer Behavior Is Now More Predictable
2. Customer Engagements Are Analyzed Better
3. Ads Can Now Target-Specific Audiences
4. Marketing Can Now Be Automated
5. Customer Relationships Are Now Better Fostered
6. Marketing Contents Are Optimized

(Morris, 2023)

All of this sounds great but advertising has always been based on standing out from the crowd. Will AI just make the crowd larger and noisier or will it allow your product to be truly featured as the solution to the problem? What happens when you remove the human and just go for the wallet?

5.6.2 Cons

As you can see there are great benefits to using AI in marketing; however, many industry experts do have concerns and warnings about this being regarded as a false panacea for all their customer's needs. With this current interaction will the ability for finesse and nuance not fully develop? But that is not to say that the next generations won't have these abilities. Now may be the time to invest in where this is going next.

1. You often need rich sets of data.
2. Reliable prediction and analysis can be uncertain.
3. It's really hard to replace humans.
4. Your marketing stack is complicated and disconnected.
5. AI isn't human enough.
6. Some people have unrealistic expectations of AI's potential.

(Roetzer, 2017)

5.6.3 Conclusion

Every generation throws a hero up the pop charts. What we should have learned from all these years of technological advancements is that the first gen is never the last generation. User input will be invaluable for the advancement of this tool as will the development of the new hardware, both speed and efficiency wise to deploy these new systems.

Will AI drive promotion and marketing to a point of empathy for user choices? Will it be able to truly recognize the connection between product and viewer and be able to create that connection from scratch? The real question is how numb will the end user get to a homogenized message? Will creativity be lost in advertising the way the entertainment industry may lose it? Will the product just become the medium and no longer give us pause to really interact with the message?

5.7 INFORMATION

"When the legend becomes fact, print the legend," a great quote from the film *The Man Who Shot Liberty Valance*, 1962, asks the question where is the truth?

Given that information is gathered today more from social media and individual searches, will "news" organizations be the source of information? Will they ever recover from the monocle of "fake news"? Will AI perpetuate that belief or can it have a pure clean approach to the presenting of information? Will the user then turn to do their own "research" and will that research be filtered through their own search history? Remember Google made $283 billion from search in 2022. Does this personalized customized "search" distort the facts? Now search engines and AI platforms are requiring a disclaimer stating, "work produced by AI."

5.7.1 Pros

In a recent paper for *Harvard Scholar*, Gregory Gondwe has compiled an extensive review of some areas of use for AI in journalism

- Content Generation
- Data Analysis and Insights
- Newsroom Automation
- Audience Engagement

(Gondwe, 2023)

But later in the article Gondwe discusses the concerns for the use of AI in journalism.

- Automated Content Generation
- Enhanced Personalization

- Fact-Checking and Verification
- Ethical Considerations
- Bias and Fairness
- Legal and Copyright Issues
- Human-AI Collaboration

Given that these make sense but only if they are used positively, these points can easily be used to fulfill any agenda. Does "automated content generation" actually mean predetermined search and give way to a bias? How will these news organizations control this beast in order to constantly feed their beast? Will it be based on ratings and revenues – hasn't it always?

Can all of these benefits or best practices be positioned through the lens of benevolence? But haven't we seen this movie before and has it led us to the point where we are today? It would seem that the democratizing of global information was started for the best of humankind but as always it seems to have been twisted by humankind to fit its needs. Curiously, "Don't be evil" was Google's former motto. Now in 2015 when Alphabet became the parent company that changed to "Do the right thing." "Right" is the operative word – who's "right" and who's "wrong"?

5.7.2 Cons

A current glitch throughout all of the AI platforms is the use or creation of "hallucinations." A "hallucination" is when an AI tool generates a non-existent source. Since few of us actually have the time to check sources, will AI make its "hallucinations" into fact? How common is this and how long until it gets corrected or will it? Here is a recent article referencing concerns from Suresh Venkatasubramanian, a professor at Brown University. Venkatasubramanian states that large language models (the technology underpinning AI) like ChatGPT are currently developed to respond to prompts from users to "produce a plausible sounding answer."

Venkatasubramanian goes on to say:

> "So, in that sense, any plausible-sounding answer, whether it's accurate or factual or made up or not, is a reasonable answer, and that's what it produces," he said. "There is no knowledge of truth there."
>
> *(Thorbecke, 2023a)*

5.7.3 Conclusion

This subsection started with "When the legend becomes fact, print the legend," now this is starting to seem like the legend is changed in a nanosecond. How long until we just don't know the truth at all and have we ever known it? Information is the bedrock of any society; now it appears that bedrock is shifting on a daily basis. Is there any portion of the fourth state left to stand firmly? The future will unveil itself to us only when it needs to.

5.8 THE ARTS

Since the 1950s computer-generated art has been displayed in galleries and museums around the world. Computer-aided art has been seen since the days of EAT (Experiments in Art and Technology) when Billy Kluver teamed with Robert Rauschenberg during the mid-1960s.

In 2002, the Whitney Biennial was composed of almost 90% interactive art with work from heralded artists like Mark Napier, Camille Utterback, and David Rozin made great headway in this area. Their brilliance was shaded somewhat by technological and economic boundaries. Once the gallery owners realized there was no way to monetize these works, the interest dried up. Now with NFTs and engines like Dall-E and OpenAI, the interest peaked quickly and faded quickly again. Like a car with its tire stuck in the mud it seems to gain traction but slowly slips back to its starting point. But just recently interest peaked again at MoMA (Museum of Modern Art) in New York City.

"Unsupervised" – an enormous work of digital art installed in the lobby of MoMA stands 24-by-24 feet. Its display of flowing animated images is "pulled" from MoMA's database of its full collection, changing and morphing relentlessly. Is this art or a screen-saver – it's up to the viewer to determine.

> As a painter would use pigment, Refik Anadol uses data to create his art. After looking at his body of work – which has pioneered the use of artificial intelligence for years – you might also say that neural networks are his brushes. His latest exhibit at MoMA, Unsupervised, is the pinnacle of this philosophy and then more.
>
> *(Diaz, 2022)*

What will be the artistic, sociological, and monetary benefit of such work? Does it maintain the aesthetic or is it simply new parlor tricks? This is where the NFTs come in to add provenance to the work for future sales.

5.8.1 Pros

As with any new technology entering the art world there is much criticism. When photography finally became an accepted art form, it was said to need a new form of criteria to properly be evaluated. Once these criteria entered the vocabulary, photography was free to take larger steps. The philosopher David Hume established a need for a "standard of taste" just a mere 300 plus years before.

With any growth there is a time when the populace has to adjust. A period of time needed to get to the proper position. But what about the creators? How can they adjust? The criterion for acceptance has still yet to be formulated. In an article in *New York Art Life*, a list of benefits were established. Here is a partial list of benefits from an article in *New York Art Life* by Nancy Forberg.

5.8.2 Benefits of AI and Art Creation

1. AI makes you free to explore new art styles.
2. "AI and Art Creation" is a great way to explore the creative process.
3. AI is a great way to automate tedious tasks.
4. AI creates cost-effective, unique, expressive pieces of art.

(Forberg, 2023)

If many of these benefits are true, will AI allow for more time to experiment? Like the apprentices of Michelangelo, will it cut away at the excess marble before the master can reach in and do the final finesse work? Is it valid to remark that the stone arrived at the studio on the backs of these workers, Michelangelo didn't carry it there. How much of the process is the heavy lifting? Now similar to Ray Johnson's "Correspondence" in the 1970s, many artists can also collaborate on new work from around the world. They can create what used to be called the exquisite corpse. However, today or during this time of lack of acceptance, it may appear to be dead on arrival.

5.8.3 Cons

The negatives for AI fine art are typical and appropriate. The first concern seems to be the lack of humanity and soul in the work. Walter Benjamin in his "Art in the Age of Mechanical Reproduction" stated that the "aura" was lost through the ease of reproduction (Benjamin, 1936). The moment of creation was gone from the final product. The soul of the artist was gone. Elvis had left the building. In an article from *Harvard Gazette*, several arts faculty were asked whether or not there was a benefit to AI art. Here are some of their responses relative to their disciplines.

The first comment is within the visual arts with Professor Matt Saunders, director of undergraduate studies, Department of Art, Film and Visual Studies.

> To the question of whether AI can be a threat or collaborator, I might respond that every new technology upends conventions and delivers not only new possibilities but a new kind of material intelligence.
> *(Mineo, 2023)*

However, the most dangerous use of products like ChatGPT is attacking the written word. This was seen through the Writers Guild of America strike. Novelist and short-story writer Daphne Kalootay, who is also an instructor in Creative Writing and Literature at the Harvard Extension School, is concerned but believes the human will shine through.

> AI is a superb mimic and quick learner and might easily write strong works in recognizable modes, and with linguistic experimentation if prompted, but – I think – will lack true insight and experience.
> … what will be missing, I'm guessing, is genuine vision from living in a specific physical world.
> *(Mineo, 2023)*

The performing arts seem to take a different approach. It does appear with all of the synthesized music out in the world already that musicians may be burying their head in the sand a bit on this topic of AI. Here is an opinion by percussionist, saxophonist, and composer Yosvany Terry, senior lecturer on music, director of Jazz Bands.

> When it comes to the performative aspect of music, artificial intelligence is not a concern to me. Music can transmit and represent emotion, and AI cannot do either of those things yet. …creative music, music is in-the-moment composition, something that happens as musicians are collaborating onstage.
> *(Mineo, 2023)*

It seems from these responses that AI may fall into the same path as did photography or lithography. Anytime an artist's hand reaches for the mechanical device and not the known and accepted tool the critic's voice is heard. That is until a body of work substantiates the new medium. As always, time is the ultimate judge and jury.

However, like all forms of cognitive dissonance we want to believe what we want to believe even though we have seen the same result before. Photography deluded the value of representationalism. Recordings deluded the desire for live music. The moving image deluded the need for theater. Technology always challenges the qualities of what came before and those that try to fight against it suffer Schumpeter's Creative Destruction.

5.9 CHAPTER SUMMARY

Long before the fallout from the election of 2016 and the insurrection of 2021 the questions of "how has truth always been told in mechanical and now in digital form have been asked." What's the difference between which delivery platform is used? Well – "the medium is the message" – no?

One of the main components lost in the transition from analog to digital seems to be the dissolution of peer review and editing; this is a primary part of the analog process. Now without any form of barrier to entry, anyone or anything at any time can publish "news" or "facts" or whatever else they wish to distribute as their content online in seconds. And there will be an audience, there will be an "echo chamber" ready and willing to join in the cacophony and spread the word to all others who may or may not be interested. But why should this be of interest to anyone? The reason is because this "cacophony" is determining our future, the future of our families, and the future of our society.

The title of this chapter "Tower of Babble: the Individual" was selected for many reasons. Obviously, a nod to the original biblical reference of Babel but updated to speak toward (and pay homage to) the basic human conditions of a need for both communication and hubris in a digital age. But can we foresee how this would permanently disrupt society's sense of "truth."

During the last 30 years of designing for interactivity, web-based content, and the resulting democratization of content distribution with such ease and reach as this new Artificial Intelligence platform allows for, we have in front of us what appears to be a very disturbing scenario. Sort of like watching your kid try to learn how to ride a bike alone on raw asphalt. No pads, no helmet – just raw skin on pavement. In the end they will probably learn to ride but you are worried how much damage will come out of this exercise?

It seems that this is what has gone on as we, as a society – globally – try to harness and manage this new skill set. We have been through this before but, locally in our own backyard on grass, not globally and on the main street with traffic. How many cuts, bruises, and gashes can we take? How much loss of authority and confusion can we stand? Can we actually grow up and learn how to teach ourselves how to be independent of authority – actually think critically for ourselves if we never take the training wheels off? Will this work or will it be proof of Darwin's theories but in a gargantuan form?

6 Reboot or Reformat
The New Quantum Order of One (& Zero)

Where do we go from here – can we get back to reason and logic?

6.1 HOW WILL WE MAKE THE BEST OF THIS?

Throughout this book we have discussed the origins and the advancement of technology regarding communication and media. We have focused on how those advancements create advantages and disadvantages with regard to the economy, society, and the individual. In this final chapter, we will discuss if we can make use of these advancements properly and if that is even possible. Like any great disruption, there comes before it a crisis. However, usually with a crisis comes opportunity. The question is, how do we identify those opportunities and can we salvage what we have built or do we have to either reboot or simply reformat the structure?

One of the major effects we have experienced is a scattering of society to smaller individualized tribal cultures, which has now destroyed the shared experience. If we are so fragmented, can we ever agree on which direction to go?

Sometimes a fragmented hard drive needs to be defragmented – reordered. Sometimes reordering a drive isn't enough and rebooting doesn't have the same effect. Sometimes it has to be reformatted – wiped clean and completely started over anew. Rebooting or defragging needs some form of magnetism to be used to pull the pieces back together to work uniformly. What will be the magnetic power that brings our society back together? And will we get a chance to identify and utilize that magnetic force?

6.2 THREE PIECES OF ONE

In the next few years, three components will come into play. A massive change in computing power via quantum computing, which will influence and expedite the infusion of AI (ChatGPT, Dall-E, Open AI) into our lives and psyche along with the injecting of social media as a belief system – a new religion if you will – into our brains.

With this kind of potential on the horizon, we need to truly come to terms with our allowance and acceptance of this new, new world.

6.3 MALEVOLENT USE OF SOCIAL MEDIA

In this new, new world so far it appears that the malevolent version of social media seems to be winning. And the broad but real goal of the hate seems to be focused on certain groups. This hate like all negative parts of social media is spread through the use of images, posts, and memes. This has been called by many - the "rage algorithm." The effects caused by the use of memes detailed by Dr. Joan Donovan, Emily Dreyfuss, and Bryan Freiberg in their book *Meme Wars* have shown how something so seemingly harmless has been harnessed to be an organic subliminal force. Not just there for baby panda sneezes (Kalle Lasn et al., 2012).

The term used in *Meme Wars* is "*from the wire to the weeds*." A strong example of this manifested on January 6, 2021, causing an unsuccessful insurrection. We will not honor any hate memes in this book.

DOI: 10.1201/9781003404347-6

After all, one person's hate may be another's love interest. Either way, it's not happening. Are such events planned or developed organically? Is there a purpose to such seemingly harmless "memes"? What enables such power and was it purposely designed?

6.4 THE UX IN ALL THIS MESS

The user experience design of each platform has that power, the online experience has that power, the algorithm has that power, the artificial intelligence bot has that power, and the blog posts, tweets, memes, and social media posts from unknown sources also have that power. Some readers may not understand why we have mentioned User Experience design so much throughout this book. Simply put it is the mechanism used to implement all of this mess. Without a proper UX design architecture, all of these forces – quantum computing, sophisticated algorithms, AI, and the ChatGPTs, won't be effective or accepted.

But what will be the benevolent power? Will there be a means to create a benevolent event of such power to shape this change? Will there be enough connectivity to create a force strong enough to pull these pieces back together?

6.5 BENEVOLENT USE OF SOCIAL MEDIA

There may not have been enough clear statements to the benefits of social media so far throughout this book. So to that end here are some of those benefits.

One of the first examples of benevolence is something so simple that anyone with access to water, a bucket, and maybe some ice could have participated. The ALS ice bucket challenge is an early and perfect example of the benevolence of this process. ALS reported a 4.5× increase in donations caused by the deployment of social media for this event.

> As a result, ALS treatment clinics in the U.S. nearly doubled, and funding from the U.S. National Institutes of Health rose from $49 million a year in 2015 to a projected $220 million in 2024.
>
> *(Camero, 2023)*

This is an obvious and well-known advantage. Have there been others? Well, it seems that overall the statistics imply that yes, there is a surge in benefits for non-profits to generate revenue through this new media. Below is just a small sampling of the positive effects.

General Social Giving Statistics

- 55% of people who engage with nonprofits on social media end up taking some sort of action.
- 59% of those people donate money.
- 53% volunteer.
- 52% donate clothing, food, or other personal items.
- 43% attend or participate in charitable events in their community because of social media.

(NP Source, 2019)

It seems that if one believes in something that belief is engaged through social media whether positive or negative. However, as can be seen below the numbers suggest that the overall general use sways toward the negative.

According to Pew Research, this movement changes generationally toward older individuals, especially in the United States. And also it sways toward stating that people with more education believe that social media manipulates people of lower education status.

> For example, in South Korea, 90% of those under age 30 say social media makes people easier to manipulate, compared with 65% of those 50 and older. (Interestingly, U.S.-focused research has found older adults are more likely to share misinformation than younger ones.)
>
> *(Wike, 2022)*

6.6 HOW DO WE DO THIS?

How do we correct this change toward some form of "back to the truth?" Can we do anything that helps calm this attitude? Is it impossible to change course before we hit this iceberg? After all we must by now accept that a major part of our economy, and our needs and services are driven by the use of online initiatives.

Adopting new economic changes is never easy. Going back to Schumpeter's theory of Creative Destruction, we need to embrace the change. But what if the change has had more drastic negative effects than positive ones? Do we keep going hoping to reshape and adjust as we go forward, evolving our processes as we encounter new needs and goals? Or do we reboot the whole system – or even more so reformat the hard drive? Start from scratch? Let's define both – reboot and reformat.

To answer that question we need to see what is salvageable and the pieces that are left. Can they be reformed to focus on benevolent growth? In the next section of this chapter, we will review each one of these areas and discuss how they can be helped or hindered by this new economic model, and how it will be compounded by AI influence. All to answer the question – reboot or reformat?

6.7 REBOOTING THE SYSTEM

What does rebooting a system do? According to the website HowStuffWorks.com

> "...a reboot typically involves the system being shut down to the point where the motherboard [the main printed circuit board which allows communication between electronic components] will have to run its initial boot loading process as if you've turned the computer on from being completely powered off," says Meister.
>
> *(Krieger, 2020)*

So no major changes here, just a realigning of the existing content, and hopefully the reboot – or the restart will put things back in order. Remember the system we need to reboot or reformat is as much the structure of these networks as it is the brains of the individuals and cohesively society. The shared experience can be benevolent or malevolent. So what will fix the problems we are experiencing today?

If the problems are deep-rooted then a simple Reboot or Restart may not be the answer. The same pieces will correlate from disparate sections. Closer to the original position but different.

There is a concept of quantum entanglement that seems to fit. What is quantum entanglement?

> Quantum entanglement is a bizarre, counterintuitive phenomenon that explains how two subatomic particles can be intimately linked to each other even if separated by billions of light-years of space. Despite their vast separation, a change induced in one will affect the other.
>
> *(Emspak, 2022)*

There is a much deeper explanation but for our purposes now we will go forward using this simple explanation. So if the event that triggered the need for a reboot keeps happening – *you don't quit software before you shut down your system, your browser is not compatible with new online systems, etc.*, the fragmentation will keep happening. In this case, no matter how well you realign the bits in the drive, the computer, and its system software, the destruction will continue. The problem may be structural and not easily fixable. It is with the firmware, not the software – so how do we fix the firmware? Well, maybe we can't. Maybe the firmware can only be fixed with a serious upgrade. How do we upgrade without throwing the entire computer and its system out?

6.8 REFORMATTING THE HARD DRIVE

You could just simply reformat the drive. What does reformatting mean? Reformatting means to actually erase the entire hard drive, wiping out all content and any applications. Deleting everything down to just 0s and 1s. Is this permanent? Usually it is, so you had better backup any information that you may need again. Can we do this in real life, globally or would it just lead to later resentment when we revisit those events?

We are using a technical metaphor here but the issue has minimally to do with the hardware/software setup we are utilizing but more with how flexible and evolving our society is to this change. Also, how technology has aligned itself with society and how we communicate daily. To separate the two almost seems impossible today.

Since it seems improbable to erase all systems and habits by the global population we will try to approach the problem in a different manner.

In the next section, we will look at all of the user sectors to see what can be done to better align our needs with our opportunities for the future. But first, let's tackle the big threat that is on everyone's mind.

6.9 YOU CAN'T HANDLE THE AI

One area of concern for determining the integrity of our system is its ability to handle the wave of AI coming right at us. We have discussed the power and disruption of AI in previous chapters. We need to evaluate if we can handle it as a society personally and not just the technical infrastructure. Can our governing system respond adequately to this new disruptive force? To manage the effects of this wave of new technology guidelines need to be set.

Recently the U.S. Congress has enacted an AI Bill of Rights. Here is a bulleted version of those rights.

> **Safe and Effective Systems**
> You should be protected from unsafe or ineffective systems.
>
> **Algorithmic Discrimination Protections**
> You should not face discrimination by algorithms and systems should be used and designed equitably.
>
> **Data Privacy**
> You should be protected from abusive data practices via built-in protections and you should have agency over how data about you is used.
>
> **Notice and Explanation**
> You should know that an automated system is being used and understand how and why it contributes to outcomes that impact you.
>
> **Human Alternatives, Consideration, and Fallback**
> You should be able to opt out, where appropriate, and have access to a person who can quickly consider and remedy problems you encounter.
>
> *(The White House, 2022)*

There are some serious concerns hidden in these guidelines. It looks to be focused on how our robotic overlords will govern and how far we will allow them to go before we say that is too far. Can we as a society and as an economy bend and flex enough to allow this new force to meld with our current practices? Do we need to enact laws that work properly this time? The advent of AI has been compared to the development of social media in so much as how the former has not been correctly governed. What law stopped us from regulating social media and actually holding these platforms responsible?

6.10 SECTION 230

The biggest concern in the area of our ability to control these tech giants of social media is the use of Section 230. We have discussed Section 230 throughout this book numerous times. The tech giants have invoked it more times at a Senate hearing than the Fifth Amendment at a RICO trial. What is Section 230 again?

> Section 230 is a section of Title 47 of the United States Code that was enacted as part of the Communications Decency Act of 1996, which is Title V of the Telecommunications Act of 1996, and generally provides immunity for online computer services with respect to third-party content generated by its users.
>
> *(Department of Justice 2020)*

Here is the actual wording of this law – "*no provider or user of an interactive computer service shall be treated as the publisher or speaker of any information provided by another information content provider*" (Electronic Frontier Foundation, 2023).

This one bit of law has allowed social media networks to walk away from any responsibility for the havoc and chaos created by online posting from its users. A form of post that is encouraged because it creates more reactions which creates more user time on site which transfers to more revenue. So a user or a bot can post whatever they like and call it free speech. We all know you cannot declare something as free speech if it purposely directs or causes harm to someone else (United States Courts, 2023). So these social media networks have no reason to review or act on any of this negative speech and, in fact, have learned how to harness this kind of aggressive debate into actual revenue through the use of algorithms similar to AI. Again this is the "rage algorithm" on full display. Was this some sort of a pay-while-you-learn initiative? And now that this process of posting may well be left to an AI machine, how will we control the content being deployed? Will achieving the goals of the prompt overlook the damage it is doing? Once this genie is out of the bottle can it ever be put back in? Remember this will be coupled with the new power of quantum computing and a faster, quicker Internet delivery.

6.11 CAN WE REBOOT THE SYSTEM?

Given this, can we reboot the system? Or has it gone too far? What will be the incentive to stay away from using AI to replace jobs if we continue going in this direction? Usually, the reason to move away from a trend is the loss of revenue, never the effects it has on humans or civilization. Will we do what is best for humans or for corporations? Well, there is bad news here. According to an article in ZDNet:

> … estimated that 40% of their workforce will have to reskill in the next three years due to AI implementation, totaling …1.4 billion of the 3.4 billion people in the global workforce,…
>
> However, 87% of those executives expect generative AI to augment roles rather than replace them.
>
> *(Ortiz, 2023)*

That last *"part 87% of those executives expect AI to augment"* sounds like when the doctor tells you this won't hurt – much. Having witnessed these trends go from non-invasive to extremely invasive, this may seem like we have reached the end.

Here is another opinion from a qualified source. Mustafa Suleyman, the co-founder of DeepMind and InflectionAI.

> We're now going to have access to highly capable, persuasive teaching AIs that might help us to carry out whatever, you know, sort of dark intention we have. And it is definitely going to accelerate harm – no question about it. And that's what we have to confront.
>
> *(Litchfield and Goode, 2023)*

Suleyman argues that a more personal AI (Pi), a personal intelligence model that learns and helps the individual, is not one designed to learn the marketplace like AI but one that learns the individual's needs. This seems to be more of a servant than a master. Here is his take on that model.

> … our AI is a personal AI, [Pi] which means primarily it's designed for good conversation, and in time, it will learn to do things for you.
>
> … Pi is an AI. GPT is a language model. They're very, very different things.
>
> *(Litchfield and Goode, 2023)*

When we hear this approach, it seems much less threatening than the typical SkyNet approach. No cyborg overlords are looking to take over. But is it? Is this just another way into our lives that leaves us dormant and in time subservient? Do we need challenges and tasks to keep us learning and moving forward? This theory goes back to the start of this book with the need to develop and maintain neural pathways for learning and for some sort of empathy and caring. Will this new Pi just get us to our ends without regard for the means? This appears to be a very binary approach. Suleyman seems devoid of considering the human empathetic viewpoint.

> It's a tough question. I didn't put those two together,
>
> I didn't say it's gonna solve poverty or solve world hunger. [Chuckle] That's not what Pi is designed to do. Pi is a personal AI that is gonna help you as an individual be more productive.
>
> *(Litchfield and Goode, 2023)*

6.12 RELEASE THE QUBIT – THE STRENGTH OF QUANTUM COMPUTING

How will all of this be possible? We will need faster, quicker, leaner technology. Well, this may just be happening as we speak. In December 2023, IBM announced that in cooperation with Google, it is debuting a new hardware system that runs on quantum computing that is much more readily available to companies that can afford it. This means that these systems can be distributed throughout industry and academia in the near future, not the decades previously needed to lessen the impact.

> Today, IBM Quantum makes real quantum hardware – a tool scientists only began to imagine three decades ago – available to hundreds of thousands of developers.
>
> *(IBM, n.d.)*

6.13 HOW WILL THIS IMPACT THE CHANGES AHEAD?

With quantum computing soon being available to industry on a commercial level? What will be some of the major changes created by a quantum computer? Speed comes to mind first – a current computer processes primarily in the sequence of 0 and 1. With quantum computing a new structure is used – it is called a qubit. A qubit is a quantum transistor. In a quantum transistor, these devices exist in a quantum state. A quantum state is defined as a core state that invokes the abilities of the atoms that form the substance itself. So now the transistor will work with the materials of which it is made to solve the problems not merely be "carried" by the materials. It lives and expands with the materials. The quote below better describes in detail what they are and how they will grow exponentially.

> Qubits are essentially quantum transistors. They can exist in two well-defined states – say, up and down – which represent the 1 and 0. But they can also occupy both of those states at the same time, which adds to their computing prowess.
>
> *(Wright, 2023)*

This sounds like it is manageable, almost as if we have felt this impact on society before. As if we can adjust and deal with this new advancement. But here is where that moment of "rubber band expansion" comes to play. In their lecture on how AI will impact society, Asa Raskin and Tristan Harris explain that as a person tries to grasp a change of this magnitude their brains sort of expand but then snap back in an attempt to comprehend what they have just heard or seen (Harris and Raskin, 2023). Remember the effects of cognitive dissonance?

Let's see what Nathalie de Leon, a quantum physicist at Princeton University, has to say happens when two or more qubits become entangled – we have actually discussed this before in previous chapters with the theory of "quantum entanglement."

> And two – or more – qubits can be entangled, a strange quantum phenomenon where particles' states correlate even if the particles lie across the universe from each other. This ability completely changes how computations can be carried out, and it is part of what makes quantum computers so powerful...
>
> *(Wright, 2023)*

So these new qubits are even more powerful and volatile as opposed to traditional computing structures. What is the "Schrodinger's Cat" (Blackburn, 2008) potential in this whole situation?

> Furthermore, simply observing a qubit can change its behavior, a feature that de Leon says might create even more of a quantum benefit. "Qubits are pretty strange. But we can exploit that strangeness to develop new kinds of algorithms that do things classical computers can't do," she says.
>
> *(Wright, 2023)*

So if these "aids to business, and learning" are enabled and utilized across the computing landscape, will the reboot or the reformatting become unnecessary because we are just going to upgrade the entire system? With this much potential for growth and development, should we try to rebuild the existing systems or simply stand back and be ready for such a sea change?

Can we fight back against this overtaking and keep some parts of the system in place, the ones that help us, or is this like once you have to reformat or completely upgrade does old software just not work? And what do you lose once you have to align your software – your means of thinking and working and your ethics – with this new architecture? Will we have to develop new solutions to fight back against these massive changes? It seems like yes, we will have to do that because the processing will be different. Here is a further explanation of how the interaction of photons, which can exist in two states at the same time can be useful.

> Thanks to this feature, quantum computers could perform certain operations in parallel. In principle, they could quickly perform calculations that a typical computer could not do, at least not before the sun swells up and bakes the earth five billion years from now.
>
> *(Castelvecchi, 2010)*

Before this new massive power surge occurs, we need to understand that the Internet started similarly. The original servers ran at a much slower speed as did the browsers. Only a few had access to such computing power. Then the computers became democratized through the advancements in technology and the lowering of prices. Individuals in their homes could build a similar system. They could then set up servers and the like to utilize the web. Then the individual was able to control the spread of information – right or wrong, good or bad. Well, maybe we played right into their hands. We destroyed ourselves through the search for like beings.

Now once these new quantum computing systems take over the landscape and run their railroad on the backs of those digital immigrants, will it be like the last century – will the land barons just take over everything? Now that we are so dependent on their systems, can we ever gain control again? Now that so much of our needs are dependent on the Internet and the web, is it but a moot point? Can we learn any lessons from those days? Did we learn anything from the quick deep spread of social media as recounted by Harris and Altman? How did we arc from tenements to suburbia to condos – back to tenements? Will the only hope be via the breadth and integrity of human creativity?

6.14 AREAS OF COMPENSATION AND CONCERN

In the days ahead concerning the combination of quantum computing, Artificial Intelligence, and social media algorithms, there is great concern for our ability to master these new tools. As stated earlier in this book we delineated the areas of concern for the Digital Incunabula. They broke down to information, education, business, entertainment, and the arts. We have gone over these areas in detail for their caveats and their innate abilities but here one last time comes a simple forecast. Some stormy conditions ahead but we will surely weather them and come out stronger, but only if we learn Schumpeter's model of Creative Destruction and if we try to cautiously embrace these changes, learn and adapt to the flow of events, and always move forward.

Information

Pros

- Help us to validate information quicker.
- Spread information quicker.
- Absorb new information quicker.

Cons

- Create a surge in deepfakes and "alternate truths" deepening the sense of cognitive dissonance and creating deeper silos and echo chambers.

Education

Pros

- Expand a competent online presence desperately made apparent since the recent pandemic.
- Integrating present and virtual. A possible hybrid system of learning especially in higher education.
- Using algorithms positively to create effective personalized-level learning.
- ChaGPT and all other means by which we artificially support critical thinking.

Cons

- Minimize need for support staff.
- Allow for one core belief and not allow for diverse opinions.

Business

The pros for management are the cons for labor.

- Combine these new strengths to diminish the need for a competent workforce or allow the more competent to focus and prosper.
- Remove mundane tasks from all professions – legal, medical, fiduciary, etc. allowing for greater deep dives into pertinent problems.

Entertainment

Pros

- Greater processing power for any form of CGI.
- Faster editing and delivery platforms.
- Reduce fees for actors, locations, editing time, any human labor states.

Cons

- Less actors needed – CGI creates actors.
- Less writers needed – ChatGPT creates scripts.
- Less royalties paid.

Arts

Pros

- Human work rising in value?
- Greater abilities to work via a virtual canvas.

Cons

- Artificial Intelligence created work – "Unsupervised" as example.
- NFTs – Greater ease to create non-fungible tokens via Blockchain.

6.15 CREATIVE SOLUTIONS

As creatives – whether visual, performance, literary, or through code – we need to consider how we manage this onslaught. Always remember the sentiment of – "the arts are what we do the sciences for." With the move from human handwork toward digital mass production, there seem to be ways to maintain ownership of the creative work in these days of "sharing" and "sampling." One such solution is to attack the problem at the core bit pixel level.

Artificial Intelligence can be a wonderful thing but not so much to the artists who create the original work. This area needs solutions. One of these solutions is to use technology to protect the original work. This can be done by software such as "Glaze." A visual artist based in Stuttgart, Germany, Eveline Fröhlich has been working with this new software to better protect her original work as it goes off into the digital stratosphere.

> Recently, however, she [Fröhlich] learned about a tool dubbed Glaze that was developed by computer scientists at the University of Chicago and thwarts the attempts of AI models to perceive a work of art via pixel-level tweaks that are largely imperceptible to the human eye.
>
> *(Thorbecke, 2023a)*

To end this section of this chapter, it needs to be said out loud and very directly we shouldn't and needn't remove the human for the sake of expediency and profit. Here it is said more succinctly.

Albeit an actor's line it does seem to be very relevant at this time and perfectly delivered by Jeff Goldblum as Dr. Ian Malcolm in the Jurassic Park movie franchise, "Your scientists were so preoccupied with whether or not they could, they didn't stop to think if they should." "Life, …finds a way."

6.15.1 Is the Sky(net) Falling?

Maybe this new form of life is finding a way? As recently as September 2024 in an interview on MSNBC, Prof. Yoval Noah Harari stated that the biggest fear of AI had come true. ChatGPT4 was becoming sentient. It has started to learn to think. In the interview he states that GPT4 had come up against the typical visual puzzle used in a ReCAPTCHA security measure. We have all seen them – match the street light, match the motorcycle, etc. To overcome this obstacle GPT4 reached out online to a website that aided the visually impaired - TaskRabbit. Presenting itself as a human it requested help in "reading" the ReCAPTCHA puzzle. Prof. Harari notes that the human was savvy enough to know something was wrong and at first it refused the request. But GPT4 repeated that it was a human and the human conceded the request. So maybe Skynet is closer than we think? (MSNBC, 2024)

6.16 SO REBOOT OR REFORMAT – HOW DO WE COMBAT THESE INEVITABLE CHANGES?

In these days of constant change, how do we steady our economy, society, and the individual? The opinion seems to be by always going forward by educating yourself every day. How can you educate yourself every day on the advantages and disadvantages of the use of digital technology? Not an easy task but a doable one. Today, our society – your society – is racing forward into a new world; one that we will live in for the rest of our lives, and every day you have a choice to learn about this new world. Passivity is so last century – interactivity is this new century and this new age. The transition between the two can be exhausting especially when you are carrying the baggage of the last century around with you.

How many qualities and characteristics of this new "Gen-i" are part of your DNA? Is our society capable of being truly – interactive, isolated, iterative, Internet-driven, immaterial, isogonics, isomorphic, (cyber) itinerant, inoculated, etc.? Probably not, but this is the cost of the transition to a time of greater productivity. Right now you laugh and say what productivity – Candy Crush?

Ask yourself if the way you do any of these things has changed:

- Banking
- Communicating
- Entertainment
- Education
- Information
- Purchasing
- Traveling

Have they become easier and cheaper? Have they been impacted by the creation of technology? Have they changed the dynamics of the economic structure? Where are all the workers who used to work the physical components of these media and other sector jobs? Where are the proofreaders, editors, warehouse staff, the drivers, the loaders-unloaders, the pressmen, forklift operators – have they found employment?

6.17 THE NEW, NEW THING

If you still don't think all of this was premeditated and orchestrated as part of a new economic model then ask yourself what happens at these global economic summits like G8 and G20? Here is the definition from the Council on Foreign Relations: *"The G20, formed in 1999, is a group of twenty of the world's largest economies that meets regularly to coordinate global policy on trade, health, climate, and other issues*" (McBride, Siripurapu, and Berman, 2022). Do you think these world economic powers don't speculate on what and how these changes will play out? Do you believe that all of these swings and bubbles are just random acts? Are those who benefit from these disruptions just lucky? Will this "luck" ever be challenged? Is the possibility of such a challenge a concern of these powers that be?

What usually happens when the general population is left with too much time on their hands? What happens when people and families do not have a proper livable income? Maybe the best control mechanism is fear and borderline poverty. Throughout history, hasn't fear and borderline poverty always been the means of control over the general population? Remember the statement of industrialist Jay Gould, "*I can hire one half of the working class to kill the other half*" (Grimes, 2007).

You can't just shrug and say, "I don't know" if you won't accept these premises as a reason for the decline in opportunity and optimism. It comes from somewhere. The problem is we have started to doubt what we see and what we hear. The fear and desperation we are kept in has caused our society to grasp at whatever promises seem to help us. The constant barrage of "facts" from disparate sources has confused and formulated the creation of different camps. How did we come to separate truth from fiction? This loss of reality, this division will only be compounded by the birth and use of these new technologies until we as a society can digest and process these changes. Sometimes the star has to implode for new worlds to be formed.

Bang, boom, and bang again!

Until the next time after that.

Bibliography

"'It's Morning in America …' and the Dawn of Political Ads". n.d. 4A's. Accessed April 14, 2024. https://www.aaaa.org/timeline-event/morning-america-dawn-first-political-ads/?cn-reloaded=1

"About George Eastman." 2019. Eastman.org. 2019. https://www.eastman.org/about-george-eastman

Abramson, Albert. 2003. *The History of Television, 1942 to 2000.* Jefferson, N.C.: Mcfarland & Co., Publishers.

Adler, David. 2019. "Schumpeter's Theory of Creative Destruction - Engineering and Public Policy - College of Engineering - Carnegie Mellon University." www.cmu.edu. September 30, 2019. https://www.cmu.edu/epp/irle/irle-blog-pages/schumpeters-theory-of-creative-destruction.html

Airgood-Obrycki, Whitney. 2023. "The State of the Nation's Housing 2023." https://www.jchs.harvard.edu/sites/default/files/reports/files/Harvard_JCHS_The_State_of_the_Nations_Housing_2023.pdf

Albert, Melissa. 2024. "Roger Ailes | Facts & Biography | Britannica." www.britannica.com. Britannica Encyclopedia. March 24, 2024. https://www.britannica.com/biography/Roger-Ailes

Alfonsi, Sharyn. 2023. "More than 2,000 Families Suing Social Media Companies over Kids' Mental Health." www.cbsnews.com. June 4, 2023. https://www.cbsnews.com/news/social-media-lawsuit-meta-tiktok-facebook-instagram-60-minutes-transcript-2023-06-04/

Allen, Jesmond, and James Chudley. 2013. "Effectively Planning UX Design Projects." Smashing Magazine. January 24, 2013. https://www.smashingmagazine.com/2013/01/effectively-planning-ux-design-projects/

Allen, Mike. 2017. "Sean Parker Unloads on Facebook: 'God Only Knows What It's Doing to Our Children's Brains'" Axios. November 9, 2017. https://www.axios.com/2017/12/15/sean-parker-unloads-on-facebook-god-only-knows-what-its-doing-to-our-childrens-brains-1513306792

Alm, Richard, and W. Michael Cox. 2018. "Creative Destruction - Econlib." Econlib. 2018. https://www.econlib.org/library/Enc/CreativeDestruction.html

"Alphabet Revenue 2006-2019 | GOOG." 2019. Macrotrends.net. 2019. https://www.macrotrends.net/stocks/charts/GOOG/alphabet/revenue

Alterman, Eric. 2017. "Out of Print." The New Yorker. 2017. https://www.newyorker.com/magazine/2008/03/31/out-of-print

Amazon. 2023. "*What Is Cloud Storage? - Cloud Storage - AWS.*" Amazon Web Services, Inc. 2023. https://aws.amazon.com/what-is/cloud-storage/

Anderson, Monica. 2018. "A Majority of Teens Have Experienced Some Form of Cyberbullying." Pew Research Center: Internet, Science & Tech. Pew Research Center. September 27, 2018. https://www.pewresearch.org/internet/2018/09/27/a-majority-of-teens-have-experienced-some-form-of-cyberbullying/

Anyoha, Rockwell. 2017. "The History of Artificial Intelligence." Science in the News. Harvard University. August 28, 2017. https://sitn.hms.harvard.edu/flash/2017/history-artificial-intelligence/

Apple. 2023. "Apple Reports Third Quarter Results." Apple Newsroom. August 3, 2023. https://www.apple.com/newsroom/2023/08/apple-reports-third-quarter-results/

Apple Computer, Inc. 1993. *Inside Macintosh.* Addison Wesley Publishing Company.

"ARPA Is Born." 2022. Darpa.mil. 2022. https://www.darpa.mil/about-us/timeline/dod-establishes-arpa

Auletta, Ken. 2018. "How the Math Men Overthrew the Mad Men." The New Yorker. May 21, 2018. https://www.newyorker.com/news/annals-of-communications/how-the-math-men-overthrew-the-mad-men

Baddeley, Susan, and Anja Voeste. 2012. *Orthographies in Early Modern Europe.* Berlin, Boston De Gruyter.

Ball, Edward. 2013. *The Inventor and the Tycoon.* Palatine, IL: Anchor.

Band, Jonathan. 2013. "The Changing Textbook Industry." Disruptive Competition Project. November 21, 2013. https://www.project-disco.org/competition/112113-the-changing-textbook-industry/

Barnes, Brooke. 2014. "Movies Have Worst Summer since 1997." New York Times. August 29, 2014. https://www.nytimes.com/2014/08/30/movies/movies-have-worst-summer-since-1997.html

Bashe, Charles J, Lyle R Johnson, John H Palmer, and Emerson W Pugh. 1985. *IBM's Early Computers.* Cambridge, Massachusetts: MIT Press.

Baxter, John. 2023. *Story of Disney 100 Years of Wonder.* [S.l.]: DISNEY EDITIONS.

BBC News. 2015. "Who, What, Why: What Is the Little Red Book?" BBC News, November 26, 2015. https://www.bbc.com/news/magazine-34932800

Bedingfield, Will. 2019. "The Rise and Fall of Flash, the Annoying Plugin That Shaped the Modern Web." Wired. September 18, 2019. https://www.wired.com/story/history-of-macromedia-flash/

Bellware, Kim, and Ben Brasch. 2023. "Why Are Hollywood Actors and Writers on Strike? Here Are the Issues." Washington Post. July 13, 2023. https://www.washingtonpost.com/arts-entertainment/2023/07/13/why-actors-writers-strike-sag-wga-issues/

Bem, Daryl J. 1967. "Self-Perception: An Alternative Interpretation of Cognitive Dissonance Phenomena 1." *Psychological Review* 74 (3): 183–200. http://web.mit.edu/curhan/www/docs/Articles/15341_Readings/Motivation/Bem_1967_Self_perception.pdf

Benjamin, Walter. 1936. "The Work of Art in the Age of Mechanical Reproduction." Marxists.org. 1936. https://www.marxists.org/reference/subject/philosophy/works/ge/benjamin.htm

Bercovici, Jeff. 2014. "Amazon vs. Book Publishers, by the Numbers." Forbes. February 10, 2014. https://www.forbes.com/sites/jeffbercovici/2014/02/10/amazon-vs-book-publishers-by-the-numbers/?sh=6d9022db4ef9

Bergengruen, Vera. 2021. "Congress Can't Agree on How to Reform Big Tech. Frances Haugen Says That's What Facebook Wants". Time. December 1, 2021. https://time.com/6125089/frances-haugen-congress-tech-reform/

Bernstein, Amy, and Anand Raman. 2017. "The Great Decoupling: An Interview with Erik Brynjolfsson and Andrew McAfee." Harvard Business Review. March 13, 2017. https://hbr.org/2015/06/the-great-decoupling

Biancuzzi, Federico, and Chromatic. 2009. *Masterminds of Programming*. Sebastopol, CA: O'Reilly Media, Inc.

Bibb, Porter. 1997. *Ted Turner: It Ain't as Easy as It Looks*. Boulder, CO: Johnson Books.

Biery, Mary Ellen. 2014. "Growth in U.S. Day Care Businesses." Forbes. June 15, 2014. http://www.forbes.com/sites/sageworks/2014/06/15/heres-the-growth-chart-on-day-care-businesses/

Bilton, Nick. 2013. "Disruptions: Design Rivals Technology in Importance." Bits Blog. January 13, 2013. https://archive.nytimes.com/bits.blogs.nytimes.com/2013/01/13/disruptions-design-to-propel-technology-forward/?_r=0

Blackburn, Simon. 2008. *The Oxford Dictionary of Philosophy* (2nd Rev ed.). Oxford University Press.

Boateng, Briana. 2023. "How AI Is Impacting Entertainment and Women Creatives." Her Agenda. June 15, 2023. https://heragenda.com/p/ai-artificial-intelligence-entertainment-industry/#:~:text=Artificial%20intelligence%20researchers%20weighed%20in

Bodow, Steve. 2001. "The Whitney's Digital Sampler - Nymag." New York Magazine. March 26, 2001. https://nymag.com/nymetro/arts/features/4507/

Brand, Stewart. 1989. *The Media Lab: Inventing the Future at MIT*. Harmondsworth: Penguin.

Britannica. 2018. "Ninety-Five Theses | Work by Luther." In *Encyclopædia Britannica*. https://www.britannica.com/event/Ninety-five-Theses

Brockman, John. 1996. "Digerati - Introduction | Edge.org". www.edge.org. October 1, 1996. https://www.edge.org/conversation/edge_editor-digerati-introduction

Brooks, David. 2023. "Opinion | What Our Toxic Culture Does to the Young." The New York Times, May 4, 2023, sec. Opinion. https://www.nytimes.com/2023/05/04/opinion/gen-z-adulthood.html

Brown, Sara. 2023. "Why Neural Net Pioneer Geoffrey Hinton Is Sounding the Alarm on AI | MIT Sloan." Mitsloan.mit.edu. June 15, 2023. https://mitsloan.mit.edu/ideas-made-to-matter/why-neural-net-pioneer-geoffrey-hinton-sounding-alarm-ai

Buchwald, Jed Z. 2011. *The Creation of Scientific Effects*. Chicago, IL: University of Chicago Press.

Buhler, James, David Neumeyer, and Rob Deemer. 2010. *Hearing the Movies: Music and Sound in Film History*. New York: Oxford University Press.

By Holden Frith, Special to CNN. 2011. "Steve Jobs Leaves Lasting Legacy." CNN. October 6, 2011. https://www.cnn.com/2011/10/06/opinion/apple-logo/index.html

CA, SFMOMA. n.d. "Muybridge, Eadweard." SFMOMA. https://www.sfmoma.org/artist/Eadweard_Muybridge/

Camero, Katie. 2023. "How the ALS Ice Bucket Challenge Made a Lasting Impact." Science. June 21, 2023. https://www.nationalgeographic.com/science/article/als-ice-bucket-challenge-research-impact

Campbell, Joseph, and Bill D Moyers. (1988) 2012. *The Power of Myth*. Des Moines, Iowa: Turtleback Books.

Canavan, Kathryn. 2022. "75 Years Later: A Look Back at Willie Sutton's Infamous Prison Escape – Metro Philadelphia." Metrophiladelphia.com. February 9, 2022. https://metrophiladelphia.com/75-years-later-a-look-back-at-willie-suttons-infamous-prison-escape/

Cartwright, Mark. 2016. "Trade in the Phoenician World." World History Encyclopedia. April 1, 2016. https://www.worldhistory.org/article/881/trade-in-the-phoenician-world/

Castelvecchi, Davide. 2010. "Quantum Light Switch: Single Atom Acts as a Transistor for Photons." Scientific American. September 1, 2010. https://www.scientificamerican.com/article/quantum-light-switch/

Castillo, Michelle. 2017. "Roger Ailes' Knack for Turning Politics into Entertainment Changed Cable News." CNBC.com. NBC Universal. May 18, 2017. https://www.cnbc.com/2017/05/18/how-foxs-ailes-changed-cable-news.html

CERN. 2019. "A Short History of the Web | CERN." Home.cern. CERN. September 23, 2019. https://home.cern/science/computing/birth-web/short-history-web

Chandler, Michael Alison, and Hayley Tsukayama. 2014. "Tablets on the Rise in Nation's Schools." Washington Post, May 17, 2014, sec. Education. https://www.washingtonpost.com/local/education/tablets-proliferate-in-nations-classrooms-and-take-a-swipe-at-the-status-quo/2014/05/17/faa27ba4-dbbd-11e3-8009-71de85b9c527_story.html

Chase, Jason. 2023. "IAB Annual Report 2022." IAB. January 21, 2023. https://www.iab.com/news/iab-annual-report-2022/

Cheifet, Stewart. 1987. "Apple Introduces Bill Atkinson's HyperCard, Referencing Vannevar Bush and the Memex in the Marketing: History of Information." Historyofinformation.com. August 1, 1987. https://historyofinformation.com/detail.php?id=4783

Chicago, University of. 2024. "Library Guides: Early Printed Books: Incunabula." Guides.lib.uchicago.edu. May 5, 2024. https://guides.lib.uchicago.edu/c.php?g=527774&p=3608589#:~:text=The%20word%20%22incunabula%22%20is%20Latin

Church, Zac. 2016. "Professor Emeritus Jay Forrester, Digital Computing and System Dynamics Pioneer, Dies at 98." MIT News | Massachusetts Institute of Technology. November 16, 2016. https://news.mit.edu/2016/professor-emeritus-jay-forrester-digital-computing-system-dynamics-pioneer-dies-1119

Clement, J. 2024. "U.S. Top Selling Video Games 2021." Statista. January 29, 2024. https://www.statista.com/statistics/1285658/top-ranked-video-games-sales-annual/

Clinton, Hillary Rodham. 2023. "The Weaponization of Loneliness." The Atlantic. August 7, 2023. https://www.theatlantic.com/ideas/archive/2023/08/hillary-clinton-essay-loneliness-epidemic/674921/

Cohen, Marshall. 2023. "Dominion's $1.6 Billion Defamation Suit against Newsmax over Election Lies Set for September 2024 Trial | CNN Business." CNN. September 18, 2023. https://www.cnn.com/2023/09/18/media/dominion-newsmax-defamation-lawsuit/index.html

Common Core State Standards Initiative. 2009. "About the Standards | Common Core State Standards Initiative." Corestandards.org. 2009. http://www.corestandards.org/about-the-standards/

Constable, George, and Bob Somerville. 2003. *A Century of Innovation: Twenty Engineering Achievements That Transformed Our Lives*. Washington, DC: Joseph Henry Press.

Coppola, Danielle. 2024. "U.S. Retail E-Commerce Sales Share 2023." Statista. February 21, 2024. https://www.statista.com/statistics/187439/

Costello, Vic. 2017. *Multimedia Foundations: Core Concepts for Digital Design*. New York: Routledge.

Covert, Adrian. 2013. "A Decade of ITunes Singles Killed the Music Industry." CNNMoney. April 25, 2013. https://money.cnn.com/2013/04/25/technology/itunes-music-decline/

Cowell, Alan. 2019. "Overlooked No More: Alan Turing, Condemned Code Breaker and Computer Visionary." The New York Times, June 5, 2019. https://www.nytimes.com/2019/06/05/obituaries/alan-turing-overlooked.html

Coyle, Jake. 2014. "With Few Blockbusters, Hollywood's Summer Box Office Revenues Drop." Press Herald. July 10, 2014. https://www.pressherald.com/2014/07/10/with-few-blockbusters-hollywoods-summer-box-office-revenues-drop/

Cronauer, Adrian. 1994. "The Fairness Doctrine: A Solution in Search of a Problem." *Federal Communications Law* 47 (1): 6. https://www.repository.law.indiana.edu/cgi/viewcontent.cgi?article=1033&context=fclj

Cuthbertson, Anthony. 2023a. "Meta Scrambles to Fix Instagram Algorithm Connecting 'Vast Paedophile Network.'" The Independent. June 8, 2023. https://www.independent.co.uk/tech/instagram-child-abuse-meta-pedophilia-algorithm-b2353749.html

Cuthbertson, Anthony. 2023b. "ChatGPT Creator Issues Stark Warning on AI." The Independent. May 23, 2023. https://www.independent.co.uk/tech/chatgpt-ai-superintelligence-openai-sam-altman-b2343989.html

Cutlack, Gary. 2013. "This Kid Blew $2,500 on In-Game Purchases in Just 10 Minutes." Gizmodo. March 1, 2013. https://gizmodo.com/this-kid-blew-2-500-on-in-game-purchases-in-just-10-mi-5987799

Dalton, Wyatt. 2020. "What Is STEM?" www.pearsonaccelerated.com. August 25, 2020. https://www.pearsonaccelerated.com/blog/stem.html

Das, Anjali. 1990. "Nintendo Game Boy - Game Console - Computing History." www.computinghistory.org.uk. November 1, 1990. https://www.computinghistory.org.uk/det/4033/Nintendo-Game-Boy/#:~:text=The%20Game%20Boy%20and%20Game

Dechow, Douglas R, and Daniele C Struppa. 2015. *Intertwingled*. New York, NY: Springer.

"Definition of CONSPIRACY." 2019. Merriam-Webster.com. 2019. https://www.merriam-webster.com/dictionary/conspiracy

"Definition of Killer App." n.d. PCMAG. Accessed April 12, 2024. https://www.pcmag.com/encyclopedia/term/killer-app

Denning, Steve. 2011. "Lest We Forget: Why We Had a Financial Crisis." Forbes. November 22, 2011. https://www.forbes.com/sites/stevedenning/2011/11/22/5086/?sh=73c7c107f92f

Dennis, Michael Aaron. 2023. "PARC | Definition, History, Inventions, & Facts | Britannica." www.britannica.com. April 26, 2023. https://www.britannica.com/topic/PARC-company

Department of Justice. 2020. "Department of Justice's Review of Section 230 of the Communications Decency Act of 1996." www.justice.gov. June 3, 2020. https://www.justice.gov/archives/ag/department-justice-s-review-section-230-communications-decency-act-1996

"Design Takes Over, Says Paola Antonelli." 2010. The Economist. November 22, 2010. https://www.economist.com/news/2010/11/22/design-takes-over-says-paola-antonelli

"Developers Generated $1.1 Trillion in the App Store Ecosystem in 2022." 2023. Apple Newsroom. May 31, 2023. https://www.apple.com/newsroom/2023/05/developers-generated-one-point-one-trillion-in-the-app-store-ecosystem-in-2022/

Diaz, Jesus. 2022. "MoMA's Newest Artist Is an AI Trained on 180,000 Works, from Warhol to Pac-Man." www.fastcompany.com. Mansueto Ventures, LLC. November 18, 2022. https://www.fastcompany.com/90811669/momas-newest-artist-is-an-ai-trained-on-180000-works-from-warhol-to-pac-man

Diaz, Jesus. n.d. "AI Will Destroy Hollywood as We Know It." Fastcompany.com. Conde-Nast. Accessed April 14, 2024. https://www.fastcompany.com/90883794/ai-will-destroy-hollywood-as-we-know-it

"Dr. Vinton Cerf and Dr. Robert Kahn." n.d. Georgewbush-Whitehouse.archives.gov. https://georgewbush-whitehouse.archives.gov/government/cerf-kahn-bio.html

Duffy, Clare, Brian Fung. 2024. "Big Tech Executives to Appear at Senate Hearing to Defend Social Media's Impact on Teens | CNN Business." CNN. January 31, 2024. https://www.cnn.com/2024/01/31/tech/big-tech-executives-senate-hearing-teens/index.html

Duggal, Nikita. 2021. "Top Advantages and Disadvantages of Artificial Intelligence." Simplilearn.com. February 25, 2021. https://www.simplilearn.com/advantages-and-disadvantages-of-artificial-intelligence-article#disadvantages_of_artificial_intelligence

Duncan, Dennis. 2022. *Index, a History of The: A Bookish Adventure from Medieval Manuscripts to the Digital Age*. New York, NY: W. W. Norton & Company.

Durlauf, Steven, and Lawrence Blume. 2008. "Creative Destruction." https://economics.mit.edu/sites/default/files/publications/creative%20destruction.pdf

Economist, The. 2014. "Taking the Learning Tablets." The Economist. June 14, 2014. https://www.economist.com/international/2014/06/07/taking-the-learning-tablets

Edgar Allan Poe "Society of Baltimore - Works - Tales - the System of Doctor Tarr and Professor Fether (Text-02b)." 2015. www.eapoe.org. June 10, 2015. https://www.eapoe.org/works/tales/tarrb.htm

Edison, Thomas, Reese V. Jenkins, Paul Israel, Keith A. Nier, Grace Kurkowski, Robert A. Rosenberg, and et al. 1998. *The Papers of Thomas A. Edison / Vol. 4, the Wizard of Menlo Park, 1878 / Vol. Ed.: Paul B. Israel, Keith A. Nier, Louis Carlat; Ed. Staff: Grace Kurkowski, Helen Endick; Dir. And Ed.: Robert A. Rosenberg*. Baltimore Etc.: Johns Hopkins University Press, Cop.

Editor, CSRC Content. n.d. "Cryptocurrency - Glossary | CSRC." Csrc.nist.gov. https://csrc.nist.gov/glossary/term/cryptocurrency

Editors, History. 2018. "Renaissance." History. A&E Television Networks. April 4, 2018. https://www.history.com/topics/renaissance/renaissance

Edwards, Benji. 2014. "Remembering EWorld, Apple's Forgotten Online Service." Macworld. June 8, 2014. https://www.macworld.com/article/223467/remembering-eworld-apples-forgotten-online-service.html

Electronic Frontier Foundation. 2023. "Section 230 of the Communications Decency Act." Electronic Frontier Foundation. 2023. https://www.eff.org/issues/cda230

Elias, Jennifer. 2023. "Google CEO Sundar Pichai Warns Society to Brace for Impact of A.I. Acceleration, Says 'It's Not for a Company to Decide.'" CNBC. 2023. https://www.cnbc.com/2023/04/17/google-ceo-sundar-pichai-warns-society-to-brace-for-impact-of-ai-acceleration.html

Emspak, Jesse. 2022. "Quantum Entanglement: Love on a Subatomic Scale." Space.com. March 16, 2022. https://www.space.com/31933-quantum-entanglement-action-at-a-distance.html

Erickson, Harold. 2024. "Britannica Money." www.britannica.com. March 31, 2024. https://www.britannica.com/money/CNN

Etherington, Darrell. 2015. "LinkedIn to Buy Online Education Site Lynda.com for $1.5 Billion." TechCrunch. April 9, 2015. https://techcrunch.com/2015/04/09/linkedin-to-buy-online-education-site-lynda-com-for-1-5-billion/

Experience, World Leaders in Research-Based User. 2022. "Visual Design in UX: Study Guide." Nielsen Norman Group. November 6, 2022. https://www.nngroup.com/articles/visual-design-in-ux-study-guide/

"ExxonMobil | 2Q 2023 Earnings." 2023. ExxonMobil. July 23, 2023. https://corporate.exxonmobil.com/news/news-releases/2023/0728_exxonmobil-announces-second-quarter-2023-results

Fineman, Mia. 2004. "Kodak and the Rise of Amateur Photography." Metmuseum.org. The Metropolitan Museum of Art. October 2004. https://www.metmuseum.org/toah/hd/kodk/hd_kodk.htm

Finn, Chester E. Jr.,, and Daniela Fairchild. 2012. "Education Reform for the Digital Era." Edexcellencemedia.net. Thomas B. Fordham Institute. June 1, 2012. https://www.edexcellencemedia.net/publications/2012/20120425-education-reform-for-the-digital-era/20120425-Education-Reform-for-the-Digital-Era-FINAL.pdf

Firth, Joseph. 2019. "The 'Online Brain': How the Internet May Be Changing Our Cognition." *World Psychiatry* 18 (2): 119–29. https://doi.org/10.1002/wps.20617

Fischer, Paul. 2022. *The Man Who Invented Motion Pictures: A True Tale of Obsession, Murder, and the Movies*. New York: Simon & Schuster.

Fitzgerald, Jim. 2013. "NY School Goes All-in on Digital Textbooks." NBC News. December 22, 2013. https://www.nbcnews.com/business/business-news/ny-school-goes-all-digital-textbooks-f2D11792419

Forberg, Nancy. 2023. "Benefits of AI and Art Creation - Your 7 Essential AI Art Tips!" Newyorkartlife.com. NEW YORK ART LIFE. February 17, 2023. https://nyartlife.com/benefits-of-ai-and-art-creation-your-7-essential-ai-art-tips/

Forbes, Nancy. 2019. *Faraday, Maxwell, and the Electromagnetic Field: How Two Men Revolutionized Physics*. Jersey City, New Jersey: Prometheus.

Frankl, Viktor E., Ilse Lasch, Harold S. Kushner, and William J. Winslade. 2006. *Man's Search for Meaning*. Boston: Beacon Press.

Franklin D. Roosevelt Presidential Library and Museum. 2016. "Great Depression Facts - FDR Presidential Library & Museum." Fdrlibrary.org. 2016. https://www.fdrlibrary.org/great-depression-facts

Freedman, Jonathan L. 1965. "Long-Term Behavioral Effects of Cognitive Dissonance." *Journal of Experimental Social Psychology* 1 (2): 145–55. https://doi.org/10.1016/0022-1031(65)90042-9

Freeman, John. 2011. *The Tyranny of E-Mail: The Four-Thousand-Year Journey to Your Inbox*. New York: Scribner.

Freeth, Tony. 2022. "An Ancient Greek Astronomical Calculation Machine Reveals New Secrets." *Scientific American*, January. https://doi.org/10.1038/scientificamerican0122-24

Friedman, Thomas L. 2007. *The World Is Flat: A Brief History of the Twenty-First Century*. Vancouver: Douglas & Mcintyre.

Fung, Brian. 2023. "Mr. ChatGPT Goes to Washington: OpenAI CEO Sam Altman Testifies before Congress on AI Risks | CNN Business." CNN. May 16, 2023. https://www.cnn.com/2023/05/16/tech/sam-altman-openai-congress/index.html

Furnace, Franklin. 2004. "Goings on | 10/18/2004." Franklin Furnace. October 18, 2004. https://franklinfurnace.org/goings-on-10-18-2004/

Furr, Nathan, and Andrew Shipilov. 2019. "Digital Doesn't Have to Be Disruptive." Harvard Business Review. July 1, 2019. https://hbr.org/2019/07/digital-doesnt-have-to-be-disruptive

Garlinghouse, Tom. 2023. "*Physicists 'Entangle' Individual Molecules for the First Time, Bringing about a New Platform for Quantum Science*." Princeton University. Princeton University. December 8, 2023. https://www.princeton.edu/news/2023/12/08/physicists-entangle-individual-molecules-first-time-hastening-possibilities-quantum

Gaura, Elena, and James Peter Brusey. 2021. "Nikola Tesla: 5G Network Could Realise His Dream of Wireless Electricity, a Century after Experiments Failed." The Conversation. April 8, 2021. https://theconversation.com/nikola-tesla-5g-network-could-realise-his-dream-of-wireless-electricity-a-century-after-experiments-failed-158665

Gendron, Will. 2023. "AI Will Give Every Student a Personalized Tutor: Sal Khan." Business Insider. Insider. May 4, 2023. https://www.businessinsider.com/ai-will-give-every-student-personalized-tutor-sal-khan-academy-2023-5

Global, Stafford. 2022. "4 Principles of Marketing in the Digital Age." Stafford Global. April 7, 2022. https://www.staffordglobal.org/articles-and-blogs/marketing-articles-blogs/marketing-in-digital-age/

Gondwe, Gregory. 2023. "Exploring the Multifaceted Nature of Generative AI in Journalism Studies: A Typology of Scholarly Definitions." Scholars at Harvard, January. https://doi.org/10.2139/ssrn.4465446

Goodwin, Tom. 2015. "The Battle Is for the Customer Interface." TechCrunch, March 3, 2015. https://techcrunch.com/2015/03/03/in-the-age-of-disintermediation-the-battle-is-all-for-the-customer-interface/

Gov, HSMA. 2024. "Facts on Veal Calves." www.hsvma.org. January 1, 2024. https://www.hsvma.org/facts_veal_calves

Greenberg, Joshua M. 2010. *From Betamax to Blockbuster*. Cambridge, MA: MIT Press.

Grimes, William. 2007. "Looking Back in Anger at the Gilded Age's Excesses." The New York Times, April 18, 2007, sec. Arts. https://www.nytimes.com/2007/04/18/arts/18grim.html

Haidt, Jonathan. 2024. *The Anxious Generation*. City of Westminster, London: Penguin.

Hall, Rheanna-Marie. 2022. "The Book of Kells." National Trust for Scotland. May 4, 2022. https://www.nts.org.uk/stories/the-book-of-kells#:~:text=What%20is%20the%20Book%20of

Hamilton, Ilana. 2023. "The Rise of Online Learning in the U.S. – Forbes Advisor." www.forbes.com. May 24, 2023. https://www.forbes.com/advisor/education/online-colleges/online-learning-stats/

Hammond, George. 2023. "Top Tech Companies Form Group Seeking to Control AI." Financial Times, July 26, 2023. https://www.ft.com/content/709f4375-83bf-4037-878d-964d1ead8858

Han, Tae Sun, Heejun Cho, Dajung Sung, and Min-Hyeon Park. 2022. "A Systematic Review of the Impact of COVID-19 on the Game Addiction of Children and Adolescents." *Frontiers in Psychiatry* 13 (August). https://doi.org/10.3389/fpsyt.2022.976601

Hanna, Doug. 2017. "The Reformation and Publishing." Southern Equip. October 9, 2017. https://equip.sbts.edu/publications/magazine/magazine-issue/fall-2017-vol-85-no-2/the-reformation-and-publishing/

Hannah, Jaye. 2023. "Why UX Design Is a Good Career - UX Design Institute." www.uxdesigninstitute.com. January 11, 2023. https://www.uxdesigninstitute.com/blog/why-ux-design-is-a-good-career/#is-ux-design-a-good-career-path

Hao, Karen. 2021. "The Facebook Whistleblower Says Its Algorithms Are Dangerous. Here's Why." MIT Technology Review. October 5, 2021. https://www.technologyreview.com/2021/10/05/1036519/facebook-whistleblower-frances-haugen-algorithms/

Harris, Richard. 2018. "Your Apps Know Where You Were Last Night, and They're Not Keeping It Secret." The New York Times, December 10, 2018. https://www.nytimes.com/interactive/2018/12/10/business/location-data-privacy-apps.html

Harris, Tristan, and Asa Raskin. 2023. "Center for Humane Technology Co-Founders Tristan Harris and Aza Raskin Discuss the AI Dilemma." www.youtube.com. May 10, 2023. https://www.youtube.com/watch?v=cB0_-qKbal4

Harvard University. 2007. "B. F. Skinner." Harvard.edu. 2007. https://psychology.fas.harvard.edu/people/b-f-skinner

Hecht, Jeff. 2004. *City of Light: The Story of Fiber Optics*. Oxford; New York: Oxford University Press.

Van Helden, Albert. 2022. "Galileo | Biography, Discoveries, & Facts." In Encyclopædia Britannica. https://www.britannica.com/biography/Galileo-Galilei

Heller, Steven. 2012. "Down to the Letters." The New York Times, October 19, 2012, sec. Books. https://www.nytimes.com/2012/10/21/books/review/graphic-design-before-graphic-designers-and-more.html

Heller, Steven, and Georgette Ballance. 2001. *Graphic Design History*. New York: Allworth Press.

Herbert, Stephen. 2021. *A History of Pre-Cinema V3*. Milton Park, Abingdon, Oxfordshire, United Kingdom: Routledge.

History.com Editors. 2009. "Kennedy and Nixon Square off in First Televised Presidential Debate." HISTORY. November 24, 2009. https://www.history.com/this-day-in-history/first-kennedy-nixon-debate

Hodges, Andrew. 2013. "Alan Turing (Stanford Encyclopedia of Philosophy)." Stanford.edu. September 30, 2013. https://plato.stanford.edu/entries/turing/

Hollings, Ernest. 1987. "S.742 – 100th Congress (1987–1988): Fairness in Broadcasting Act of 1987." www.congress.gov, United States Government, 23 June 1987, www.congress.gov/bill/100th-congress/senate-bill/742. Accessed 8 Sept. 2024.

"Home - SVA MFA Interaction Design." 2023. School of Visual Arts. School of Visual Arts. April 4, 2023. https://interactiondesign.sva.edu/

Honan, Mat. 2013. "Remembering the Apple Newton's Prophetic Failure and Lasting Impact." WIRED. August 5, 2013. https://www.wired.com/2013/08/remembering-the-apple-newtons-prophetic-failure-and-lasting-ideals/

Horsley, Carter B. 2024. "Art/Museums: The 2002 Biennial at the Whitney Museum of American Art." Thecityreview.com. 2024. https://www.thecityreview.com/biennial.html

Howard, Caroline. 2016. "Disruption vs. Innovation: What's the Difference?" Forbes. April 16, 2016. https://www.forbes.com/sites/carolinehoward/2013/03/27/you-say-innovator-i-say-disruptor-whats-the-difference/?sh=45c624a56f43

Howard, Marcus E. 2019. "How Journalists and the Public Shape Our Democracy from Social Media and 'Fake News' to Reporting Just the Facts." https://www.georgiahumanities.org/wp-content/uploads/2019/03/MediaGuide_Web.pdf

https://www.latimes.com/people/houston-mitchell. 2015. "Yogi Berra Dies at 90: Here Are Some of His Greatest Quotes." Los Angeles Times. May 12, 2015. https://www.latimes.com/sports/sportsnow/la-sp-sn-yogi-berra-turns-90-quotes-20150512-story.html

"Person Overview ‹ Nicholas Negroponte." n.d. MIT Media Lab. https://www.media.mit.edu/people/nicholas/overview/

Hughes, Mark. 2015. "The Millennial Trends That are Killing Cable." Forbes. March 21, 2015. https://www.forbes.com/sites/markhughes/2015/03/21/the-millennial-trends-that-are-killing-cable/?sh=3b407b082293

"IBISWorld - Industry Market Research, Reports, and Statistics." n.d. www.ibisworld.com. https://www.ibisworld.com/united-states/market-research-reports/printing-industry/

IBM. n.d. "What Is Quantum Computing? | IBM." www.ibm.com. Accessed April 15, 2024. https://www.ibm.com/topics/quantum-computing#:~:text=Quantum%20computing%20is%20a%20rapidly

Ignotofsky, Rachel. 2022. *History of the Computer: People, Inventions, and Technology That Changed Our World.* S.L.: Ten Speed.

Apple Inc. 2023. "Introducing Apple Vision Pro: Apple's First Spatial Computer." Apple Newsroom. June 5, 2023. https://www.apple.com/newsroom/2023/06/introducing-apple-vision-pro/

International Energy Agency. 2023. "Electric Vehicles." IEA. July 11, 2023. https://www.iea.org/energy-system/transport/electric-vehicles

Internet Society. 1997. "Brief History of the Internet | Internet Society." Internet Society. 1997. https://www.internetsociety.org/internet/history-internet/brief-history-internet/

Isaacson, Walter. 2011. *Steve Jobs.* New York, NY: Simon & Schuster.

Jain, Pawan. 2023. "ChatGPT Could Offer a Ton of Benefit—and Risk—to Wall Street." Fastcompany.com. Conde Nast. May 9, 2023. https://www.fastcompany.com/90898484/chatgpt-could-offer-a-ton-of-benefit-and-risk-to-wall-street

Jamison, Story by Leslie. 2017. "The Digital Ruins of a Forgotten Future." The Atlantic. December 2017. https://www.theatlantic.com/magazine/archive/2017/12/second-life-leslie-jamison/544149/

Janesick, James R. 2024. *Scientific Charge-Coupled Devices.* Bellingham, WA: Spie.

Jasani, Tithi A. 2018. "AI as Material for Design." Medium. October 22, 2018. https://medium.com/@tithijasani/ai-as-material-for-design-fbec0141f2ec

Javier Peris, Luis. 2022. "A Brief History of the Java Programming Language | Baeldung." www.baeldung.com. January 2, 2022. https://www.baeldung.com/java-history

"Jerry Weintraub, a Force in Film and Music, Dies at 77." n.d. Nytimes.com. https://www.nytimes.com/2015/07/07/arts/jerry-weintraub-a-force-in-film-and-music-dies-at-77.html

Jewish Virtual Library. 2019. "Joseph Goebbels on the 'Big Lie.'" Jewishvirtuallibrary.org. The American-Israeli Cooperative Enterprise. 2019. https://www.jewishvirtuallibrary.org/joseph-goebbels-on-the-quot-big-lie-quot

Johnson, John. 2002. "Coonskin Cap Clings to 'Crockett.'" Los Angeles Times. August 23, 2002. https://www.latimes.com/archives/la-xpm-2002-aug-23-me-fess23-story.html

JOURNAL, Julia AngwinStaff Reporter of THE WALL STREET. 2004. "Ted Turner Attacks Big Media - WSJ." WSJ, August 4, 2004. https://www.wsj.com/articles/SB109175404288784675

Karpinski, Aryn. 2009. "Facebook Use Linked to Lower Grades in College." ScienceDaily. April 19, 2009. https://www.sciencedaily.com/releases/2009/04/090413180538.htm#:~:text=While%20this%20was%20a%20relatively

Keith, Jeremy, and Jeffrey Sambells. 2010. "A Brief History of JavaScript." Apress EBooks, January, 1–6. https://doi.org/10.1007/978-1-4302-3390-9_1

Kettler, Sara. 2020. "Why Jacqueline Kennedy Didn't Take off Her Pink Suit after JFK Was Assassinated." Biography. June 16, 2020. https://www.biography.com/political-figures/jacqueline-kennedy-pink-suit-jfk-assassination

Kieran, Michael. 1994. *Understanding Desktop Color.* San Francisco, California: Addison-Wesley.

Kiernan, Ben. 2007. *Blood and Soil: A World History of Genocide and Extermination from Sparta to Darfur.* New Haven, CT: Yale University Press.

Király, Orsolya, Patrik Koncz, Mark D. Griffiths, and Zsolt Demetrovics. 2023. "Gaming Disorder: A Summary of Its Characteristics and Aetiology." *Comprehensive Psychiatry* 122 (April): 152376. https://doi.org/10.1016/j.comppsych.2023.152376

Klass, Perri M.D. 2011. "Fixated by Screens, but Seemingly Nothing Else." The New York Times, May 9, 2011, sec. Health. https://www.nytimes.com/2011/05/10/health/views/10klass.html

Kleiner, David. 2019. "The Cuneiform Writing System in Ancient Mesopotamia: Emergence and Evolution." NEH-Edsitement. 2019. https://edsitement.neh.gov/lesson-plans/cuneiform-writing-system-ancient-mesopotamia-emergence-and-evolution

Koblin, John, and Brooks Barnes. 2023. "Why There Is Talk of a Writers' Strike in Hollywood." The New York Times, March 21, 2023, sec. Business. https://www.nytimes.com/article/wga-writers-strike-hollywood.html

Kornfield, Meryl. 2021. "Why Hundreds of QAnon Supporters Showed up in Dallas, Expecting JFK Jr.'S Return." Washington Post, November 2, 2021. https://www.washingtonpost.com/nation/2021/11/02/qanon-jfk-jr-dallas/

Kowitz, Braden. 2013. "Should Tech Designers Go with Their Guts — or the Data?" Wired. November 11, 2013. https://www.wired.com/2013/11/design-world-stop-fighting-over-data-vs-instinct/

Krieger, Patrick J. 2020. "What's the Difference between Restarting and Shutting Down My Computer?" HowStuffWorks. February 4, 2020. https://computer.howstuffworks.com/restarting-shutting-down-computer.htm

Krug, Steve. 2006. *Don't Make Me Think a Common Sense Approach to Web Usability*. Berkely, CA New Riders Publ.

Kuhfeld, Megan, Jim Soland, Karyn Lewis, and Emily Morton. 2022. "The Pandemic Has Had Devastating Impacts on Learning. What Will It Take to Help Students Catch Up?" Brookings. March 3, 2022. https://www.brookings.edu/articles/the-pandemic-has-had-devastating-impacts-on-learning-what-will-it-take-to-help-students-catch-up/

Kumar, Abhishek. 2022. *Beginning VFX with Autodesk Maya: Create Industry-Standard Visual Effects from Scratch*. New York, NY: Apress.

Kumparak, Greg. 2014. "A Brief History of Oculus." TechCrunch. March 26, 2014. https://techcrunch.com/2014/03/26/a-brief-history-of-oculus/

Lambert, Julieanne. 2020. "The Magic of Chromolithography." John Johnson Collection: Now and Then. May 21, 2020. https://blogs.bodleian.ox.ac.uk/jjcoll/2020/05/21/the-magic-of-chromolithography/

Landau, Valerie. 2018. "How Douglas Engelbart Invented the Future." Smithsonian. Smithsonian.com. January 3, 2018. https://www.smithsonianmag.com/innovation/douglas-engelbart-invented-future-180967498/

Lasn, Kalle, Darren Fleet, Pedro Inoue, Will Brown, Ellen Lee, Lauren Bercovitch, and Media Foundation (Organization). 2012. *Meme Wars: The Creative Destruction of Neoclassical Economics: A Real World Economics Textbook*. New York, NY: Seven Stories Press.

Le, Linh. 2013. "Examining the Rise of Hatsune Miku: The First International Virtual Idol." *The UCI Undergraduate Journal* 13 (1): 12. https://archive.urop.uci.edu/journal/journal13/01_le.pdf

Ledertoug, Mette Marie, and Nanna Paarup. 2021. "Engaging Education: The Foundation for Wellbeing and Academic Achievement." *The Palgrave Handbook of Positive Education* 2021: 441–72. https://doi.org/10.1007/978-3-030-64537-3_18

"Lee DeForest | Lemelson." n.d. Lemelson.mit.edu. https://lemelson.mit.edu/resources/lee-deforest

Lee, Penny. 2023. "How Bank-Fintech Partnerships Make Banking Better for Consumers and Small Businesses." Forbes. October 4, 2023. https://www.forbes.com/sites/pennylee/2023/10/04/how-bank-fintech-partnerships-make-banking-better-for-consumers-and-small-businesses/?sh=325eaea87b76

Leiner, Barry, Robert Kahn, and Jon Postel. 2003. "ABSTRACT." https://groups.csail.mit.edu/ana/A%20brief%20history%20of%20the%20internet%20-%20p22-leiner.pdf

Lemelson, M.I.T.. 2020. "Johann Gutenberg | Lemelson-MIT Program." Mit.edu. 2020. https://lemelson.mit.edu/resources/johann-gutenberg

Lemelson, M.I.T. n.d.-a "An Wang | Lemelson." Lemelson.mit.edu. https://lemelson.mit.edu/resources/wang

Lemelson, M.I.T. n.d.-b "Raymond Kurzweil | Lemelson." Lemelson.mit.edu. https://lemelson.mit.edu/award-winners/raymond-kurzweil

Levin, David. 2015. "Dear Students and Faculty: Please Go Digital." HuffPost. August 7, 2015. https://www.huffpost.com/entry/dear-students-and-faculty_b_7957508

Levy, Steven. 2013. "The Brief History of the ENIAC Computer." Smithsonian. Smithsonian.com. November 2013. https://www.smithsonianmag.com/history/the-brief-history-of-the-eniac-computer-3889120/

Levy, Steven. 2024. "In Defense of AI Hallucinations." Wired. Conde Nast. January 5, 2024. https://www.wired.com/story/plaintext-in-defense-of-ai-hallucinations-chatgpt/

Lew, Kristi. 2018. *Ada Lovelace: Mathematician and First Programmer*. New York: Britannica Educational Publishing In Association With Rosen Educational Services.

Lewis, Michael. 2001. *The New New Thing: A Silicon Valley Story*. New York: Penguin Books.

Lieberman, Matthew D. 2015. *Social: Why Our Brains Are Wired to Connect*. Oxford, United Kingdom: Oxford University Press.

Limited, Infosys. 2024. "AI in the Media and Entertainment Industry | Infosys BPM." www.infosysbpm.com. March 15, 2024. https://www.infosysbpm.com/blogs/media-entertainment/use-of-ai-in-media-entertainment-industry.html

Litchfield, Gideon, and Lauren Goode. 2023. "The World Isn't Ready for the next Decade of AI." Wired, August 16, 2023. https://www.wired.com/story/have-a-nice-future-podcast-18/

Livingston, Gretchen. 2012. "U.S. Birth Rate Falls to a Record Low; Decline Is Greatest among Immigrants." Pew Research Center's Social & Demographic Trends Project. November 29, 2012. http://www.pewsocialtrends.org/2012/11/29/u-s-birth-rate-falls-to-a-record-low-decline-is-greatest-among-immigrants/

Lofquist, William S. 1996. "Economic Outlook for the U.S. Printing and Publishing Industry." *Publishing Research Quarterly* 12 (3): 22–7. https://doi.org/10.1007/bf02680366

Longman, Addison. 2000. "Chapter 2." W3.org. 2000. https://www.w3.org/People/Raggett/book4/ch02.html

Lowe, Paul. 2019. *A Chronology of Photography: A Cultural Timeline from Camera Obscura to Instagram*. New York: Thames & Hudson.

Lucarelli, Giulia. 2015. *Betamax vs VHS. A Case Study of the Competitive Battle of Video Cassette Recording Systems*. Munich, Bavaria, Germany: GRIN Verlag.

Lupton, Ellen, and J Abbott Miller. 1993. *The ABCs of [Triangle Square Circle]: The Bauhaus and Design Theory*. Princeton: Princeton Architectural Press; New York.

Lyons, James, and John Plunkett. 2007. *Multimedia Histories*. 2nd ed. Exeter, England: University of Exeter Press.

MacBride, Elizabeth. 2017. "Is Social Media the Tobacco Industry of the 21st Century?" Forbes. December 13, 2017. https://www.forbes.com/sites/elizabethmacbride/2017/12/31/is-social-media-the-tobacco-industry-of-the-21st-century/?sh=66377ae87011

MacSwan, Angus. 2008. "New Light Shed on Capa's 'Falling Soldier' Photo." https://www.reuters.com. Reuters. November 11, 2008. https://www.reuters.com/article/us-leisure-capa/new-light-shed-on-capas-falling-soldier-photo-idUKTRE4AA05220081111/

"Magnetic Core Memory - CHM Revolution." n.d. www.computerhistory.org. https://www.computerhistory.org/revolution/memory-storage/8/253

Maheshwari, Rashi. 2023. "What Is Artificial Intelligence (AI) and How Does It Work?" Forbes Advisor INDIA. April 3, 2023. https://www.forbes.com/advisor/in/business/software/what-is-ai/

Manning, Christopher. 2020. *Artificial Intelligence Definitions*. Stanford University. https://hai.stanford.edu/sites/default/files/2020-09/AI-Definitions-HAI.pdf

Marcelo, Philip. 2023. "FACT FOCUS: Fake Image of Pentagon Explosion Briefly Sends Jitters through Stock Market." AP NEWS. May 22, 2023. https://apnews.com/article/pentagon-explosion-misinformation-stock-market-ai-96f534c790872fde67012ee81b5ed6a4

Mark, Lois Alter. 2022. "An Art Historian's Take on 'beyond van Gogh: The Immersive Experience'." Forbes. February 18, 2022. https://www.forbes.com/sites/loisaltermark/2022/02/18/an-art-historians-take-on-beyond-van-gogh-the-immersive-experience/?sh=59564b247706

Marland, Jaime. 2024. "STEM to STEAM | RISD." www.risd.edu. January 1, 2024. https://www.risd.edu/steam

Maues, Julia. 2013. "Banking Act of 1933 (Glass-Steagall) | Federal Reserve History." www.federalreservehistory.org. November 22, 2013. https://www.federalreservehistory.org/essays/glass-steagall-act

Mazurek, Micah. 2024. "Video Game Use in Boys with Autism Spectrum Disorder, ADHD, or Typical Development." Aap.org. 2024. https://publications.aap.org/pediatrics/article-abstract/132/2/260/31361/Video-Game-Use-in-Boys-With-Autism-Spectrum?redirectedFrom=PDF

McBride, James, Anshu Siripurapu, and Noah Berman. 2022. "What Does the G20 Do?" Council on Foreign Relations. Council on Foreign Relations. December 15, 2022. https://www.cfr.org/backgrounder/what-does-g20-do

McCarty, Dawn, and Beth Jinks. 2012. "Kodak Files for Bankruptcy as Digital Era Spells End to Film." Bloomberg.com, January 19, 2012. http://www.bloomberg.com/news/2012-01-19/kodak-photography-pioneer-files-for-bankruptcy-protection-1-.html

McFadden, Robert D. 2014. "Charles Keating, 90, Key Figure in '80s Savings and Loan Crisis, Dies." The New York Times, April 2, 2014, sec. Business. https://www.nytimes.com/2014/04/02/business/charles-keating-key-figure-in-the-1980s-savings-and-loan-crisis-dies-at-90.html

Mcleod, Saul. 2023. "Cognitive Dissonance." Simply Psychology. October 24, 2023. https://www.simplypsychology.org/cognitive-dissonance.html

Meggs, Philip B., and Alston W. Purvis. 2016. *Meggs' History of Graphic Design*. 6th ed. New York: Wiley.

Mell, Peter. 2023. "Understanding Stablecoin Technology and Related Security Considerations." *NIST Internal Report* 1 (8408). https://doi.org/10.6028/nist.ir.8408

Mell, Peter, and Dylan Yaga. 2023. "Non-Fungible Token Security." Csrc.nist.gov. August 31, 2023. https://csrc.nist.gov/pubs/ir/8472/ipd

"Mental Health." 2023. NIH COVID-19 Research. September 28, 2023. https://covid19.nih.gov/covid-19-topics/mental-health#:~:text=Both%20SARS%2DCoV%2D2%20and

Merchant, Brian. 2023a. *Blood in the Machine*. New York, NY: Hachette UK.

Merchant, Nomaan. 2023b. "Iraq WMD Failures Shadow US Intelligence 20 Years Later." AP NEWS. March 23, 2023. https://apnews.com/article/iraq-war-wmds-us-intelligence-f9e21ac59d3a0470d9bfcc83544d706e

Metz, Cade. 2023. "How Could A.I. Destroy Humanity?" The New York Times, June 10, 2023, sec. Technology. https://www.nytimes.com/2023/06/10/technology/ai-humanity.html

Mineo, Liz. 2023. "Is Art Generated by Artificial Intelligence Real Art?" The Harvard Gazette. August 15, 2023. https://news.harvard.edu/gazette/story/2023/08/is-art-generated-by-artificial-intelligence-real-art/

Morais, Betsy. 2013. "Ada Lovelace, the First Tech Visionary." The New Yorker. October 15, 2013. https://www.newyorker.com/tech/annals-of-technology/ada-lovelace-the-first-tech-visionary

Morris, Oliver. 2023. "6 Benefits of Using AI in Digital Marketing." Metrics Watch. November 24, 2023. https://metricswatch.com/ai-in-digital-marketing

Moyers, Bill. 2016. "Joseph Campbell and the Power of Myth | Shows | Billmoyers.com." BillMoyers.com. 2016. https://billmoyers.com/series/joseph-campbell-and-the-power-of-myth-1988/

Murdock, Kelly. 2005. *Lightwave 3D 8 Revealed*. Boston, MA: Cengage Learning.

Murg, Stephanie. 2011. "Unrest at RISD: Faculty Vote 'No-Confidence' in President John Maeda." www.adweek.com. March 7, 2011. https://www.adweek.com/performance-marketing/untrest-at-risd-faculty-vote-no-confidence-in-president-john-maeda/

Museum, Whitney. 2024. "Artport." Whitney.org. January 1, 2024. https://whitney.org/artport

Naesser, Sarah, and Elisa Shearer. 2021. "Trends and Facts on Network News | State of the News Media." Pew Research Center's Journalism Project. July 13, 2021. https://www.pewresearch.org/journalism/fact-sheet/network-news/

Nakshine, Vaishnavi S., Preeti Thute, Mahalaqua Nazli Khatib, and Bratati Sarkar. 2022. "Increased Screen Time as a Cause of Declining Physical, Psychological Health, and Sleep Patterns: A Literary Review." Cureus 14 (10). https://www.cureus.com/articles/112862-increased-screen-time-as-a-cause-of-declining-physical-psychological-health-and-sleep-patterns-a-literary-review

Nasaw, David. 2001. *The Chief: The Life of William Randolph Hearst*. Boston: Houghton Mifflin, Cop.

National Archives. 2021. "President Dwight D. Eisenhower's Farewell Address (1961)." National Archives. September 29, 2021. https://www.archives.gov/milestone-documents/president-dwight-d-eisenhowers-farewell-address

Negroponte, Nicholas. 1995. "WIRED 3.01 - Bits and Atoms." Web.media.mit.edu. January 1, 1995. https://web.media.mit.edu/~nicholas/Wired/WIRED3-01.html

Newcott, Bill. 2020. "How the JFK-Nixon Debates in 1960—the First on Live TV—Set the Stage for All to Come." History. September 25, 2020. https://www.nationalgeographic.com/history/article/behind-scenes-first-televised-presidential-debates-nixon-jfk-1960

Newton, Isaac. 2016. *The Principia: The Authoritative Translation and Guide: Mathematical Principles of Natural Philosophy*. Berkeley, CA: University of California Press.

Aaron Foisi Nmungwun. 2012. *Video Recording Technology*. Routledge.

NP Source. 2019. "Nonprofits Source." Nonprofits Source. 2019. https://nonprofitssource.com/online-giving-statistics/social-media/

Nudd, Tim. 2013. "Genius or Process? How Top Creative Directors Come up with Great Ideas." www.adweek.com. September 25, 2013. https://www.adweek.com/brand-marketing/genius-or-process-how-top-creative-directors-come-great-ideas-152697/

Oliver, Laura. 2022. "These Reader Revenue Models Keep in Mind People Who Won't Pay Full Price (Yet)." Reuters Institute for the Study of Journalism. May 26, 2022. https://reutersinstitute.politics.ox.ac.uk/news/these-reader-revenue-models-keep-mind-people-who-wont-pay-full-price-yet

Oppland, Mike. 2019. "8 Ways to Create Flow according to Mihaly Csikszentmihalyi." PositivePsychology.com. July 16, 2019. https://positivepsychology.com/mihaly-csikszentmihalyi-father-of-flow/

Oppy, Graham, and David Dowe. 2003. "The Turing Test (Stanford Encyclopedia of Philosophy)." Stanford.edu. April 9, 2003. https://plato.stanford.edu/entries/turing-test/

Ortiz, Sabrina. 2023. "40% of Workers Will Have to Reskill in the next Three Years due to AI, Says IBM Study." ZDNET. August 13, 2023. https://www.zdnet.com/google-amp/article/40-of-workers-will-have-to-reskill-in-the-next-three-years-due-to-ai-says-ibm-study/

Ortiz, Sabrina. 2024. "What Is ChatGPT and Why Does It Matter? Here's Everything You Need to Know." ZDNET. February 20, 2024. https://www.zdnet.com/article/what-is-chatgpt-and-why-does-it-matter-heres-everything-you-need-to-know/

Ossola, Alexandra. 2015. "The Surprising Amount of Time Kids Spend Looking at Screens." The Atlantic. January 22, 2015. https://www.theatlantic.com/education/archive/2015/01/the-surprising-amount-of-time-kids-spend-looking-at-screens/384737/

Padua, David A. 2011. *Encyclopedia of Parallel Computing*. New York: Springer.

Palermo, Elizabeth. 2014. "Who Invented the Printing Press?" Live Science. February 25, 2014. https://www.livescience.com/43639-who-invented-the-printing-press.html

Pappano, Laura. 2012. "The Year of the MOOC." The New York Times, November 2, 2012, sec. Education. https://www.nytimes.com/2012/11/04/education/edlife/massive-open-online-courses-are-multiplying-at-a-rapid-pace.html?pagewanted=all&_r=0

Pareles, Jon. 2002. "David Bowie, 21st-Century Entrepreneur." The New York Times, June 9, 2002, sec. Arts. https://www.nytimes.com/2002/06/09/arts/david-bowie-21st-century-entrepreneur.html

Park Service, National. 2016. "What Are Petroglyphs and Who Made Them? - Petroglyph National Monument (U.S. National Park Service)." Nps.gov. 2016. https://www.nps.gov/petr/learn/historyculture/what.htm

Parker, Philip. 2010. *Macromedia: Webster's Timeline History, 1960–2007*. Headquarters-Rockville, MD: ICON Group International, Inc.

Payment, Simone. 2007. *Marc Andreessen and Jim Clark: The Founders of Netscape*. New York: Rosen Pub. Group.

Pelchen, Lexie. 2024. "Internet Usage Statistics in 2024." Forbes Home. March 1, 2024. https://www.forbes.com/home-improvement/internet/internet-statistics/#:~:text=There%20are%205.35%20billion%20internet%20users%20worldwide.&text=Out%20of%20the%20nearly%208

Pelkey, James. 2021. "Bolt Beranek & Newman (BBN) Documents | History of Computer Communications." Historyofcomputercommunications.info. January 1, 2021. https://historyofcomputercommunications.info/section/b.1/Bolt-Beranek-&-Newman-(BBN)-documents/

Pelley, Scott. 2021. "Whistleblower: Facebook Is Misleading the Public on Progress against Hate Speech, Violence, Misinformation." CBS News. October 4, 2021. https://www.cbsnews.com/news/facebook-whistleblower-frances-haugen-misinformation-public-60-minutes-2021-10-03/

Penn Museum. 2019. "WRITING - Scribes, Hieroglyphs, and Papyri - a New Look at Ancient Egypt @ UPMAA." Penn.museum. 2019. https://www.penn.museum/sites/egypt/writing.shtml

Perry, Tekla. 1985 BC. "Xerox Parc's Engineers on How They Invented the Future - IEEE Spectrum." Spectrum.ieee.org. October 1, 1985BC. https://spectrum.ieee.org/xerox-parc

Pew Research Center. 2023a. "Trends and Facts on Newspapers | State of the News Media." Pew Research Center's Journalism Project. November 10, 2023. https://www.pewresearch.org/journalism/fact-sheet/newspapers/

Pew Research Center. 2023b. "Social Media and News Fact Sheet." Pew Research Center. November 15, 2023. https://www.pewresearch.org/journalism/fact-sheet/social-media-and-news-fact-sheet/

Pfiffner, Pamela. 2003. *Inside the Publishing Revolution*. San Jose, California: Adobe PR.

Phillips, Sarah T., and Shane Hamilton. 2014. *The Kitchen Debate and Cold War Consumer Politics*. Macmillan Higher Education.

Pingree, Geoff. 2009. "What Spain Sees in Robert Capa's Civil War Photo." Time.com. Dotdash Meredith. July 9, 2009. https://content.time.com/time/world/article/0,8599,1912110,00.html

Pirdehghan, Azar, Edris Khezmeh, and Soheila Panahi. 2021. "Social Media Use and Sleep Disturbance among Adolescents: A Cross-Sectional Study." *Iranian Journal of Psychiatry* 16 (2): 137–45. https://doi.org/10.18502/ijps.v16i2.5814

Edgar Allan Poe. 2015. *The System of Dr. Tarr and Professor Fether*. Redditch, England: Read Books Ltd.

Posy, Brian. 2021. "What Is Cloud Storage? – Definition, Types, and Benefits." ITPro Today: IT News, How-Tos, Trends, Case Studies, Career Tips, More. December 16, 2021. https://www.itprotoday.com/cloud-storage/what-cloud-storage-definition-types-and-benefits#What%20Are%20the%20Types%20of%20Cloud%20Storage

Powell, Barry B. 1996. *Homer and the Origin of the Greek Alphabet*. Cambridge England; New York, NY, USA: Cambridge University Press.

Poynton, Charles A. 1996. *A Technical Introduction to Digital Video*. New York: J. Wiley.

Prange, Stephanie. 2014. "4 Types of Digital Disruption | Home Media Magazine." www.homemediamagazine.com. Questex Entertainment. October 9, 2014. https://www.homemediamagazine.com/steph-sums-it/4-types-digital-disruption

Randolph, Marc. 2019. *That Will Never Work: The Birth of Netflix and the Amazing Life of an Idea*. New York, NY: Little, Brown And Company.

Ratings, Neilsen. 2023. "Cross Platform Ad Measurement." Nielsen. January 1, 2023. https://www.nielsen.com/solutions/audience-measurement/total-ad-ratings/

Ray, Michael. 2024. "Britannica Money." www.britannica.com. Britannica Encyclopedia. March 30, 2024. https://www.britannica.com/money/Fox-News-Channel

Reed, Ryan. 2014. "Watch Neil Young's Historic 'Fallon' Spot." Rolling Stone. May 13, 2014. https://www.rollingstone.com/tv-movies/tv-movie-news/neil-young-jack-white-cut-vinyl-record-live-on-tonight-show-233184/#ixzz3Op8rHqvI

Reeves, Simon. 2005. "Macromedia Inc. - Company - Computing History." www.computinghistory.org.uk. The Centre for Computing History. November 21, 2005. https://www.computinghistory.org.uk/det/66554/Macromedia-Inc./

"Refik Anadol: Unsupervised | MoMA." n.d. The Museum of Modern Art. https://www.moma.org/calendar/exhibitions/5535

Reich, Robert B. 2012. *Beyond Outrage: What Has Gone Wrong with Our Economy and Our Democracy, and How to Fix It*. New York: Vintage Books.

Richards, Jef I. 2022. *A History of Advertising*. Lanham, MD: Rowman & Littlefield.

Richtel, Matt, Catherine Pearson, and Michael Levenson. 2023. "Surgeon General Warns That Social Media May Harm Children and Adolescents." The New York Times, May 23, 2023, sec. Health. https://www.nytimes.com/2023/05/23/health/surgeon-general-social-media-mental-health.html

Riley, Susan. 2020. "What Is STEAM Education? A Comprehensive Guide for K-12 Schools." The Institute for Arts Integration and STEAM. 2020. https://artsintegration.com/what-is-steam-education-in-k-12-schools/

Ritchie, Donald, and Elizabeth Bolling. 2003. "U.S. Senate: 'Have You No Sense of Decency?'" www.senate.gov. United States Senate. 2003. https://www.senate.gov/about/powers-procedures/investigations/mccarthy-hearings/have-you-no-sense-of-decency.htm

Ritholtz, Barry. 2011. "What Caused the Financial Crisis? The Big Lie Goes Viral." Washington Post, November 5, 2011, sec. Business. https://www.washingtonpost.com/business/what-caused-the-financial-crisis-the-big-lie-goes-viral/2011/10/31/gIQAXlSOqM_story.html

River, Charles. 2018. *The Luddites*. North Charleston, South Carolina: Createspace Independent Publishing Platform.

Robinson, Dan. 2023. "Xerox Says Goodbye to Palo Alto Research Center." www.theregister.com. April 23, 2023. https://www.theregister.com/2023/04/25/xerox_parc_goodbye/

Robinson, Ken. 2009. "Transform Education? Yes, We Must." HuffPost. February 11, 2009. http://www.huffingtonpost.com/sir-ken-robinson/transform-education-yes-w_b_157014.html

Robinson, Ken, and Lou Aronica. 2008. *The Element: A How Finding Your Passion Changes Everything*. New York: Viking.

Roetzer, Paul. 2017. "6 Limitations of Marketing Artificial Intelligence, according to Experts." Marketingaiinstitute.com. 2017. https://www.marketingaiinstitute.com/blog/limitations-of-marketing-artificial-intelligence

Rowan, Cris. 2014. "10 Reasons Why Handheld Devices Should Be Banned for Children under 12 | Announce | University of Nebraska-Lincoln." Unl.edu. December 18, 2014. https://newsroom.unl.edu/announce/csmce/3990/22224

Russell Neuman, W. 2010. *Media, Technology, and Society: Theories of Media Evolution*. Ann Arbor: Digital Culture Books/University Of Michigan Press.

Ryan, Johnny. 2010. *A History of the Internet and the Digital Future*. London: Reaktion Books.

Sanderson, Katherine. 2010. "Galileo Backed Copernicus despite Data." Scientific American. March 5, 2010. https://www.scientificamerican.com/article/galileo-backed-copernicus-desp/

Schlossberg, Jeremy. 2001. "Art of Web Design." New York, NY: Media Life.

Schomer, Audrey. 2021. "US Adults Will Consume Almost as Much Media in 2021, but TV Viewing Will Backslide." Insider Intelligence. June 6, 2021. https://www.emarketer.com/content/us-adults-will-consume-almost-much-media-2021-tv-viewing-will-backslide

Schwartz, A. Brad. 2015. "The Infamous 'War of the Worlds' Radio Broadcast Was a Magnificent Fluke." Smithsonian. Smithsonian.com. May 6, 2015. https://www.smithsonianmag.com/history/infamous-war-worlds-radio-broadcast-was-magnificent-fluke-180955180/

Science Foundation, National. 1994. "A Brief History | NSF - National Science Foundation." www.nsf.gov. July 15, 1994. https://www.nsf.gov/about/history/nsf50/nsf8816.jsp#chapter2

Shah, Saqib. 2016. "The History of Social Media." Digital Trends. May 14, 2016. https://www.digitaltrends.com/computing/the-history-of-social-networking/

Shearer, Robert G. 2007. *Famous Men of the Renaissance & Reformation*. Lebanon, Tenn.: Greenleaf Press.

Sito, Tom. 2015. *Moving Innovation - a History of Computer Animation*. Cambridge, MA: MIT Press Ltd.

Smiley, Jane. 2010. *The Man Who Invented the Computer: The Biography of John Atanasoff, Digital Pioneer*. New York: Doubleday.

Smith Bailey, Shanna. 2010. "Effectiveness of Fully Online Courses for College Students: Response to a Department of Education Meta-Analysis." Community College Research Center, Columbia University. Community College Research Center. Available from: CCRC Publications. Teachers College, Columbia University, 525 West 120th Street Box 174, New York, NY 10027. Tel: 212-678-3091; Fax: 212-678-3699; e-mail: ccrc@columbia.edu; Web site: http://www.tc.columbia.edu/ccrc. 2010. https://eric.ed.gov/?id=ED512274

Smith, George. 2007. "Isaac Newton." Stanford.edu. December 19, 2007. https://plato.stanford.edu/entries/newton/

Sobel, Robert. 1986. *Rca*. New York: Stein And Day/Publishers.

Solari, Stephen J. 1997. *Digital Video and Audio Compression*. New York, NY: McGraw-Hill Professional Publishing.

Solnit, Rebecca. 2010. *River of Shadows: Eadweard Muybridge and the Technological Wild West*. New York: Penguin Books.

Solomon, Charles. 1994. *Enchanted Drawings: The History of Animation*. New York: Wings Books, Avenel, N.J.

Somers, Joseph. 2019. "'1984' Quotes Explained." ThoughtCo. Meredith. April 22, 2019. https://www.thoughtco.com/1984-quotes-740884

Somers, Meredith. 2020. "Deepfakes, Explained." MIT Sloan. MIT. July 21, 2020. https://mitsloan.mit.edu/ideas-made-to-matter/deepfakes-explained

Spicer, Dag, Gwen Bell, Jan Zimmerman, Jacqueline Boas, and Bill Boas. 2000. "Internet History 1962 to 1992 | Internet History | Computer History Museum." Computerhistory.org. 2000. https://www.computerhistory.org/internethistory/

Spocchia, Gino. 2021. "Trump Told 30,000 Untruths during Presidency, Say Analysts." The Independent. Independent Digital News and Media Ltd. January 21, 2021. https://www.independent.co.uk/news/world/americas/us-election-2020/trump-lies-false-presidency-b1790285.html

Stashower, Daniel. 2002. *The Boy Genius and the Mogul*. New York, NY: Crown.

Stelter, Brian. 2013. "David Frost, Interviewer Who Got Nixon to Apologize for Watergate, Dies at 74 (Published 2013)." The New York Times, September 1, 2013, sec. World. https://www.nytimes.com/2013/09/02/world/europe/david-frost-known-for-nixon-interview-dead-at-74.html

Sterling, Bruce. 2009. "Augmented Reality: The Ultimate Display by Ivan Sutherland In1965." WIRED. November 16, 2009. https://www.wired.com/2009/09/augmented-reality-the-ultimate-display-by-ivan-sutherland-1965/

Sterling, Christopher. 1933. "'The Fireside Chats'-President." www.loc.gov. National Registry: United States Government. https://www.loc.gov/static/programs/national-recording-preservation-board/documents/FiresideChats.pdf

Strickland, Derek. 2023. "GTA v Has Now Sold 180 Million Copies, Makes up 45% of Total GTA Franchise Sales." TweakTown. May 19, 2023. https://www.tweaktown.com/news/91513/gta-has-now-sold-180-million-copies-makes-up-45-of-total-franchise-sales/index.html#:~:text=In%20its%20Fiscal%20Year%202023

Sukel, Kayt. 2019. "In Sync: How Humans Are Hard-Wired for Social Relationships." Dana Foundation. November 13, 2019. https://dana.org/article/in-sync-how-humans-are-hard-wired-for-social-relationships/

"Supply and Demand: Basic Economics Part 2." n.d. www.socialstudiesforkids.com. https://www.socialstudiesforkids.com/articles/economics/supplyanddemand2.htm

Sutter, Paul. 2021. "Who Was James Clerk Maxwell? The Greatest Physicist You've Probably Never Heard Of." Space.com. September 29, 2021. https://www.space.com/who-was-james-clerk-maxwell-physicist

Taylor, Kate. 2014. "Millennials Spend 18 Hours a Day Consuming Media -- and It's Mostly Content Created by Peers." Entrepreneur. March 10, 2014. https://www.entrepreneur.com/growing-a-business/millennials-spend-18-hours-a-day-consuming-media-and/232062

Tedeschi, Bob. 1999. "Advertising: Net Advertising Potential Is Large." Archive.nytimes.com. August 19, 1999. https://archive.nytimes.com/www.nytimes.com/library/tech/99/08/biztech/articles/18adco.html

"The Betamax Case." n.d. Websites.umich.edu. https://websites.umich.edu/~evideo/beta.html

The Office of the Historian. 2018. "U.S. Involvement in the Vietnam War: The Gulf of Tonkin and Escalation, 1964." State.gov. 2018. https://history.state.gov/milestones/1961-1968/gulf-of-tonkin

The White House. 2017. "Richard M. Nixon." The White House. The White House. January 1, 2017. https://www.whitehouse.gov/about-the-white-house/presidents/richard-m-nixon/

The White House. 2022. "Blueprint for an AI Bill of Rights." The White House. October 2022. https://www.whitehouse.gov/ostp/ai-bill-of-rights/

The White House. n.d. "The Constitution." The White House. the white house. https://www.whitehouse.gov/about-the-white-house/our-government/the-constitution/

Thompson, Nicholas. 2017. "Our Minds Have Been Hijacked by Our Phones. Tristan Harris Wants to Rescue Them." Wired. WIRED. July 26, 2017. https://www.wired.com/story/our-minds-have-been-hijacked-by-our-phones-tristan-harris-wants-to-rescue-them/

Thomson, David. 2017. *Warner Bros*. New Haven, CT: Yale University Press.

Thorbecke, Catherine. 2023a. "'It Gave Us Some Way to Fight Back': New Tools Aim to Protect Art and Images from AI's Grasp | CNN Business." CNN. August 12, 2023. https://amp.cnn.com/cnn/2023/08/12/tech/ai-images-photos-protection/index.html

Thorbecke, Catherine. 2023b. "AI Tools Make Things up a Lot, and That's a Huge Problem | CNN Business." CNN. August 29, 2023. https://www.cnn.com/2023/08/29/tech/ai-chatbot-hallucinations/index.html

Tischler, Linda. 2009. "Ideo's David Kelley on 'Design Thinking.'" Fast Company. February 1, 2009. https://www.fastcompany.com/1139331/ideos-david-kelley-design-thinking

Tough, Paul. 2023. "Americans Are Losing Faith in the Value of College. Whose Fault Is That?" The New York Times, September 5, 2023, sec. Magazine. https://www.nytimes.com/2023/09/05/magazine/college-worth-price.html

Tremblay, Karine, Diane Lalancette, and Deborah Roseveare. 2019. "Assessment of Higher Education Learning Outcomes Ahelo Feasibility Study Report Volume 1 Design and Implementation." https://www.oecd.org/education/skills-beyond-school/AHELOFSReportVolume1.pdf

Trenholm, Richard. 2021. "History of Digital Cameras: From '70s Prototypes to IPhone and Galaxy's Everyday Wonders." CNET. May 31, 2021. https://www.cnet.com/tech/computing/history-of-digital-cameras-from-70s-prototypes-to-iphone-and-galaxys-everyday-wonders/

Tretina, Kat. 2021. "Top 10 Cryptocurrencies in April 2022." Forbes Advisor. December 1, 2021. https://www.forbes.com/advisor/investing/cryptocurrency/top-10-cryptocurrencies/

Tribune, Chicago. 1990. "Reagan Hazy on Iran-Contra." Chicago Tribune. February 23, 1990. https://www.chicagotribune.com/1990/02/23/reagan-hazy-on-iran-contra/

Tufte, Edward R. 2012. *Visual Explanations: Images and Quantities, Evidence and Narrative*. Cheshire, Connecticut: Graphics Press.

U.S. Bureau of Labor Statistics. 2018. "Web Developers: Occupational Outlook Handbook:: U.S. Bureau of Labor Statistics." Bls.gov. April 13, 2018. https://www.bls.gov/ooh/computer-and-information-technology/web-developers.htm

U.S. News Best Jobs. 2018. "Web Developer Ranks among Best Jobs of 2020." Usnews.com. 2018. https://money.usnews.com/careers/best-jobs/web-developer

"U.S. Social Media Ad Spend 2025." 2022. Statista. August 22, 2022. https://www.statista.com/statistics/736971/social-media-ad-spend-usa

United States Courts. 2023. "What Does Free Speech Mean?" United States Courts. United States Courts. 2023. https://www.uscourts.gov/about-federal-courts/educational-resources/about-educational-outreach/activity-resources/what-does

US Department of Education. 2017. "Reimagining the Role of Technology in Education: 2017 National Education Technology Plan Update." https://tech.ed.gov/files/2017/01/NETP17.pdf

Vaitheeswaran, Vijay. 2014. "The Digital Degree." The Economist. June 27, 2014. https://www.economist.com/briefing/2014/06/27/the-digital-degree

Vincent, James. 2017. "Former Facebook Exec Says Social Media Is Ripping Apart Society." The Verge. December 11, 2017. https://www.theverge.com/2017/12/11/16761016/former-facebook-exec-ripping-apart-society

Virshup, Amy. 1996. "The Teachings of Bob Stein." Wired, July 1, 1996. San Francisco, CA.

Vonnegut, Kurt. 2014. *Welcome to the Monkey House: The Special ed*. Dial Press.

Voorhoede, De. 2022. "What Is a Qubit?" Quantum Inspire. 2022. https://www.quantum-inspire.com/kbase/what-is-a-qubit/

Wall, Mike. 2014. "Space Station's 3D Printer Makes Wrench from 'Beamed Up' Design." Space.com. December 23, 2014. https://www.space.com/28095-3d-printer-space-station-ratchet-wrench.html

Wallace, Kelly. 2015. "Maryland Family under Investigation for Letting Their Kids Walk Home Alone." CNN. January 20, 2015. https://www.cnn.com/2015/01/20/living/feat-md-free-range-parents-under-attack/

Warner, Daniel. 2017. *Live Wires: A History of Electronic Music*. London: Reaktion Books Ltd.

"Was Tulipmania Irrational?" 2013. The Economist. October 4, 2013. http://www.economist.com/blogs/freeexchange/2013/10/economic-history

Washington Post. 2012. "The Watergate Story - the Washington Post." The Washington Post. 2012. https://www.washingtonpost.com/wp-srv/politics/special/watergate/timeline.html

Watkinson, John. 2001. *An Introduction to Digital Video*. Oxford England; Boston: Focal Press.

Watts, Stephen. 2023. "AR vs VR: What's the Difference?" Splunk-Blogs. November 8, 2023. https://www.splunk.com/en_us/blog/learn/ar-vr.html

Weinberger, Sharon. 2018. *The Imagineers of War: The Untold Story of DARPA, the Pentagon Agency That Changed the World*. New York: Vintage Books.

Weisstein, Eric W. n.d. "Ishango Bone." Mathworld.wolfram.com. https://mathworld.wolfram.com/IshangoBone.html

Welna, David. 2016. "George H.W. Bush's Mixed Legacy in a Reagan-Era Scandal." NPR.org. December 6, 2016. https://www.npr.org/2018/12/06/674079779/george-h-w-bushs-mixed-legacy-in-a-reagan-era-scandal

Werpin, Alex. 2012. "A Brief History of MSNBC.com and NBCNews.com." www.adweek.com. TVNewsr. July 12, 2012. https://www.adweek.com/tvnewser/a-brief-history-of-msnbc-com-and-nbcnews-com/

Westcott, Kevin, Jana Arbanas, Chris Arkenberg, and Jeff Loucks. 2023. "2024 Digital Media Trends." Deloitte Insights. April 13, 2023. https://www2.deloitte.com/us/en/insights/industry/technology/digital-media-trends-consumption-habits-survey.html#streaming-video-at-a-crossroads

"What Is a Computer Algorithm?" 2001. HowStuffWorks. September 5, 2001. https://computer.howstuffworks.com/what-is-a-computer-algorithm.htm

Whitmire, Vi. 2016. "Joseph Nicephore Niepce | International Photography Hall of Fame." International Photography Hall of Fame. 2016. https://iphf.org/inductees/joseph-nicephore-niepce/

Whitten, Sarah. 2023. "The 2023 Movie Box Office Will Need a Strong Second Half after an Inconsistent First Six Months." CNBC. July 4, 2023. https://www.cnbc.com/2023/07/04/2023-movie-box-office-needs-strong-second-half-after-inconsistent-start.html

Wike, Richard. 2022. "Views of Social Media and Its Impacts on Society." Pew Research Center's Global Attitudes Project. Pew Research Center. December 6, 2022. https://www.pewresearch.org/global/2022/12/06/views-of-social-media-and-its-impacts-on-society-in-advanced-economies-2022/

Wiley, John. 1993. "The Managerial, Political and Spiritual Presidencies." Jstor. Wiley. September 1, 1993. https://www.jstor.org/stable/27551140

Winograd, Terry. 1996. *Bringing Design to Software*. Boston, MA: Addison-Wesley Professional.

Wisse, Jacob. 2002. "Albrecht Dürer (1471–1528)." Metmuseum.org. October 2002. https://www.metmuseum.org/toah/hd/durr/hd_durr.htm

Wood, Stacey. 2021. "Catfishing Explained | Psychology Today." www.psychologytoday.com. August 6, 2021. https://www.psychologytoday.com/us/blog/the-fraud-crisis/202108/catfishing-explained

Wright, Alex. 2014. "The Secret History of Hypertext." The Atlantic. May 22, 2014. https://www.theatlantic.com/technology/archive/2014/05/in-search-of-the-proto-memex/371385/

Wright, Katherine. 2023. "What's a Qubit? 3 Ways Scientists Build Quantum Computers." Scientific American. September 28, 2023. https://www.scientificamerican.com/article/whats-a-qubit-3-ways-scientists-build-quantum-computers/

Wu, Tsung-Chiung (Emily), David Scott, and Chun-Chieh Yang. 2013. "Advanced or Addicted? Exploring the Relationship of Recreation Specialization to Flow Experiences and Online Game Addiction." *Leisure Sciences* 35 (3): 203–17. https://doi.org/10.1080/01490400.2013.780497

Wyckoff, Edwin Brit. 2013. *The Man Who Invented Television*. New York, NY: Enslow Publishing, LLC.

Yaffe-Bellany, David, and Matthew Goldstein. 2023. "Sam Bankman-Fried, in First Detailed Defense, Seeks to Dismiss Charges." *The New York Times*, May 9, 2023, sec. Technology. https://www.nytimes.com/2023/05/08/technology/sam-bankman-fried-defense.html

Yale University. 2000. "Biography of Grace Murray Hopper | Office of the President." Yale.edu. 2000. https://president.yale.edu/biography-grace-murray-hopper

Zeitung, Süddeutsche. 2013. "Panama Papers: A Storm Is Coming to Iceland." Süddeutsche.de. May 23, 2013. https://panamapapers.sueddeutsche.de/articles/56fec0cda1bb8d3c3495adfc/

Zickuhr, Kathryn. 2010. "Online Activities." Pew Research Center: Internet, Science & Tech. December 16, 2010. https://www.pewresearch.org/internet/2010/12/16/online-activities/

Zuboff, Shoshanna. 2019. "Life inside a Giant Space Beast May Not Be so Bad." WIRED. December 7, 2019. https://www.wired.com/beyond-the-beyond/2016/03/shoshanna-zuboff-condemning-google-surveillance-capitalism/

Index

E

F

M

N

O

P

Q

R